Digital Audio Editing

Digital Audio Editing

Correcting and Enhancing Audio in Pro Tools, Logic Pro, Cubase, and Studio One

Simon Langford

NEW YORK AND LONDON

First published in 2014
by Focal Press,
70 Blanchard Rd Suite 402
Burlington, MA 01803

Simultaneously published in the UK
by Focal Press,
2 Park Square, Milton Park, Abingdon, Oxon OX14 4RN.

Focal Press is an imprint of the Taylor & Francis Group, an informa business.

© 2014 Taylor & Francis

The right of Simon Langford to be identified as author of this work has been asserted by him in accordance with sections 77 and 78 of the Copyright, Designs and Patents Act 1988.

All rights reserved. No part of this book may be reprinted or reproduced or utilized in any form or by any electronic, mechanical, or other means, now known or hereafter invented, including photocopying and recording, or in any information storage or retrieval system, without permission in writing from the publishers.

Notices
Knowledge and best practice in this field are constantly changing. As new research and experience broaden our understanding, changes in research methods, professional practices, or medical treatment may become necessary.

Practitioners and researchers must always rely on their own experience and knowledge in evaluating and using any information, methods, compounds, or experiments described herein. In using such information or methods, they should be mindful of their own safety and the safety of others, including parties for whom they have a professional responsibility.

Product or corporate names may be trademarks or registered trademarks and are used only for identification and explanation without intent to infringe.

Library of Congress Cataloging in Publication Data
Langford, Simon.
Digital audio editing : correcting and enhancing audio in Pro Tools, Logic Pro, Cubase, and Studio One / Simon Langford.
    pages cm
ISBN 978-0-415-82958-8 (paper back)
1. Digital audio editors. 2. Pro Tools 3. Logic (Computer file) 4. Cubase 5. Studio One (Computer file) I. Title.
 ML74.3.L36 2014
 780.285'65—dc23
                                                        2013023380

ISBN: 978-0-415-82958-8 (pbk)
ISBN: 978-0-203-51289-0 (ebk)

Typeset in Giovanni
Project Managed and Typeset by: diacriTech

Printed in Canada by Marquis

Bound to Create

You are a creator.

Whatever your form of expression — photography, filmmaking, animation, games, audio, media communication, web design, or theatre — you simply want to create without limitation. Bound by nothing except your own creativity and determination.

Focal Press can help.

For over 75 years Focal has published books that support your creative goals. Our founder, Andor Kraszna-Krausz, established Focal in 1938 so you could have access to leading-edge expert knowledge, techniques, and tools that allow you to create without constraint. We strive to create exceptional, engaging, and practical content that helps you master your passion.

Focal Press and you.

Bound to create.

> We'd love to hear how we've helped you create. Share your experience:
>
> **www.focalpress.com/boundtocreate**

Focal Press
Taylor & Francis Group

Contents

ACKNOWLEDGMENTS xi

INTRODUCTION
　What Exactly Is Audio Editing? 1
　Book Format 3

CHAPTER 1　Audio Editing 101 5
　Editing History: Tape 5
　Editing History: Digital Tape and Hard Disk-Based 6
　Editing History: DAW 9
　The Here and Now 12

CHAPTER 2　Different Aims of Audio Editing 15
　Corrective Editing versus Creative Editing 15
　Restorative Editing 17
　The Role of the Editor 20

SECTION 1　Corrective Editing 23

CHAPTER 3　Cutting, Copying, Pasting, and Moving 25
　Why Start Here? 25
　The Basics 25
　Start at the Beginning 27
　Tightness of Edits 27
　"Zero-Crossing" Edits 28
　"Visual" Editing versus "Audible" Editing 30
　Multiple-Track Editing 32
　Potential Problems 33
　Hands On 35
　　　Introduction 35
　　　Logic 35
　　　Pro Tools 37
　　　Studio One 40
　　　Cubase 43

CHAPTER 4　Fades and Cross-Fades 47
　Differences between Fades and Cross-Fades 47
　Different Fade Shapes 47
　　　Linear 48
　　　Logarithmic 49
　　　Exponential 50
　　　S-Curve 50
　Different Fade Shapes: Summary 51
　Using Fades 52
　Using Cross-Fades 54
　Phase Cancellation 54
　Periodic Patterns 56

Hands On	57
Introduction	57
Logic	58
Pro Tools	60
Studio One	62
Cubase	64
CHAPTER 5 Level Control	69
Should This Be a Part of the Editing Process?	69
Destructive versus Nondestructive Edits	69
Different Types of Level Control	72
Destructive Edits	72
Static Nondestructive Edits	73
Variable Nondestructive Edits	74
Automatic Nondestructive Edits	75
Hands On	76
Introduction	76
Logic	76
Pro Tools	78
Studio One	80
Cubase	82
CHAPTER 6 Tonal Matching	85
What Is Tonal Matching?	85
CHAPTER 7 Comping and Alternate Takes	89
The Search for Perfection	89
Putting It All Together	90
Making the Grade	91
Tonal Matching	93
Pitch Adjustments	94
Timing Changes	96
Comping Made Easy	97
Hands On	99
Introduction	99
Logic	99
Pro Tools	102
Studio One	105
Cubase	108
CHAPTER 8 Multi-Track Comping	113
Ideal-World Multi-Track Comping	113
Real-World Multi-Track Comping: Spill Issues	115
Real World Multi-Track Comping: Timing Issues	117
How Can We Fix It?	118
Frequency-Dependent Gating	119
Expand Your Options	120
Hands On	122
Introduction	122
Logic	122
Pro Tools	125
Studio One	127
Cubase	130

CHAPTER 9 Transient Detection 133
Why Use Transient Detection? 133
Simple Level-Based Detection 135
Advanced Level-Based Detection 136
Spectral Detection Techniques 137
Possible Applications and Limitations 138
Hands On 140
    Introduction 140
    Logic 140
    Pro Tools 142
    Studio One 145
    Cubase 147

SECTION 2 Creative Editing 151

CHAPTER 10 Beat-Mapping and Recycling 153
Introduction 153
Beat-Mapping 153
Quality Limitations of Beat-Mapping 155
Recycling 156
Extended Uses of Recycled Files 158
More Creative Uses 159
    Beat-Mapping 159
    Recycling 161
Quick and Easy Multi-Samples 163
Rhythm from Anything 163
Hands On 164
    Introduction 164
    Logic 164
    Pro Tools 166
    Studio One 169
    Cubase 172

CHAPTER 11 Drum Replacement 177
Using Recycled Loops 177
Dedicated Drum Replacement Tools 179
Drum Replacement Tools in Use 181
Hands On 184
    Introduction 184
    Logic 185
    Pro Tools 187
    Studio One 190
    Cubase 192

CHAPTER 12 Time-Stretching 195
Introduction 195
Why Is It So Difficult? 197
The Additive Approach 198
The Granular Approach 199
How We Use It and When We Use It 201
Time-Based Stretches 202
Tempo-Based Stretches 203

Length-Based Stretches	204
Filling in the Gaps	204
"Manual" Time-Stretching	206
Hands On	207
Introduction	207
Logic	207
Pro Tools	209
Studio One	212
Cubase	215

CHAPTER 13 Elastic Audio (Time) 219

Definition of Elastic Audio	219
Different Algorithms	220
Advantages and Disadvantages Compared to Regular Time-Stretching	221
Quantization of Audio Files	222
Auto-Conform and Follow Tempo	223
The Best of Both Worlds	224
Hands On	225
Introduction	225
Logic	225
Pro Tools	228
Studio One	231
Cubase	233

CHAPTER 14 Pitch-Shifting 239

Introduction	239
A Brief History of Pitch-Shifting	239
Analogue	239
Digital	240
Computer-Based	241
Different Approaches to Pitch-Shifting	244
Size Matters	245
Formants	247
Uses of Pitch-Shifting	250
Hands On	252
Introduction	252
Logic	253
Pro Tools	255
Studio One	257
Cubase	260

SECTION 3 Restorative Editing 263

CHAPTER 15 Editing in the Third Dimension 265

Introducing the Idea of Spectral Editing	265
Taking Things Further	268
FFT Analysis and Waterfall Plots	269
Spectrograms	271

CHAPTER 16   Spectral Editing 275
   Introduction 275
   Tools of the Trade 276
   Corrective Spectral Editing. 278
      Attenuation. 278
      Copying and Pasting 279
      Clicks and Crackle 282
      Noise and Hum 285
   Creative Spectral Editing 288
      Compression. 288
      De-clipping 290

CHAPTER 17   Applications of Audio Restoration. 293
   The Recording Studio and Beyond 293
   Beyond Restoration 295

CHAPTER 18   Demixing. 299
   What Exactly Is Demixing?. 299
   Is That Even Possible?. 300
   How Does It Work? 302
   What Are the Applications? 304
   Legalities of Possible Uses 307
   Fair Use. 308
   Moral Issues. 309
   It's Not All Bad 310

CHAPTER 19   Thinking Outside the Box. 311
   Sound Design 311
   Using and Abusing Audio Editing Techniques 312

INDEX. 317

Acknowledgments

The irony is not lost on me that this book about editing of one form actually started out life roughly 50% longer than it ended up! It was very much an exercise in restraint and refinement that led from that initial form to the final product here. I have to thank everyone in my life for their patience and understanding while I have been deeply ensconced in the process of refining things down to where they finished up. However, special thanks must (of course) go to those who played a bigger part in the whole process.

I would first like to thank Meagan and Anais and all the other members of the Focal Press team who have worked with me in getting the ideas out of my head and onto paper. Writing any book is a labor of love, and the author will naturally have a vested interest. But the dedication of those whose sole aim is to bring knowledge to the world and empower others through that knowledge amazes me. To all of you, whether I have dealt with you personally or not, I give my deepest gratitude.

Then I need to thank my long-time collaborator, proofreader, and friend Alex Sowyrda. You, Alex, probably know nearly as much as I do about audio editing after reading through my original and painfully long version of this book, so I owe you my thanks once again.

I also need to thank the many, many people who have, over the years that I have been involved in the music industry, shared their expansive knowledge with me. I wasn't born knowing any of the information that I have included in this book, and it is only through the kindness and generosity of others and their willingness to share their own knowledge that any of this was possible. You all know who you are and what you have taught me, so this book is as much about your vast experience as it is about mine.

I also need to give my thanks to Rodney Orpheus at PreSonus, Emma Clifton at Avid, and the teams at Yamaha/Steinberg and Apple for their assistance in making this book happen.

And finally, given that this entire book is dedicated to the concept of making the best version possible of something, I have to give my personal thanks to the two people who drive and motivate me to be the best possible version of myself. I dedicate this book to Ben and Hayley, without whom I wouldn't have had the courage or belief to even get started on this long journey. My life simply wouldn't be as bright, as colorful, or as memorable without either of you, and I certainly wouldn't be the person I have become today.

Introduction

WHAT EXACTLY IS AUDIO EDITING?

Welcome to *Digital Audio Editing: Correcting and Enhancing Audio in Pro Tools, Logic Pro, Cubase, and Studio One*. During the course of this book, I will guide you through the principles and practice of a number of regularly used audio editing techniques, as well as a few less-frequent ones, and will wrap things up with a look at some of the more advanced and unusual editing techniques.

I would also like to point out, right at the very beginning, that the audio editing process can often cross over into the production process and even the "sound design" process. Many of the techniques that we will be looking at in this book could be seen as the fundamentals of sound design as well, and it is very tempting to branch off into both of these areas as an extension of what we will be looking at. However, in an effort to try to keep things focused and manageable, we should probably stick to the definition that audio editing involves taking sounds and recordings that we already have and making the best version of them for the purpose they have to fulfill. I would view sound design more as the process of creating new sounds, either from scratch or by heavily manipulating existing sounds or recordings, and would view production more as the process of combining the edited audio recordings into a finished piece.

The simplest analogy that comes to mind is that of building a house. We could consider audio editing as actually making the bricks as straight and uniform and correct as we can. These bricks are arranged and put in place by the builder (mix engineer) under the direction of the architect (producer) who has the blueprints for the building. If the bricks aren't straight and uniform, then the builder's job becomes far more difficult.

So with that in mind, let's ask the key question: What exactly is audio editing? I think that everybody would think of the obvious things such as cutting, pasting, cross-fades, and track "comping," but there are many more ways to edit audio. As a general rule, we can put all these editing methods under one of three main headings: corrective, creative, and restorative. We will look at each of these in more detail as the book progresses, but, as a taster, I would suggest that the most common corrective tasks include the previously mentioned cutting, copying, pasting, and comping, while the most common creative tasks would include time-stretching and pitch manipulation. Restorative tasks are an area where technology seems to be changing at a ridiculous pace, and we now have the ability to do things that were simply unimaginable until fairly recently. We can carry out spectral repairs and manipulate individual sounds within a complex mix and further manipulate "mixed" recordings in some truly mind-bending ways. In fact, we can change pretty much any aspect of audio recordings in one form or another, some more successfully and easily than others, obviously.

There are a couple of things to consider at this point, and both of these relate to assessing the true need for these advanced techniques. The things we are able to do with audio recordings today are breathtaking, and having these abilities and techniques available to us gives us a great deal of freedom when working with audio files. The problem, as I see it, is that having this much power to manipulate the recordings can lead us down the path of "let's fix it in post-production." I don't think that this is necessarily a healthy attitude to have. Of course there are situations when we, as audio editors, will be presented with a collection of files and expected to sort out the mess that we have been given, and that is when all the techniques presented in this book will become invaluable to us. But equally there might be times when the knowledge of what is possible after the recording has been completed will make those actually doing the recording a little lazy.

Let's look at the relatively straightforward situation of working with vocals. We can compile a "master" take from a number of different takes in order to not be reliant on the singer actually delivering a (subjectively) "perfect" take. We can correct the pitch if singers are a little off-key, and we can correct the timing if they are a little loose in that respect. We can even control the dynamics of their performance to some degree and perhaps even the tone of their voice, although the plug-ins and software that allow us to do this will give us only a very limited range to change before the effects become very noticeable and, in my opinion, damaging. Armed with this knowledge, it wouldn't be a huge leap for the people involved in the recording (including the singer, the producer, and the engineer) to simply make do and rely on fixing it later. While I am sure that this doesn't happen very often, I can understand this from a financial point of view, because hiring a studio is usually far more expensive in terms of hourly rate than having somebody fix things later. But to look at it this way would be a major mistake, in my opinion. If at all possible we should always strive to have the very best take/performance that we can to work with. We can achieve great things with audio editing, and we can make an average performance or recording sound pretty damn good! But with that in mind, imagine what we could do with a take or performance that is already very good!

In essence, what I am trying to get across is that audio editing shouldn't be seen as a way to cut corners at the recording stage. In the early days of recording, it simply wasn't possible to do anything like the level of editing that we do now, and it was crucial to get the best performances right from the very beginning. I don't think that, looking back, many of us would condemn the quality of the actual performances in older recordings. Yes, we can clearly hear that the technical quality of the recordings might not be up to today's standards, but many would argue that there was a certain magic in those recordings that isn't always present today, and much of that can be put down to getting the very best performance from the band or artist. Of course there are situations where it isn't possible to be quite so selective (live-band recordings, location news reports, and things of nature that are one-off events that occur "in the moment"), and in those situations we can be glad that we can use all the tools we have today to clean up and polish rough edges that might be present.

BOOK FORMAT

As mentioned above, the book is divided into three main sections: Corrective Editing, Creative Editing, and Restorative Editing. Each of these sections will then be divided into chapters, in which each chapter will look at a particular aspect of the editing process. The main body of each chapter will be largely theoretical in nature and will discuss the concepts and ideas in general, but this will be followed by a "Hands On" section at the end of each chapter, which will look at ways in which the concepts discussed can be achieved in each of the four digital audio workstations (DAWs) we will be covering. Space limitations prevent a full breakdown of how to do everything that is covered in the each chapter, but there will be an additional content on the accompanying website that will provide additional "Hands On" sections as well as other background information.

When specific DAW commands are referenced in the book, they will use colored, bold text and will follow one of two formats. Keyboard commands will be shown, such as **Cmd[Mac]/Ctrl[PC] + V**, which indicates that all the keys indicated (Cmd or Ctrl and V in this case) should be pressed at the same time. In some cases the keys required will be the same for both Mac and PC, in which case only a single combination will be shown. Menu commands will be shown as a series of menu and sub-menu names eventually leading to a menu item. The format will be along the lines of **Edit > Cut/Insert Time > Insert Silence Between Locators**, which would mean to go to the *Edit* menu, then the *Cut/Insert Time* sub-menu, and then choose the *Insert Silence Between Locators* item.

4 Book Format

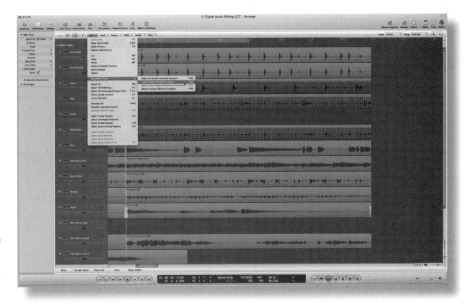

FIGURE I.1
When a menu shortcut is included in the text, it will list the main menu name first, then any sub-menus, and finally the menu item required.

So now that we have covered that, let's start our journey by going back to the drawing board and seeing how all this started. As I mentioned above, this flexibility that we have to manipulate the recordings that we are working with is only a comparatively recent development, and in order to fully appreciate the options that we have available to us today, I think it might be prudent to start by rewinding back to the era of multi-track tape recorders.

CHAPTER 1
Audio Editing 101

EDITING HISTORY: TAPE

It was only with the advent of multi-track tape recorders that any kind of meaningful editing became possible. Audio editing (in the *very* loosest definition of the term) was technically possible with earlier mono or stereo tape recordings, but any editing would have been limited to very simple repeating of sections or removal of sections of the whole recording. When multi-track tape recorders came into the world (courtesy of a company called Ampex in the 1950s) a few more options became available to the creative (and patient) engineer. On the surface there doesn't seem to be any more flexibility, because editing of a multi-track tape in the way it had previously been done still involved cutting sections of the entire tape (across all the tracks at once) to do the edits. The advantage with a multi-track tape is that individual tracks could be copied to another tape and then copied back over after the edits had been done—so it was time-consuming, but definitely possible.

With the required patience, it was possible to copy sections from other places within the track and paste them, although that would have been accomplished by recording the sound or section in question out onto another tape machine and then rerecording it back onto the original multi-track tape at the appropriate place. This was far from easy, and, even though the two tape machines could possibly have been synchronized through the use of a time code recorded on to each of them, it would still have been a very difficult process, and there certainly wouldn't have been the sub-millisecond accuracy that we have now. Nonetheless, it was the first of many steps toward where we are today.

In addition to the copy and paste abilities of multi-track tape, there was also the option to process individual tracks or sounds using a variety of studio equipment. One of the corrective aspects of audio editing involves level and dynamics control. Multi-track tape recording allowed the engineers and producers to have control over the relative balance of the tracks *after* they had been recorded for the first time; in addition, individual tracks could be processed with compressors and other outboard equipment to allow for changes to the dynamics as well as level control. Things that we take massively for granted these days and can accomplish in seconds with the flick of a wrist and a few well-placed mouse clicks may well have taken many minutes, if not more back, in the early days of multi-track recording. This kind of level control could be seen as a part of mixing or production, but there are times when it will fall under the remit of the editor, so it was worth mentioning here.

"Comping" (short for *compiling*) became a realistic possibility as well, because a performer could record multiple takes onto adjacent tracks, and then the best sections of each performance could be noted down and the relevant tracks soloed (or channel faders on a mixing desk moved) at the appropriate time, and the resulting compiled track could be recorded on to another track on the tape—often called bouncing—or perhaps even another tape machine. Because the timing of the changes wasn't always accurate, it is most likely that the changes between takes would have happened between lines of a song or perhaps, at best, between words. Nowadays we can quite easily edit between different takes, even in the middle of a single word or note and, with a little work, make the resulting edit sound utterly smooth and natural.

A further development of this concept of comping is in the ability to change individual drum hits within a recording if the drums have been recorded with a multiple microphone setup. There are many limitations to this, which we will look at in detail later on, but, with care, it is possible to remove an errant drum hit within the context of the whole recording, which simply wouldn't be possible with a mono or stereo "mixed" drum-kit recording. Obviously this became possible only with the advent of multi-track tape (and in particular multi-track tape machines with a good number of tracks available) and is something that could prove very useful in fine-tuning a recording.

Looking at this very short summary of the editing possibilities of multi-track tape, it becomes clear that a lot of what we would consider to be the "corrective" editing processes that we use today could, in fact, be done with the medium of multi-track tape. What should also be clear, however, is that many of these processes were more of an art (and not always a 100% accurate one) rather than a science, and a great deal of skill would be required in order to achieve decent results, and, even if that skill level was attained, it was still a time-consuming process. The next stage of evolution of recording (and therefore editing) came with the advent of digital audio, and it was with this development that things started to become more accurate, reliable, and repeatable, and required less technical skill and time.

EDITING HISTORY: DIGITAL TAPE AND HARD DISK–BASED

The idea of recording sound digitally actually goes back much earlier than most people think. The very first digital audio recording experiments can be traced back as far as the late 1960s and early 1970s, which predates the 1981 introduction of the CD format by over a decade. Just a year after the unveiling of the CD, Sony debuted its DASH digital multi-track systems, which offered twenty-four or even forty-eight tracks of digital audio recorded onto half-inch wide tape. These systems were, naturally, staggeringly expensive but were totally groundbreaking and paved the way for the systems that would follow later and at a much more affordable price. In use, these Digital Audio Stationary Head (DASH) systems (also available from Studer and Teac Audio Systems Corporation America [TASCAM]) were very much like a conventional analog tape–based multi-track in that, through the use of some very clever encoding and error correction, the tape could be physically edited and spliced in the same way that people were used to doing with analog systems. In fact, that was pretty much the only way you could edit with these systems. So in this regard they served as a very easy transition into digital recording. Most of the editing techniques that editors already used could still be applied. The only thing that was different was that it was a digital signal being recorded rather than an analog one.

One fundamental advantage of digital recording is that there is a much lower noise level on recorded signal, which means that multiple tracks layered up would be a great deal cleaner and clearer than a comparable analog recording. Another advantage is that, in the event that a digital track needed to be recorded or bounced to another track or another recorder altogether, assuming that the signal was kept digital during the process and not converted back to analog and then to digital again, there would be no loss of quality and no increase in background noise levels. One final advantage that is worth mentioning, and which could prove very important if trying to copy and paste sounds from different parts of the recording, is the fact that digital recordings have no inherent "wow" or "flutter." Both of these terms relate to minor (or sometimes major) fluctuations in the speed of the tape as the recording is being made or played back. This is problematic, because an analog audio recording on tape has a direct relationship between the pitch of the sound and the speed of the tape. When there are minor variations in the speed of the tape during recording or playback, the pitch of the recording will vary with the changes in tape speed, resulting in a "wobbling" effect. With a digital system, this simply doesn't apply. You can still change the pitch and tempo of a digital recording by adjusting the playback frequency to something different than the sampling frequency of the recording, but at least every part of every track will be affected by the same amount (and there will be no wow or flutter) which allows for much more consistency when cutting, pasting, and moving audio among different parts of the same recording.

It would take a different format entirely to start the transition to what we know today, and that came in the form of using hard drives as the recording medium. Although this is often thought of as a development that came after digital multi-track tape, the very first hard-disk multi-track recorder was actually a software add-on for the NED Synclavier II system and was released in 1982, the same year that the Sony DASH system was introduced. Because of the cost and scarcity of this system, tape-based digital systems were the ones that really pushed forward the digital revolution in multi-track recording. Though behind the scenes the work continued, and then, in 1993, the Otari RADAR (Random Access Digital Audio Recorder) was introduced. The key to its success, and the subsequent move to hard-disk recording, lies in the name: Random Access.

This one aspect of the RADAR system would become a huge selling point and would revolutionize the way that audio recording and editing (more so the latter) were carried out. Tape-based systems are essentially linear recording systems in both input and output. Sounds are recorded from a start point to an end point and then are played back (physical splicing and editing aside) from that same start point to the same end point. Hard-disk recording as offered by the RADAR system was slightly different. Of course the recording would still be linear in input—a recording would start at a particular point and continue on for a set amount of time—but the linearity of the output wasn't required any more. This *non-linear editing* process allowed editors sample-accurate editing and the ability to select, cut, copy, paste, and move sections of audio without any physical process at all, and, best of all, this was the start of the ability to undo the edits if they either weren't done correctly or simply didn't have the desired effect.

The RADAR system had many limitations, though. While it could record up to twenty-four tracks of audio, the internal 1 GB hard drive only allowed for a total of roughly two hundred track minutes of recording (total number of minutes available over the total number of tracks) or eight minutes of twenty-four-track recording. This in itself wouldn't have been a major shock, because a typical reel of tape running at thirty inches per second would give around fifteen minutes of recording. The problem was the cost of hard drives relative to tape. A reel of two-inch tape was by no means cheap, but, in 1993, a 1 GB hard drive would have cost close to $1000, which would have been a very limiting factor. It was, though, only a matter of time before hard-drive prices began to fall to more reasonable levels; at the same time, capacities start to increase to the point where a whole drive could record much more than eight minutes.

Even though these systems were inherently more flexible than tape-based systems, there were huge costs associated with them, so tape-based digital recording wasn't quite dead just yet. The next big step in the evolution of tape-based digital multi-track recording came in 1992 with the release of the original Alesis ADAT recorder. This was a system that, at a relatively affordable price, offered eight tracks of digital recording (multiple units could be synchronized to provide up to 128 tracks if required) on S-VHS format cassettes. This may well have been an important factor in the success of the original ADAT machines. Open reel tape (as used on analog and DASH systems) could be very expensive, as well as taking up a lot of space to store. The S-VHS cassette format made the recording media for the ADAT compact and affordable, very much like the machine itself. It wasn't all positives, though, as the splicing and editing ability that both analog multi-track machines and DASH machines possessed was now gone. Given that the tape was enclosed within the cassette and the cassette itself inserted into the recorder, there was simply no way of editing in the way that people had been used to before, even if the underlying technology (as in the DASH systems) had allowed it.

Hard-disk recorders were dependent on storage space, and, in the early 1990s, this was, as we have seen, still prohibitively expensive. However, as the various storage media prices fell, the number of machines on the market increased. By the turn of the century, there were eight-, sixteen-, and twenty-four-track hard-disk recorders on the market from many of the companies familiar to studio owners, such as Akai, Fostex, Roland, TASCAM, Yamaha, and many others at prices that were far more affordable. As these "all-in-one" digital studios started to become more accessible, the ADAT/sampler setup became less attractive, and eventually the ADAT recorders were discontinued, and the move to nonlinear systems was pretty much completed. However, while these stand-alone recorders were being developed, the same technology was being harnessed in a different way to allow software sequencers running on PC or Macintosh computers to add audio-recording abilities to their already established MIDI recording, and, when this became possible, the digital audio workstation, or DAW, was born.

EDITING HISTORY: DAW

Computer-based digital audio recording, and the later change into what we now call DAW software, was a total game-changer for both the recording and editing processes. By not using any form of tape media for the actual recording, DAW software promised far greater track counts (limited only by computing power which, even at the time that DAW software first started to emerge, was already increasing exponentially), almost stupidly easy cutting and pasting, and significantly greater editing options.

Strictly speaking, the very first digital audio workstation was released in 1978 by a company called Soundstream, used hard disks (very low in capacity by today's standards) for storage, and allowed for very basic editing of the recorded audio in addition to mix-down and cross-fades. The facilities of the system would be considered arcane today, but at the time they represented a quantum leap in audio recording and editing freedom. Most PCs of the era were text-based (DOS) systems, and this factor alone contributed to DAWs' not really developing that quickly. Aside from the limitations of the technology (processing power and storage), these early systems were not overly user-friendly. You could argue that they were, in many ways, more user-friendly than working with tape, and there is some logic in that, but any systems that relied on text and numbers to define absolute positions and commands was in some ways counterintuitive to the natural and organic nature of music-making. What was needed was a better way of interacting with the underlying technological ideas of the Soundstream system.

That better way was a natural development of the move to PCs, which operated with graphical user interfaces (or GUIs). The fact that things were represented onscreen with objects and icons that were visually related to the things they represented allowed for a much more natural way to operate computers. By the late 1980s, there were many affordable computer platforms using these GUI operating systems, which had already had sequencing software packages written for them to allow the control and recording of MIDI instruments. The most common platforms in use for music sequencing at the time were the Apple Macintosh, Atari ST, Commodore, Amiga, and generic "PC." The processing power of all these machines at the time was starting to make audio recording a realistic proposition, and in 1989 a company called Digidesign released their ground-breaking SoundTools software. While this would later evolve into the now almost ubiquitous Pro Tools, this early relative was limited to only mono or stereo recording and was less of a DAW than a stereo digital recorder with some quite advanced editing features.

One of the most interesting was the FFT Window, which provided a Fast Fourier Transform view of the audio recording. This provides a three-dimensional overview of a sound that is similar to that shown by a spectrum analyzer (with frequency on the horizontal axis and amplitude on the vertical axis) but with the additional third dimension showing the change of this spectrum over time. If you imagine a series of snapshots of a spectrum analyzer display stacked up one behind the other, then that is the basic concept. This FFT view allows a more intuitive view of how a sound changes over time, but, at this period in DAW evolution, while the FFT display was a valuable addition for figuring out what needed doing, it was purely an analytical tool, and no changes could be made directly to the sound in this view. The FFT view would be used to figure out what needed to be done and where, and then the traditional tools would be used to actually make the changes. Nonetheless, it paved the way for more-advanced editing techniques later on.

It didn't take long for the technology and ideas used in SoundTools to make its way into more advanced software, and in 1990 a company called OSC (distributed by Digidesign) launched their Deck software. This was similar in functionality and principles to SoundTools but allowed four simultaneous tracks instead of the two that SoundTools was limited to. Digidesign were already well aware of the potential of the Deck software and actually licensed the core technology for use in the first version of their Pro Tools software that was released 1991. This first Pro Tools version was still limited to eight tracks, but, like so many developments in digital audio, the limitations were entirely due to the hardware that the systems were running on. Everybody realized that it wouldn't be too long before eight tracks became a realistic possibility, once the computing power and hard drive speed allowed. So Pro Tools evolved and allowed greater numbers of tracks over the years, and eventually became a very good alternative to a multi-track tape system. Later versions added in MIDI sequencing facilities as well to provide more of a one-stop studio environment. And, of course, as the processing power increased, the editing options and technologies increased and diverged to include "plug-ins" that allowed third parties to create mini software applications that ran inside of Pro Tools.

Meanwhile other companies were heading toward the same solution but from the completely opposite direction. There were two software companies that many considered to be the dominant and industry-standard ones when it came to MIDI sequencing software: Steinberg and C-Lab/Emagic. Both companies had developed software that had, over a number of revisions, become very sophisticated, and in use mirrored many of the paradigms of multi-track tape recording, only using MIDI instruments instead of audio recordings and, in doing so, retaining a greater degree of flexibility and editing freedom. But these applications lacked the ability to include any real audio recordings unless the user was prepared (and able to) record into a MIDI sampler and use the sampler effectively as an audio recording "add on." Many did just that, but it still wasn't an ideal solution, so work continued to bring native audio recording and editing functionality to the flagship products. And then, in 1991, Steinberg introduced Cubase Audio for the Macintosh (audio-capable versions of Cubase didn't arrive on the Atari platform until 1993 and on the PC platform until 1996, owing to the relative limitations of those platforms compared to the DSP-enhanced possibilities of the Macintosh platform), which was a real game-changer. For the first time there was software that incorporated all the flexibility of MIDI sequencers with the options and additional benefits of audio recording and editing. In this respect, at least, it put Steinberg well ahead of the game, as at this time Pro Tools was strictly audio-only and Logic (Emagic's direct Cubase competitor) was still MIDI only. It wasn't until 1994 that Emagic would release a version of Logic (again for the Macintosh platform first) that would include audio recording features. Another manufacturer of historical significance is Mark of The Unicorn (often known simply as MOTU), who released their Digital Performer in 1990. This was originally created as a front-end to Digidesign's Audiomedia hard-disk audio-recording platform, which added sequencing abilities to the audio features already available in Audiomedia. In doing so they added another possibility to the already rapidly growing array of options.

In addition to the extensive feature sets of these newly developed packages, there were continued advances in the quality and feature sets of two-track audio editing software. Fundamentally these offer many—if not all—of the features of the audio side of a comparable DAW but work on a single two-track (stereo) audio file at any given time. If you think of it as working with a two-track "master" audio tape or a single track of a DAW, then, conceptually, at least, you will be in the right ballpark. Additionally, they often have features not available in DAWs, as the requirements and editing paradigm are a little different. DAW software is (mostly) about real-time playback and editing, and, while they do have destructive editing capabilities, most of the editing is done in a nondestructive way. Two-track editors, on the other hand, tend to have more destructive editing options as their intended use (working on final "masters" and individual files), which lends itself more to this way of editing. Perhaps the best-known examples of these types of editors are Steinberg's Wavelab and Sony's Soundforge, although there are many other options available, including Audacity and Adobe Audition.

THE HERE AND NOW

Over the next nearly twenty years, to bring us up-to-date, DAW software has become a studio staple, and now many—possibly even *most*—studios are based on a DAW system. There are, of course, some studios that still incorporate analog multi-track machines, and there are probably a few that are all analog and have not a hint of digital trickery in sight. But, for the most part, DAW software is now the standard. Those early four-tracks systems have now expanded into systems that offer hundreds of tracks (or in some cases track numbers are limited only by processing power and hard-drive speed), which have very advanced editing facilities right out of the box and the ability to expand those through plug-ins, and which offer higher quality than the 1981 "CD Standard." There are also many more software packages on the market to choose from. There are a few that are still audio-only, but many incorporate MIDI sequencing as well. The emphasis between audio and MIDI varies between them, as does the workflow. Most are still based around the tape-recorder model, but more recently there have been more and more developments that move away from that well-known format into other, more performance-oriented setups. Ultimately, though, they all aim to do the same thing, and most of them do it very well. So what exactly *can* we do in terms of editing on modern DAW software? Let's have a brief look at some of the advantages that DAW software has brought us.

Although the recordings that we work with are still linear recordings in that they start at a certain time and continue for a certain period of time and then stop, DAW software allows far greater freedom of how to use those basic recordings than did the previous generation of digital multi-track recorders. Cutting and pasting have been possible since the advent of multi-track tape, but with each successive generation (tape, then digital, then DAW) the process has become more intuitive and much easier. With many DAW users working to a click track or clearly defined tempo, it was very easy to move things to positions that made sense in musical terms (bars and beats rather than hours, minutes, seconds, and frames). Of course, not everybody works this way, and, for those people who prefer to work with absolute time instead of bars and beats, that option is still available. In fact, if you are working with sound for picture, then it is the preferred option, as the seconds and frames will tie in with what you are seeing in the video you are working with.

In either case the cutting/copying/pasting/moving process is greatly helped by having a number of different ways of measuring time and by having the ability to have your edit point "snap" to certain subdivisions. If you are working on the bars and beats scale, then you could set your software up so that the position of any cuts you made was automatically snapped (moved to) the nearest bar (or beat, eighth note, sixteenth note, etc.), and, if you are working on a time-based scale, your cuts could be snapped to the nearest second or frame. This won't always be appropriate or even what you want, so this feature can, of course, be turned off. But it is one of a great number of ways in which DAW software has greatly streamlined the modern recording and editing work-flow and made it possible to do in minutes, or even seconds, what would not all that long ago have taken a great deal of time.

Automation is another major plus for DAW software. The idea of automation isn't really new, as there had been a number of high-end recording consoles that had offered automation and recall facilities before the advent of DAW software, but the systems that we have today go a lot further. Automation data can now be moved just as easily as the audio recordings and is often linked to the underlying audio, so that, if we move or copy a piece of audio, the automation data moves (or has the option to move) with it. Once again this is more of a work-flow development and a time-saver than anything else. It might seem like a very simple thing, but these simple cut/paste/move operations are probably carried out hundreds if not thousands of times over the course of recording, editing, and mixing a record, so, if something simple like this can save us even ten or fifteen seconds at a time, it is pretty clear that the time saved could be quite substantial.

The combination of the two of these, in conjunction with the Undo command, means that editors now have a great deal of scope to just try things out. There is no need to worry about the cost of making a mistake and ruining a reel of two-inch tape, and the range of things that we can do *nondestructively* in DAW software is greater than we could do with previous digital multi-track recorders. Having an Undo command has meant that editors are much more likely to experiment with different ways of doing things, as we have the safety of knowing that we can step back to our original version quickly and simply.

All of these benefits would, on their own, be very welcome for anybody who does any substantial amount of audio recording or editing but, above all these, the single greatest advantage of a DAW system over any other kind or recording system is the graphical overview or waveform display of the recordings. At the most fundamental level, these overviews can help us to quickly locate different parts of a recording. Obviously if the sound or instrument on the track is fairly consistent, then it won't help us at this macro level, but for other sounds that are less constant (such as a vocal), we will often be able to see different sections of the recording clearly from the periods of activity and pause. Again, this is a simple benefit that could save a lot of time, even at this level.

As we look at the overview in more detail, we will start to get more information. We will, for example, start to see the dynamics of the recording more clearly. Any unwanted "pops" and "thumps" from microphones will show up quite clearly as much higher peaks on the overview. We could also use this level of detail to give us a good indication of any overall level changes throughout the recording. These will often be very clear upon listening, so we wouldn't necessarily need a visual display of the waveform to recognize these changes, but the visual helps us not only to locate them in a potentially complex recording (locating visually rather than having to make a long list of written notes) but also can, to the more trained and experienced eye, even give a very rough indication of how significant the change is. It is doubtful that anybody could look at a waveform overview and see a dip in volume and be able to say that it was a dip of 2.4 dB and be accurate, but, with experience, you may be able to look at the waveform and estimate a drop of between 2 and 5 dB and be fairly accurate. This might not seem overly useful, but it does sometimes come in handy, and we will look at this more later.

Taking things a little further, we get to the point where we can see the actual waveform shape, and this brings us to the point where we can be truly accurate with our editing and have sub-millisecond accuracy and the means to make sure that all our edits can take place at zero-crossing points and be phase-aligned. The ability to have sub-millisecond (closer to 1/100th of a millisecond if working at a sample rate of 96 KHz) accuracy may seem unnecessary, and there is a lot of weight behind that argument. Very few people can detect timing accuracy to anything less than 5–10 ms, so this degree of accuracy isn't really necessary to preserve timing. It can be useful, however, in working with edits that need to maintain accuracy within complex sounds recorded at high sample rates.

And that brings us fully up-to-date. We are currently in a world where the tools we have at our disposal allow us to do amazing things with sound. The digital age has in many ways brought an unprecedented level of freedom for audio editors, sound designers, and engineers, as we have numerous safety measures in place that do their best to ensure that we don't ever mess up the basic recordings. This freedom can, if used properly, allow a great deal of experimentation and a spirit of adventure, which just begs us to try things and see what happens. Equally, though, the technology that we have can also lead us down a very different path and deeper into the "fix it later" approach that I mentioned earlier. Just because we *can* fix a lot of things these days doesn't necessarily mean that we *should*, and we should always strive for the best result at each stage of the process. If you are both recording engineer and editor, then you should always try to capture the very best recording that you can (both from a technical and an artistic point of view) and then build upon that quality with good editing before moving on to the next stage. If, however, you are handed less-than-perfect recordings to work with, then you can be grateful for the tools at your disposal and do the best job that you can before, again, passing it on to the next stage.

CHAPTER 2
Different Aims of Audio Editing

CORRECTIVE EDITING VERSUS CREATIVE EDITING

Each of the editing processes described in this book will fall into one (or more) of three categories: corrective editing, creative editing, or restorative editing. Sometimes the lines between the three will be very blurred, and some techniques are equally applicable to multiple areas. In fact, it is arguable that restorative editing is largely a combination of a corrective and creative process. I agree with this to some extent, but the techniques used in audio restoration are often quite specific, so I have, for the sake of clarity, chosen to separate them in this book.

The biggest difference between the first two categories is in the *intent*. The aim, or intent, of corrective editing is to solve problems that have occurred during the recording stages and to get the best possible version of an audio "part," while the aim of creative editing is to take an audio "part" and turn it into something new. Sometimes that will be a radical change, and other times it will be a subtle one, but it will always be noticeably different from the original.

Commonly used corrective processes include cutting, copying, pasting, fades and crossfades, level alterations and "comping" (the process of putting together a "composite" track from a number of alternate takes), while commonly used creative processes include beat-mapping and Recycling as well as time-stretching and the use of elastic audio–type processing to allow changes in the timing and pitch of audio files. More often than not, you will need to use a combination of many different techniques in order to achieve a particular task.

"Elastic audio" is a term first coined in Pro Tools and refers to a group of tools and techniques that allows for the changing of the pitch or tempo aspects of a recording in real-time and in a nondestructive way. While the name may vary from platform to platform, the principles are quite similar, and, in my opinion, elastic audio is a very apt name, so that is the term I will use throughout the book in the absence of an industry standard and easier way of explaining.

Corrective Editing versus Creative Editing

For example, in order to obtain a perfect vocal take, you might use elastic audio to correct a mistimed word in the middle of a chorus and then use it to correct the pitch of a slightly off-key note. Once that is done, you might think that the first line of the chorus from the second chorus was actually better than the corresponding line from the first chorus, so you would cut that line from the second chorus and replace (or paste) it in the first chorus. In order to ensure there are no clicks or pops, you would probably cross-fade (see chapter 4) overlapping audio regions. Once that is done, you might feel that, even though the majority of the chorus comes from the first chorus, there was actually a slightly better version of the last line of the chorus from one of the alternate takes, so you would use that instead. Next you might go in and make some minor level changes to some of the words to make the overall performance more consistent. All of this would give you a "comped" chorus, which you could then bounce/render to a new audio file and, if you wanted to, copy and paste to all of the relevant locations throughout the song.

Once again, though, we find ourselves at risk of doing *too much* editing. We need to decide quite early on in the editing process what the goals are. There will be times when we (or our clients) want the result to sound very "natural," and there will be other times when we want the result to sound "polished." And, of course, there is a lot of space in between those two extremes. While over-editing is possible in both corrective and creative ways, it is far easier to overdo creative editing than it is corrective editing. With corrective editing there is, of course, the temptation to copy and paste large sections of audio (whole choruses, for example) in order to give consistency. And there is also the option to comp takes together in almost ridiculous detail. Whether our quest for audio perfection drives us to those levels is something that is very personal. What is much more certain is that, in all but the most extreme situations, that level of detail simply isn't necessary in trying to create the best version.

Creative editing is something that is much more prone to overuse. It is comparatively easy to use editing to come up with recordings that are mathematically correct in both pitch and in timing, are completely consistent in terms of dynamics, and are perfect in every way. And anybody who has ever succeeded in doing this will tell you that it sounds utterly wrong! We need some "feel" in the things we create. Our brains are able to detect timing differences with sub-millisecond accuracy, and it is this ability that allows us to determine the location of sounds just by hearing them. However, while most people wouldn't *consciously* be able to determine such small periods of time, most people can consciously notice timing differences of 10 to 15 ms or more. Things can sound a tiny bit early or late when the timing is out by pretty small amounts like this. But if you asked any musician or performer to just make that first note 15 ms or so earlier, you would certainly be disappointed.

To illustrate this let's consider drummers playing a simple beat. If they want the drum to be hit at a certain point in time, then they actually need to start moving their arm or leg muscles a short time *before* they want the drum to be hit, owing to the fact that it takes time for the drum stick to travel from the raised position of their arms to the point where it hits the drum. And this isn't something we can readily calculate and act upon while we are playing. It's all about "feel." And that feel is something that we subconsciously get used to as we listen to music.

The advent of MIDI sequencing allowed us to quantize the musical timing, but that led to a lot of people feeling that sequenced music felt too "mechanical" and had no soul. Various features were then added in to MIDI sequencers to give back some of this human feel, but until very recently audio recordings simply didn't have the option to be quantized in this way with any degree of accuracy. Now that the tools exist, it is easy to find ourselves wanting to quantize things for accuracy, but, if we want to retain the feel and emotion of the original performance, this should be done sparingly and should always be done manually rather than simply hitting a Quantize button if natural results are wanted. By all means use the quantize tool (if appropriate) to snap the timings to the timing grid, but then go in and nudge things into the positions where they *feel* right rather than just looking right or being mathematically correct.

Of course it almost goes without saying that the same is true of pitch correction. It can be used in an automatic way (effectively quantizing pitch instead of time), but this should really be avoided if you want to really get the best sounding version of a vocal take. It is more time-consuming to do it manually (for both timing and pitch), but I genuinely believe that the results more than justify the time spent, especially for a lead instrument such as a vocal performance.

So in summary, sometimes (perhaps even often) more work doesn't always mean a subjectively better result. There seems to be no exact rule of what is "too much," and it very much depends on the situation: sometimes more is more but, equally, sometimes less is more. Perhaps the best approach to the question of how much in audio editing was summed up by a good friend of mine, who said that his goal was to do "enough to be noticed but not enough to be heard."

RESTORATIVE EDITING

Being able to restore older (perhaps damaged) recordings and preserve them in a modern format for the enjoyment of future generations is a very useful ability to have. In recent times, advances in software have meant that not only is much more advanced audio restoration possible, but also the techniques can be applied in more creative ways to extract individual sounds from within a complex audio file. While this isn't strictly restoration, the techniques and underlying principles are fundamentally the same as those used in restoration, so they will be looked at alongside them. There are also certain techniques that could be considered as being corrective but that, again, use techniques mainly used in restoration, so those are described here to avoid confusion and repetition.

So, first and foremost, why would we want to perform audio restoration? Since the advent of digital recording (and especially if we consider recording files at qualities higher than CD quality) there really haven't been any great problems with the quality of recordings deteriorating over time. As long as a digital audio file is readable and uncorrupted, then the quality will be exactly the same as the day it was recorded. Therefore, as long as suitable steps are taken to back up the files, there should be no issue with archiving digital recording far into the future. Perhaps in the future audio will be recorded at even higher quality than we have now, but at least the recordings made now won't have deteriorated over time.

Analog recordings, sadly, do not have quite the same life-span. Tape reels can, even if stored correctly, chemically destabilize and delaminate, making the recordings on them utterly irretrievable. Vinyl records perhaps fare better in this regard but are far from perfect. And both of these formats will suffer from mechanical damage each time they are played. And, naturally, the older the recordings are, the worse the problems can be, not only due to the length of time passed since they were recorded and any physical wear and tear but also because the quality of the media itself wasn't as good at the time the recordings were made.

Given that the earliest examples of audio recording date back to the mid-1800s, and the invention of the gramophone (which greatly expanded the market for recordings) occurred in the late 1880s, there still exist a *huge* amount of audio recordings that are decaying day by day. It is easy enough to preserve those recordings in the state they are currently in by simply playing them back (assuming suitable equipment can be found, of course) and rerecording the sound into a modern digital system. That is a great thing to be able to do, to preserve those recordings for the future, but many would love to hear them the way they originally were. After all, audio recordings capture a particular performance—a particular expression of a feeling at a defined moment in time.

As such, and especially in the case of "live" recordings, those moments can never be re-created. In the case of a classical piece, it would be easy enough to simply rerecord the piece with a new orchestra and with modern recording technology, but it wouldn't be the same as the original performance. That's not to say that it would necessarily be worse, or better, but it *would* be different. And sometimes there is a certain magic in a recording that we want to preserve—an energy that is unique. In situations like these, having the ability to take an already damaged recording and work on it to try to recapture some of the quality and fidelity of the original version is quite special.

One of the main problems with old recordings is the fluctuation of speed of the recording medium. As we mentioned briefly in the previous chapter, this is referred to by the two terms "wow" and "flutter." They both represent variations in the recording and/or playback speed of the medium. Both were actually originally conceived to describe phenomena associated with tape, but there can be similar problems—albeit with different underlying reasons—in other analog media. In order to figure out how to fix the problems, we first need to identify what each of them means and what makes it happen. Wow is a slow and cyclic variation in pitch that can be heard as a distinct up and down in pitch and tempo, while flutter is a much faster change and can sound less like a pitch or tempo variation and more like a modulation of amplitude and a subtle "choppiness" to the sound. On a tape machine, both of these are caused by the various wheels and rollers in the tape transport mechanism. With a vinyl record, this mechanism doesn't exist, but there can still be problems with wow if the center locating hole on the record isn't precisely centered.

Now that we know what the problem is we can figure ways of trying to resolve it in a digital system. Given that the pitch and tempo are directly related to the playback speed, we should be able to address this by varying the speed of the playback of the digital file. While this seems easy enough, the fundamental problem in doing this is in establishing by how much you need to vary the playback speed. There have been attempts in the past, some quite successful, to address this by using the tape bias signal that is recorded along with any audio signals. Because magnetic tape has a nonlinear response and, as a result, low-level signal fidelity can suffer, tape bias signals are added. The most common form of tape bias is AC tape bias, where an inaudible (anywhere from 40 kHz to 150 kHz) signal is added to any audio signal recorded. The effect of this is to ensure that, even when the audio signal is of a low level, the actual signal level being recorded to tape is still high enough to minimize or avoid the effects of these nonlinearities. Now because this bias signal is of a constant frequency any deviations (due to wow and flutter) can be easily measured and quantified. This variation can then be used as a control signal to modulate the playback system. However, older tapes may have degraded, meaning that the bias signal might not be clear. In addition, it is possible that the recordings were made before tape bias was common, meaning that this signal isn't even present. Clearly, though, this is applicable only to tape systems and assumes that the bias signal is clearly present. On other mediums, such as vinyl, we don't have a bias system. In addition, when an analog signal is recorded into a digital system, there is always a frequency above which no signals are recorded. This frequency, known as the Nyquist frequency, is half of the sampling frequency, so on a CD-quality (44.1 kHz) digital recording, no frequencies above 22.05 kHz will be recorded, meaning that any tape bias system would be eradicated. If, on the other hand, a 96 kHz sampling frequency was used, and if the tape bias signal as at the lower end of those used (around 40 kHz) then it might well be recorded along with the audio. So in cases where it isn't recorded, or when it wasn't in the first place, we need to find an alternative.

More recent technological advances in pitch tracking have led to the ability to correct these problems on any recording, irrespective of whether a bias signal is present or not. These new techniques work on the same principle of tracking variations in the playback speed of the original recording and then using that to generate a correction "map." However, owing to the fact that it isn't a clear, consistent pitch that is being tracked as a reference signal, the task is made much more difficult. The only software currently on the market that makes this a reality is Celemony's Capstan. The software builds upon the pitch-tracking algorithms they developed for their Melodyne pitch-correcting software and takes this extracted pitch information and then calculates a correction "map" for the audio being processed. The actual variations proposed by the software are shown as a

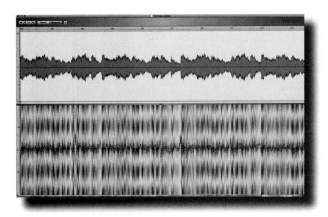

FIGURE 2.1
Celemony's Capstan uses advanced pitch-tracking technology to detect wow and flutter in recordings and allow you to correct it with ease.

line that is overlaid on top of the waveform overview. Any manual changes that you wish to make to the suggested map can be drawn in by hand, and the results can sound very realistic and natural indeed. While this technology isn't strictly editing *inside* your DAW it does represent the current state-of-the-art in audio restoration of this kind and would still most likely be carried out on the same PC or Mac that you run your DAW on, so it wouldn't be a huge interruption to your work-flow to use it.

THE ROLE OF THE EDITOR

Finally, before we delve into a detailed look at each of the processes involved in editing, let's take a moment to consider the role of the editor. By our own definition, audio editing most often involves working on improving the quality of particular individual sounds and sometimes working individual parts of a more complex recording or improving a recording as a whole. We have also stated that there are corrective, creative, and restorative forms of audio editing, but consistent through all of this is the fact that audio editing is largely a *technical* task rather than an *artistic* one. Editing tasks ordinarily have a particular technical goal attached to them: remove background noise, compile the best take, make this sound in tune, change the timing of that particular passage, get rid of that unwanted sound, and so on. While a client may simply ask for things to be made to sound "better," a good audio editor will work through a number of distinct processes in order to meet a certain technical goal.

Mixing and production, on the other hand, are much more conceptual in their targets. The end goal is usually a feeling or a mood rather than an absolute technical achievement. As such, audio editing can require a slightly different mind-set. If, like many people, you are required to be an audio editor as just one of many roles that you play, it is definitely advisable to try to separate all but the most minor of editing jobs out and either complete them before any serious creative work gets under way, or at the very least try to set aside a particular time to do them and not simply do them as you go along. Editing jobs can take a good deal of concentration if they are quite involved, and that can put a serious dent in your creativity if you are in the middle of writing, arranging, mixing, or producing.

Equally, though, you might be called upon to work on a job that involves only editing. In that case you can probably have more freedom to really get a feel for what is involved and what might be the best way to go about it rather than having to try to think on your feet and figure out the quickest way to get what you need. In my experience the "quick" route will normally get you to 80% or

90% (at the absolute most) of the quality you could potentially achieve. Depending on the context of what you are doing, the extra time may simply not be justifiable, but, on any sound or performance that will be an easily identifiable feature (as in, noticeable by somebody who isn't listening specifically for it), it is always advisable to make the extra effort. Whether time is on your side or not, though, the first step is always to identify what you have and what you want. Without a starting point and a destination, it is impossible to plan a route.

Once you have figured out what needs to be done in order to get you from where you are to where you want to be, you need to then break the process down into steps and figure out what order to do them in. Each of you will find a system that works best for you, but, as a general rule, it is advisable to leave the most potentially damaging processes until the very end. In general that would be things like time-stretching and pitch correction as well as anything involving spectral editing. This is purely so that you have the cleanest possible file to work on when you do the more complex processes. For that reason I have arranged the chapters in this book to reflect what I believe to be a naturally running order for the editing steps. Of course there will always be exceptions and exemptions, but, as a general guide, I feel that this order is a good place to start.

As I have stated early, and will undoubtedly refer to again, the main purpose of editing is to clean up, optimize, fine-tune, and polish the audio files you are working on to be the best version of themselves that they can be according to whatever criteria have been set. Therefore, in addition to all the technical skills required, you also need to have a great ear for detail, the patience to listen in minute detail, the lateral thinking to find solutions to seemingly impossible requests, the flexibility to deal with changes to what is asked of you, and the understanding that, even if you give your clients what they ask for, may not be what they actually want, because they didn't explain themselves correctly or just didn't realize exactly what was needed. The role of the editor can be a challenging one, but, if you are the kind of person who gets a great sense of achievement from doing the best that you can in whatever you do, it can be a role that offers a great deal of opportunity to do things that can really make the difference between "good" and "spectacular."

SECTION 1
Corrective Editing

CHAPTER 3
Cutting, Copying, Pasting, and Moving

WHY START HERE?

The main reason why I have decided to start this section of the book off with a look at cutting, copying, pasting, and moving is because these are the most fundamental (and therefore widely used) editing techniques in use today. This makes a lot of sense, considering the history of audio editing that we looked at in the very first chapter. But there is more to it than that, because, without getting your basic edits right in the first place, all of the other more-elaborate techniques described later will be harder to work on. Finally, cutting, copying, pasting, and moving are probably the easiest of the editing techniques to visualize in a physical sense, because they have very clear and obvious "real world" equivalents. For example, if you imagine a long strip of paper, it is very easy to visualize cutting it into sections and rearranging the order of those sections. It isn't so easy to come up with a real-world analogy for cross-fading using a similar physical and tangible idea.

It will also probably become apparent very early on that most of the emphasis is placed on audio editing in a musical context throughout this book, and many of the examples and reference will relate to situations that you might come across in dealing with editing music and vocal performances for songs. However, all of the basic techniques are equally applicable to dialogue for film and TV, location recordings, and even more diverse uses such as Foley and sound effects. My reasons for doing this are simply that editing audio for music generally (and I do mean generally, not exclusively) presents a wider set of challenges than any other audio application and is probably relevant to a greater number of people than more specialized applications. If you work in audio editing in a field other than music recording and production, please do not take offense. I don't wish to imply that what you do is any easier than editing audio for music!

So, without any further delays, let's get into things properly.

THE BASICS

The list of situations where you might use cutting, copying, pasting, and moving techniques is almost limitless. Everything from (manual) timing correction to repetition of certain sections, from creating new rhythms from existing parts to removing headphone spill or unwanted ambient noise in quiet sections—all of these and many more can make use of these most basic of techniques. In the days of multi-track tape recording, edits like these were, at least to some extent, available to the dedicated and skilled studio engineer, but things that we can seemingly do with a flick of the wrist and a few mouse clicks would have taken an almost unimaginable amount of time longer and would have been far more costly and risky.

Many of us will have decided to reuse one part of a recording in a later part of the song, the main example of this being the use of the same chorus throughout a song (perhaps with some variations or ad libs in later choruses). To a modern day DAW user, this is simply the case of selecting all the relevant vocal parts and selecting them, choosing the copy option, and then choosing the paste option to place them at the relevant place. But to multi-track tape users, it would have involved copying the required part onto another reel of tape and then finding the exact point that they wanted to make the edit by slowly running the tape spools back and forth, until they heard some kind of clear "locator" (for example, the start of a kick drum) and then physically cutting the tape with a razor blade at that point. They would then have to do the same at the end of the section they wished to copy. Then they would have to go back to the original reel and find the same two locators again—but this time in the chorus that they wished to replace—and make the physical cuts again. Once this was done, they could take the section of tape that they cut from the copy reel and physically attach it to the original reel by using sticky tape to hold the sections in place. This was a very difficult thing to do physically, because there was always a chance that, if the two sections didn't match up exactly, there could be an audible change in the sound where the tape itself wasn't perfectly aligned or an audible "thump" where there was a tiny gap or overlap in the tape at the edit point.

Potentially more problematic, however, was the fact that the edit had to be made to the whole reel of tape, which meant that all the instruments and recorded parts on the different tracks would have to be edited at exactly the same point. This could have been massively limiting, because there was always a chance that the first chorus had a wonderful guitar take on it but the vocal was less than perfect. The second chorus, on the other hand, could have had a perfect vocal but a not-so-good guitar. None of this is a problem for modern software, but, on a multi-track tape, there were only two solutions: rerecord the parts in question, or, if the studio happened to have another multi-track tape machine, perform some quite involved and elaborate redubbing of certain tracks onto a different machine before doing the edits and then dubbing them back once the edits had been completed, all the while hoping that the time code synchronization was good enough that things would end up in exactly the right place after all the messing around.

The relative complexity and time-consuming nature of doing these edits meant that experimenting with alternative arrangements or alternative takes was no simple matter, so there would have been a lot of thought given to things like arrangement prior to even commencing recording. With DAW software even complex changes to arrangements are relatively simple (especially if the track has been recorded to a click track), so there is less of a requirement to have things set in stone before the recording starts. Whether this is a good thing is up for debate, and there are very compelling arguments on both sides, but having options and flexibility is never really a bad thing, as long as it isn't used as a reason for procrastination.

START AT THE BEGINNING

Perhaps the most important tip that I can give you is one that relates to, but doesn't directly involve, the editing process.

> *Rule number one is that the editing process actually starts during the recording session.*

Wherever possible, don't let the fact that we have amazing editing tools available to make you lazy in terms of the recording itself. Sure, there may be times when you are asked to do edits on material that was recorded already, but if you are actually involved in both the recording and editing processes, then *always* strive to obtain the very best recording that you can to start with. The reason that I say this is simply because taking time at the record stage can save you time at the editing stage. This is simply because a better recording means less editing, and it could be a very quick fix to get a better recording. What will take you two minutes now could save you literally *hours* of work later on.

In terms of the actual editing itself, I would say that the most important rule is to try at all costs to avoid an attitude of "that will do." There will be times when deadlines are fast approaching, and you *have to* compromise somewhere, but, in all other situations, you should always strive to make it as good as it can be rather than just "good enough." Another potentially useful tip is a very simple one: try to avoid focusing too much on the quantization grid in your DAW. Depending on how you have your preferences set up in your DAW of choice, there is a good chance that you will have your edit points snap to a certain timing subdivision whether that be bar, beat, sixteenth note, Society of Motion Picture and Television Engineers (SMPTE) frame, or some other consistent timing reference. When you are performing audio edits, I really advise you to either turn that Snap To option off completely or, at the very least, commit to memory any key combination or modifier key (a key that is pressed and held while you are performing another action) that will temporarily override the Snap To function. There are times when using the grid is very useful, but otherwise you should use the grid only as a tool to help you get things roughly in the right, as taking the easy route and snapping to edit points will not do anything to help preserve the human feel in the recordings.

TIGHTNESS OF EDITS

The next thing I want to discuss is the "tightness" of the edits that you make when cutting and pasting. What I mean by that is, how close to the actual sound itself do you make the edit? In some situations the edit will be made in the middle of a word or note or drum hit, but in others (and this is usually an easier option) the edit point will be made in a section of silence (or, at the very least, largely reduced level) prior to the start of, or immediately following, the actual place where the edit should ideally be. Just how close this should be is a matter for debate. It is always easier to line up an audio file to where you want it (or to other audio files) if the actual sound starts at the very beginning of the region in the *Arrange* window. Equally, though, you might prefer to leave a little bit of room at the beginning or end of the section that you want to cut or copy, just to make sure that you don't cut off anything from the sound that you didn't mean to.

You should never underestimate the listener's ability to hear the tiniest things. There may be a particular edit that you have been working on, which, when listened to in isolation, doesn't sound 100% perfect but when listened to in the whole track falls under the banner of "I will probably get away with it." These kinds of things can be incredibly tempting to just pass over, but you really shouldn't, because sometimes, without actually listening for bad edits, I have heard things in tracks I have been working on that I didn't even realize weren't right. It may have been just a tiny, little sound: perhaps a "glitch" or perhaps the tail end of a sound cut ever-so-slightly short—things that, when I actually looked into it, were in fact editing mistakes and things that I never would have realized would have been as audible as they were.

"ZERO-CROSSING" EDITS

Now we come to one of the most (perhaps *the* most) important things to consider when you are making edits of any kind, and even more so if you are making edits in the middle of a word/sound/note. Even though making cuts in "silent" parts of the recording is always easier, it isn't always an option. In those situations where you have to make a cut in the middle of a sound, you should certainly aim to make your edit at phase-aligned zero-crossing points, as these will give you the very best chance of making any edits sound smooth. So what exactly is a phase-aligned zero-crossing edit?

Starting with the simpler part to explain, a zero-crossing edit is so named because you are making the edit/cut at the point where the audio waveform is at a level of "zero." In physical terms, if you imagine a speaker cone moving in and out as it plays back a sound, the zero-crossing points on the waveform of that sound would be the points where the speaker is in its "at rest" or "neutral" position. On the actual waveform display this is the point where the line representing the waveform passes through the vertical center of the waveform display (often marked by a faint horizontal line).

The reason that we try to cut the audio file at these points is simply to ensure continuity from the end of one audio region to the start of another. If we were to cut the end of one audio file at the point where the waveform line was at the very top of the waveform display, this would represent the speaker being at one extremity of its travel. If the very start of the following region (assuming that it played immediately following the previous region, with no gaps) were at the very bottom of the waveform display, this would represent the speaker being at the other extremity of its travel. Now, if we were to leave the edit like that, we would most probably hear a very obvious click or "pop" sound, because what we would be asking our speakers to do is jump from one extremity of their travel to the other in the space of one sample (anything from one 44/1000 of a second right up to one 192/1000 of a second, depending on the sampling frequency). Good zero-crossing edits aren't just a matter of finding a point where the waveform crosses the center line, though, and we will look at that next.

The phase-aligned part also makes sense when you look at it in terms of the mechanics of a loudspeaker or a vibrating guitar string (or even a vocal cord). Any oscillation in the speaker or string will be cyclical and (more or less) smooth and flowing. It might not be a simple back-and-forth from one extreme to the other, and there might well be smaller interruptions on the way, but even these will have a smooth transition from traveling in one direction to traveling in the other. The vibrating object will always go through a point of being momentarily static before changing direction. Not only that, but also, as it approaches this moment of being static, it will be decelerating. If we now move into our DAW and look at the waveform display of pretty much any acoustic source, we will see exactly this. In order to perform the best edits, we not only need to try to make the edits at the zero-crossing point but also to preserve the overall direction (upward or downward) of the waveform.

We can actually take this idea even further by not only considering zero-crossing points and waveform direction (phase) but also by looking at the point in the waveform "cycle" that we are at. Most sustaining sounds have two stages: transients and sustaining portions. The transient part of the sound is usually the "energizing" phase of the sound. This could be the point where a drum skin is struck, the sound of a bow scraping across a violin string, the sound of a plectrum plucking a guitar string, or the sound of the initial breath in a flute sound. This part of the sound is fundamental to the identity of the sound and gives it much of its character. In fact, if you were to remove all of the transient parts of a single violin note and a single trumpet note, it would be much harder to identify which was which. In fact, transients are important enough to warrant their own chapter later in the book. But at this point, we are more interested in the sustaining portion of the sound.

The reason this is important is because most instruments have a relatively steady waveform pattern that they settle into after the transient portion is out of the way. This wouldn't necessarily be something as simple as a sine wave or a sawtooth wave, and it could, in fact, be much more complex than that. But if you take a wide-enough view of the shape of the waveform, you will usually see a repeating pattern of some kind. There may well be more than one pattern: one very short cycle, one slightly longer, and one slightly longer still. This can be very useful for us if we are trying to edit together two different parts of a sustaining sound. Let's assume for a moment that you have zoomed right into the waveform display and have made sure that your edit

"Visual" Editing vs. "Audible" Editing

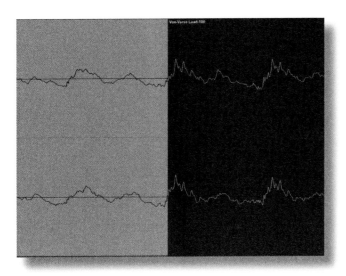

FIGURE 3.1
A good edit will not only be made at a zero-crossing point but will also be phase-aligned, as shown in the example above.

point is right on a zero-crossing point, and that you have also made sure that the phase of the waveform is consistent between the two audio regions. The next thing to do would be to zoom out a little more (in a horizontal/time sense) and now take a look at the overall waveform pattern. If you can see a repeating pattern, then it is always a good idea to try to cut the end of one audio region and the start of the following region at similar points in the waveform cycle to ensure the best possible continuity between the two regions. Of course, if you move either the start or end of a region to adjust to this new edit position, then you should always go back to check that your phase alignment and zero-crossings are still in place.

Of course there will be times when there isn't a clearly defined pattern, though, so it won't always be possible to match up the two regions exactly as you might like, but, as with so many aspects of modern music production, you should always trust your ears more than your eyes. The waveform display in your DAW is incredibly useful for things like zero-crossing edits and waveform direction or phase coherence, but you should never make a decision based solely on how the display looks on the screen. Your ears should always have the last word when it comes to critical decision making in *all* aspects of audio editing.

"VISUAL" EDITING VERSUS "AUDIBLE" EDITING

When I am actually mixing or arranging a track, I will often look in the other direction or simply close my eyes when I am trying to make a judgment call about something. To me there is something uniquely deceiving about being able to "see" a piece of music on-screen while you are working on it. Of course music has for a very long time been represented visually as notes written on a stave, but sheet music of this sort was never able to convey anywhere near as much information about the piece as a whole like a DAW can. This can cause problems when you are trying to work on the arrangement of a track, because you can get a sense of what is coming next by seeing a visual representation of the arrangement on the screen. The same is true of adjusting plug-in settings. For example, if you are working with an EQ plug-in that offers some kind of EQ curve display, it can actually guide your decisions in a way that can be counterintuitive. Even something as simple as having a numerical readout of frequencies for an EQ cut or boost can lead you to making decisions based on numbers that seem comfortable or familiar rather than what actually sounds best.

Cutting, Copying, Pasting, and Moving CHAPTER 3

This effect carries over into the audio editing world where you can, if you aren't careful, start making decisions based on what things look like rather than how they sound. As I have already mentioned, the visual side of DAWs is great for the extremely detailed aspects of editing, such as zero-crossing points, but when it comes to positioning audio regions, it is actually better to use your ears to place them. The amount of difference that this can make will depend on the type of sound you are dealing with. Sounds with sharp and clear transients, such as drums and percussion sounds and acoustic guitars, can be quite easy to place using a more visual method, as they will have a clear timing reference at the beginning of the sound, and their position will probably end up being fairly close to this theoretical grid. On the other hand, sounds such as vocals, brass, or woodwind instruments have a much slower (relatively speaking) attack, so the actual point of the waveform that you would ideally like to use as a timing reference isn't actually the very beginning of the waveform that you can see on the screen. This obviously makes it harder, if not impossible, to place these types of sounds using a purely visual method.

The other thing to consider is that aligning sounds entirely visually (when it is actually possible) might give you a great deal of accuracy in the placement of an individual sound or event, but what it will be almost impossible to do is to create something that sounds like it has a natural "human feel" to it. Even if you are working with largely sequenced and electronic material, there is still a chance that you might want something with a bit more feel to it, so it is important to realize that some placement of sounds is best done by ear in order to get things sitting exactly how you want them. By all means use the visuals to get things roughly in the right place when you are cutting/copying and pasting/moving, and use all of the little techniques that we have mentioned so far if applicable, but do be prepared to do some moving around and adjusting of the exact placement of things if you want to get the best result.

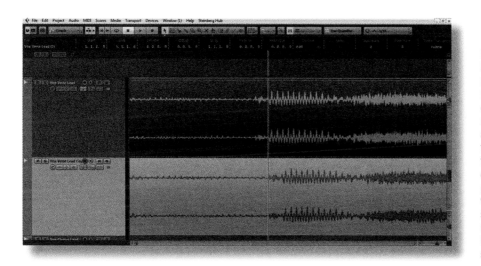

FIGURE 3.2
Sometimes you may need to move a region slightly in order for it to sound right. In the image above, the bottom track shows the transient at the start of the sound lined up perfectly to the beat, while the top track shows where the region was moved to in order for it to sound correct.

Of course this isn't an exact science, and it works better with some sounds than with others, but it's definitely worth considering if you are having trouble getting the timing right on a particular track or instrument. This is one way in which the visuals of DAWs can actually work alongside using your ears and assist the process without completely automating it.

MULTIPLE-TRACK EDITING

So far we have mainly been considering editing on single sounds or tracks, but there are a number of situations when you might have to make edits across a number of tracks at once. The first and most obvious example is a drum kit recorded with multiple close microphones. Each microphone will be aimed at a particular drum, but, even with the most directional of microphones, there is a good chance of spill from the other drums into a particular microphone. Noise gates (very useful in this context, where manual editing could be extremely difficult) can go some way toward cleaning up the sound, but they will remove spill that happens only when the drum that is supposed to be there isn't playing. It is very unlikely that we will get a completely clean signal of one drum into one mic in all but the simplest passages. So what does this mean in practice? Well, the main consideration is that we can simply move individual drum hits around in isolation on an individual track, because there is a chance that the spill into (for example) the snare drum mic at one point in the track could, if moved to another place, cause phase-cancellation issues. Equally, there could be the tail end of a low tom note in the background of a snare drum hit. In the original location this isn't a problem, because that low tom will be in the same place on the low tom mic. If, however, we move that particular snare hit to another position, there could be an audible "ghost" of the low tom note.

In situations like these, the easiest option is to make any cuts across the entire group of tracks. That doesn't mean that all the edits have to take place at exactly the same time. They should still take into account the phase-aligned zero-crossing principle that we have just spoken about, but they should all be as close as possible in time to one another. Once this is done, the copying/pasting/moving can be done with the certainty that, for that particular group of regions, there are no phasing or other issues. That's not to say that the beginning and end of the regions will match up exactly with the surrounding regions at the new location, though, so there could be further complications to deal with.

Clearly any kind of "multi-mic" recording will need to have special attention paid when you are moving things around. In the case of a drum kit, the closer the mics are to the drums they are aimed at and the more directional the mics are, the better, as it will minimize any spill and therefore minimize any of the associated problems, but, to be perfectly honest, I think that it could go without saying that any kind of editing should be done (or at least checked) in context. There is a whole chapter later on multi-track comping that looks at this in more detail and offers a few potential solutions as well, but it is worth mentioning here. There are a few other situations when I would actually consider cutting/copying multiple tracks at once, even if I hadn't previously had an issue with one or more of the tracks in question. Let me give you an example to illustrate what I am talking about here.

Let's take a typical live-band recording situation where we have the above-mentioned drum kit recorded with multiple microphones, a bass guitar, a rhythm electric guitar part, a lead electric guitar part, perhaps an electric piano part, and a lead vocal and two backing vocals. Now let's assume that we have carefully edited the drum parts, making sure that we don't have all kinds of phasing/ghost note issues, and we are now looking at the other parts that are providing the groove and feel of the track. In the second verse, we might be totally happy with the bass guitar part, but the rhythm guitar part might not feel right. It might not be an obvious timing issue; it could be something else, perhaps a dynamics issue, or just that the first verse part just felt like a better performance. So we copy over the rhythm guitar part from the first verse and place it in the second verse (for the purposes of this example, I will assume that everything was recorded to a click track, so we don't have any issues with different tempo—more on this in a moment), and, in isolation, it sounds better. But what of the bass guitar part? The performance on the bass guitar track was fine on the second verse, but now, when we copy over the guitar from the first verse, things don't seem to sit that well. The reason for this is that, with "live" instruments, the players will react to subtle changes in the groove. This is especially true if the whole band is recorded at once, because there seems to be a natural connection between the band members, and they can "read" what the others are doing, but it is also true to a lesser extent even if the parts are recorded one by one. And in our example the subtleties of the groove between bass guitar and rhythm guitar were right in the first verse, and they were right again in the second verse, even though the rhythm guitar part wasn't a great performance. But if you try to mix parts from the two different verses, things might not work together so well, so at that point you need to make a judgment call about the best way to approach things.

In an ideal world, it might be better to rerecord the parts or make sure that they were right in the first place, but you may find yourself in a situation where that isn't possible, or simply where you have been asked to do some editing and tidying up after the recording session, and a few things seem to have slipped through the net before the project came to you. Therefore, the more techniques you have at your disposal and the more options that you give yourself, the more likely you are to be able to give your clients whatever it is they are looking for, even if that isn't what you would ideally like to do or how you would ideally like to do it.

POTENTIAL PROBLEMS

In the example above, I stated that we would assume that the song was recorded to a click track in order to prevent problems with different tempos, so that's what I want to take a quick look at next. It's probably true to say that a large majority of all songs recorded today, and even those of live bands, are recorded to a click track. There are pros and cons for this, and it is a subject that is often argued a great deal from an artistic point of view, but, from the perspective of an audio editor, click-track-based recordings are clearly preferable because of the increased ease of working on them. Recording a live drummer who is playing to a click track doesn't mean that the drum parts will be robotic or mechanical—not at all. There will still be all the feel that the drummer had before, and there *will* still be tempo variations, but (if the drummer is experienced at playing to a click track) these variations should be less drastic and should also move around a "midpoint" of the click-track tempo.

What this will mean in real terms to the audio editing process is that, with a few little pushes and pulls here and there, you probably *could* copy one section of a track (for example, a chorus) for any particular instrument and paste it into a different but corresponding section. It is probably pretty rare that you wouldn't have to do any polishing or tidying up at all if you did this, but, again, it is a useful option to have if you need it.

But what if the track wasn't recorded to a click track? What if the tempo fluctuations are significant enough that you can't copy over a region from one part of the song to another? Is there nothing that you can do to fix it? Of course there is! There might be a greater or lesser amount that you can reasonably do, depending on the source material that you are working with. Single instruments on a single track are generally much more workable than "stems" or "submixes," but even then there are some things that can be done. The bad news is that a lot of it will be quite intensive manual work. There are some plug-ins and software applications that claim to be able to automatically align different takes to one another (such as Synchro Arts Vocalign), which might work in a context like this, but it is worth remembering that the more complex the source material, the less predictable the results. The only real way to get it done with certainty is to roll up your sleeves and get to work with the editing tools, cutting up the part you have copied into individual notes, words, or phrases, and then moving them into the right place by hand.

The other main problem that can occur with this copying and pasting approach is a difference in tone. This can happen with instrument recordings but is most often an issue with vocal performances. It is entirely possible that the vocal for a track was recorded over a number of hours or even a number of different days. During that time the singer's voice may have changed slightly or even not-so-slightly. Two different takes recorded a couple of days apart could have significant differences in tone. It could be something as simple as the singer having the start of a cold or a strained voice from too much singing, but these differences can cause problems when comping takes together (more in chapter 6) and when copying and pasting parts of the recording. Sometimes you might be able to get away with it if, for example, the vocals for the verses were recorded on one day and the choruses on another, because there would often be a natural change in dynamics and power between those two sections anyway, and this fact could be used to cover up the tonal change in the vocals. But if you are trying to copy and paste between different parts of the same verse or chorus with two takes that vary in tone, then the change might be too noticeable.

Cutting, Copying, Pasting, and Moving

HANDS ON

Introduction

Perhaps the most useful things to discuss for particular chapter relate not to the actual processes of cutting, copying, pasting, and moving as such but more to the process of selecting the areas that you wish to work with. Indeed, being familiar with the different methods of selecting regions or parts of regions is fundamental to most of the topics that follow, so this seems like a very good place to start with our guides to useful tips and tricks for the subjects contained in each chapter. So with that said, let's begin.

Logic

There are two fundamental ways to make selections in Logic: parametrically and visually. The fundamental difference is that the parametric selection methods work on clear criteria that you choose that are of the before/after, same/different type and are restricted to whole regions, while visual selections can be applied to whole regions or parts of regions and are completely free of restrictions as to where or when they happen. There is quite a lot of overlap between the different methods, but in general it is fair to say that the parametric methods are generally more useful on a larger scale, while the visual methods tend to be better suited to detail work.

One of the most useful parametric methods is *Select Inside Locators*, because it quickly and easily allows you to select all regions (including any MIDI regions that are present) between two defined points (the locators). This can be useful for selecting everything in a certain section of a song (a chorus, for example), but it should be noted that it will select all regions that are even in part inside the locators. What this means in practice is that, if you use this method, you might well end up with a number of regions that hang over the edges of the locator points being selected. This isn't a problem, of course, as you can always split these regions at the locator points if you needed to, but it is worth being aware of this, as it can lead to unexpected results sometimes. Also worthy of mention as a huge time-saver is the *Select Muted Regions* command. Quite often when editing, you will make a number of edits and temporarily mute regions only to then go on to something else and forget to delete the muted parts. Over time this can lead to a very confusing Arrange window, so this simple command will allow you to highlight all muted regions and either delete them en masse or at least be able to see quickly and easily where they are for further investigation.

Of the visual methods, other than simply clicking on a particular region or regions to select them, undoubtedly the most useful tool available to you is the *Marquee* tool. This tool allows you to make selections that are only a part of a region, without having to manually cut the region into parts first. Marquee selections can be adjusted once they are made by holding down the Shift key while clicking and dragging either edge of the Marquee selection (with the Marquee tool still selected). The only real negative about the Marquee tool is that only one selection can be made at once using this tool. It isn't, for example, possible to make selections in different regions at the same time using this tool.

Once you have made your Marquee selection, you have a number of work-flow-enhancing options available to you. If you then change back to the *Pointer* tool and click within the highlighted part, the region will be split at the boundaries of the Marquee selection. This is usually a far quicker way to create a separate region compared to having to use the *Scissors* tool twice. In addition, if you use the Marquee tool and then choose the *Mute* tool, clicking within the highlighted part both splits the region as already described and mutes the newly split region in one step. Similarly, if you have a Marquee selection and use the *Eraser* tool to click within the selection, the region will be split, and the highlighted sections are deleted. While none of these are things that you can't achieve any other way, they are certainly very valuable work-flow enhancements.

The Marquee tool isn't just limited to splitting regions into smaller parts, though. If you have a Marquee selection active and you start playback, only the Marquee selection will be played. This is a very quick way of auditioning a particular selection without having to split regions and then manually start playback from the correct position. As an extension of this, you could also make a Marquee selection and then choose the *Set Locators* command to quickly isolate a particular part of a region and loop around it.

Finally, you can also use the Marquee selection with track automation as an easy way of creating a parameter change for only the selected area. Normally, if you were drawing automation rather than recording it, you have to click on the automation line to create new nodes (automation points), so if you wanted to just change the automation for a particular selection, you would have to create four nodes (two on either side of the selection) and then adjust the parameter between the two innermost nodes. The Marquee tool makes this almost ridiculously easy. If you have the automation visible for the parameter that you want to change, select the Marquee tool and highlight the relevant section, and then change back to the Pointer tool and click inside the selection. Four nodes will have been automatically created for you, and you can now drag the section of automation between the innermost nodes as before. Once you start using the Marquee tool in the ways we have spoken about here, you will soon find yourself using it more and more, as it really can save a huge amount of time.

If you find yourself using the Marquee tool all the time, then there is an option available to have it appear automatically. If you go to Preferences > General > Editing (Tab), you will see an option for *Pointer Tool in Arrange Provides: Marquee Tool Click Zones*. With this option enabled, any time that you move your cursor over the bottom half of an audio region, the Marquee tool will automatically take effect. When the cursor is moved back to the top half of the region, the normal cursor will return. This can save you having to keep selecting and deselecting the Marquee tool if you use it a great deal.

Cutting, Copying, Pasting, and Moving   CHAPTER 3

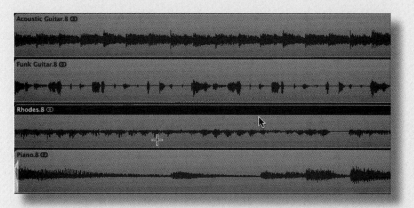

FIGURE 3.3
The Logic preferences can be set up to automatically switch between the Pointer tool and the Marquee tool. Moving the cursor to the top half of a region selects the Pointer tool, and moving to the bottom half of a region selects the Marquee tool.

The final thing to cover here, which relates to the process of making selections, is zooming. There will be some occasions when you are selecting either parametrically or are visually selecting large groups of regions, when an overview of your whole arrangement is preferable. Equally, though, if you are working with the Marquee tool or are making other detailed cuts and edits, you may wish to see only a single region (or even part of a region) on-screen in as much detail as possible. In order to do this, you need to be able to zoom in and out quite quickly and easily. Once again there are many methods, but the two that often prove to be the most useful are among the many Logic keyboard shortcuts.

The first group is used to zoom in and out one level at a time, and you simply hold down Ctrl[Mac]/Start[PC], Alt + one of the arrow keys (left and right arrows zoom out and in horizontally and up and down arrows zoom out and in vertically). I find this method preferable to the many other options, simply because it is very easy to control, as you are zooming in or out only one level each time you press a key. The other advantage of this method is that it is very repeatable and quick to find a particular zoom level that you need. The second option, which is especially useful for detailed editing work, is the *Toggle Zoom to Fit Selection or All Contents* command which, by default, is assigned to the Z key. This very useful command will zoom in both horizontally and vertically to the highest level possible in order to fit the entire contents of your current selection on-screen at once. Obviously, if this is a single region, then only that region is used to set the zoom parameters, but if you have multiple regions selected, and potentially across multiple tracks, then the zoom level will be set so that all the currently selected regions are visible, even if they are spaced very widely apart. This enables a very quick way to zoom in to the region that you are working on, and, very usefully, an additional press of the Z key will return to the zoom level that was chosen prior to the initial press.

Pro Tools

Pro Tools has a number of different editing tools that, in combination with some key commands, allow you to carry out all the basic cutting, copying, pasting, and moving that you need to. The three main tools that Pro Tools has available for these tasks are the *Trimmer* tool, the *Selector* tool, and the *Grabber* tool. Of these both the Trimmer and Grabber tool have multiple modes, which we will look at shortly. In addition there is a fourth variation, the *Smart* tool, which works as a combination of all these tools, depending on the physical position of the cursor within the region.

These tools are located at the top of the screen immediately to the left of the main song position display, and each of them can be selected by either clicking on them individually (or by clicking on the small bar above the number three to activate the Smart tool), by pressing **Esc** to cycle through the various tools, or, finally, by a direct key command: F6 or **Cmd**[Mac]/**Ctrl**[PC] + 2 for the Trimmer tool; F7 or **Cmd**[Mac]/**Ctrl**[PC] + 3 for the Selector tool; F8 or **Cmd**[Mac]/**Ctrl**[PC] + 4 for the Grabber tool, and F6 + F7, F7 + F8 or **Cmd**[Mac]/ **Ctrl**[PC] + 7 for the Smart tool.

The Trimmer tool also has three different modes, and these can be seen by clicking and holding the Trimmer tool button. The three modes are *Standard*, *TCE*, and *Loop*. Of these, the mode of most interest in the context of what we are looking at is the default Standard mode. We will look at the TCE mode in a later chapter, and the Loop mode is more geared toward arrangement issues. With this tool selected, you can truncate the start and end points of a region by placing the cursor—which will be either in the shape of an open bracket ([ ) or a closed bracket ( ]), depending on which end of the region you are working on—at the region boundary and then clicking and dragging to the desired length. The tool will automatically change from a open bracket to a closed bracket as you move the cursor past the middle point of the region. Usefully, you don't actually need to click and drag from the very beginning or end of a region. If you click anywhere in the first half of the region, the start will jump to the point where you clicked, and you can continue to adjust from there; equally, if you click in the second half, then the end will jump to the point where you clicked.

The Grabber tool also has three distinct modes: *Time*, *Separation*, and *Object*. Of these, Time and Object are superficially similar in function, in that they allow you to select entire regions and move those entire regions either along the timeline or onto other tracks. The difference between them lies in the fact that the Time mode will select a certain time period on a given track. If you select a single region, then that region will be the time period selected, but if you then hold down **Shift** while clicking on a later region on the same track, all the regions between the start of the first region and the end of the last region will be selected. Object mode, however, allows you to select regions on an ad hoc basis, and they can be on multiple tracks and at any point along the timeline, but only the regions explicitly selected shall be selected. To select multiple regions, simply hold down **Shift** while clicking on each region. The Separation mode works a little differently, in that, used on its own, it functions exactly the same way as the Time mode, but, if used following the Selector tool, it will split the region to the current selection.

The Selector tool simply allows us to make selections within a region that we can then extract to separate regions and move independently. To use the Selector tool, you simply click and drag over the region(s) that you wish to work on. By default this tool will snap to whatever resolution you have set in your *Grid*. If your Grid is set to sixteenth notes, then the Selector tool will snap the selection to the nearest sixteenth note at both the start and the end. However, if you hold down the Cmd[Mac]/Ctrl[PC] key while using the tool, the snap to grid is overridden, and the selection can be fine-tuned. Equally, if you hold down Ctrl[Mac]/Start[PC] while using the Selector tool, the selection will be snapped to the nearest bar, regardless of your Grid setting. Also of use is the fact that double-clicking within a region while using the Selector tool will select the whole region, and triple-clicking will select all regions on that track (this can also be done by clicking anywhere on the required track and pressing Cmd[Mac]/Ctrl[PC] + A).

Once you have a selection, you can either switch to the *Separation Grabber* tool to split the region and move or copy (hold down Alt while clicking and dragging to make a copy) the newly created region, or you can hit the Backspace key to delete just that selection within the region. Or, alternatively, if you just wish to split a region and create a new region from your selection, you can use Cmd[Mac]/Ctrl[PC] + E (or Edit > Separate Clip > At Selection), which will split the region at the selection boundaries. Another possibility is to use the Selector tool to highlight a selection and then use Cmd[Mac]/Ctrl[PC] + T (or Edit > Trim Clip > To Selection) to trim the region to the selection. This can be a much quicker way of working the using the Trimmer tool on both ends of the region and can also help to reduce the number of tool changes.

It is also very easy to split a region at the cursor point. Clicking in a region and pressing Cmd[Mac]/Ctrl[PC] + E will split the region at the cursor position. In addition, clicking at a particular point on the timeline at the top of the *Edit* window and then pressing Cmd[Mac]/Ctrl[PC] + E will split all of the regions whether they are selected or not in the Edit window. If you click and drag to create a selection in the timeline, then pressing Cmd[Mac]/Ctrl[PC] + E will cut all the regions at both boundaries, while pressing Cmd[Mac]/Ctrl[PC] + T will trim all the relevant regions to these boundaries.

Finally, regarding the tools, we should take a quick look at the Smart tool. As I stated earlier, this is a combination of a number of different tools, and its function is determined by the cursor position within the region. If the cursor is in the top half of the audio region, then the Selector tool (and associated cursor) is active, while in the bottom half of the region, the Grabber tool is active. However, the Grabber tool will use whatever mode was selected for the tool individually. At the left- and right-hand edges of the region, the Trimmer tool will be selected and the four corners enable to tool to create fades (more in the next chapter). As a result, the Smart tool can be the best choice if you have a lot of these types of edits to do, as it can save quite a lot of time changing between different tools.

Hands On

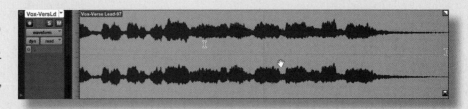

FIGURE 3.4
The Smart tool in Pro Tools adapts to the position of the cursor. When placed in the top half of the region, the Selector tool is active, while in the bottom half the Grabber tool becomes active. At the edges of the region, this changes to the Trimmer tool, and the corners allow you to create fades (top corners) or cross-fades (bottom corners).

In order to get the best out of these tools, it can often be necessary to zoom in to increase the level of visible detail and then back out to get an overview. Fortunately, in addition to the +/− buttons at the bottom right-hand corner of the screen for both the vertical and horizontal scrollbars, Pro Tools gives some very handy key commands that can be used. To zoom in or out in the horizontal sense, you can use **Cmd[Mac]/Ctrl[PC]** and/or, and each press will either double or half the amount of bars/time on-screen at once. If you have a particular region selected, then you can hold down **Alt** while clicking on the *Zoomer* tool button, and the view will be zoomed, so that the current selection fills the width of the screen. Double-clicking on the Zoomer tool button will zoom out, so that the entire length of the session is visible on the screen at once.

When it comes to vertical zooming, there are two options. You can either increase the visible size of the track, so that fewer tracks are visible, or you can zoom in (in a vertical sense) on the waveform within the region itself. To increase the size of the track as a whole, you can press **Ctrl[Mac]/Start[PC]** and the **up and down arrows**. Doing this will halve or double the vertical size of the currently selected track. However, if you press **Ctrl[Mac]/Start[PC] + Alt + up or down arrows**, then it will halve or double the vertical size of all of the tracks. Finally, pressing **Ctrl[Mac]/Start[PC] + Cmd[Mac]/Ctrl[PC] + down arrow** will zoom the currently selected track to fill the whole screen in a vertical sense.

If you have a quietly recorded part, or if you just want to have more detail on the quieter part of a recording, you can instead press **Alt + Cmd[Mac]/Ctrl[PC] + [or]**, and this will increase the vertical resolution of the waveform in the region itself. Note that this doesn't increase the volume of the actual region, even though it could look like it had done so. This is purely a zoom on the actual waveform view.

Studio One

Studio One has a very intuitive system for carrying out basic editing tasks. To select any of the main editing tools, you can either click on them (they are located at the top of the window) or you can select them using the number keys located above the QWERTY keys at the top of your keyboard (not the numeric keypad number keys). If you right-click on the background of the main *Arrange* view, then the contextual pop-up menu will display a list of the different tools (along with other basic commands), and you can select a tool that way. We will take a look at some of those tools, starting with the most versatile (and probably most used) one, but, before we do, it should be noted that the general

Cutting, Copying, Pasting, and Moving   CHAPTER 3

term "region" that I have used throughout the book is (loosely) called an event in Studio One. For the sake of continuity, I have continued the reference to regions, but, in Studio One, at least, I am referring to events.

The *Arrow* tool (press 1 to select) is very much a multifunctional tool, in that its exact use changes depending on whereabouts in a particular region you are using it. It's most broad use is in selecting regions, and clicking on any single region will select that region. If you hold down Shift while clicking, you can select multiple, non contiguous regions, and, finally, if you click on the background of the Arrange view and drag over a number of regions they will all be selected. Moving on from this, we can use the Arrow tool to resize regions. If you place the tool at a region boundary, you will see the cursor change to a vertical line with left and right arrows either side. Clicking and dragging with this cursor will adjust the region length accordingly. If you have multiple regions selected, this will adjust the length of all selected regions simultaneously. There are two other uses of the Arrow tool that we will look at in later chapters, but for now we will focus only on the uses that relate to the topics in this chapter.

The other primary tool used to make selections is the *Range* tool (press 2 to select). While the Arrow tool is great for making large-scale selections, the Range tool is designed for more specific and selective work. Using this tool, you can click and drag to select a range within a region (or regions). This range can be within a single region, over multiple regions on a single track, or it can be the same range over regions in a number of different tracks. Moving the Range tool to the edge of a range will see the cursor change to a vertical bar with left and right arrows, and, using this, you can fine-tune the size of the range. If you hold down the Shift key while using the tool, you can select multiple ranges on the same track or across different tracks. If you have multiple ranges selected, the resize part of the tool will change the size of only the range you are working on, and all others will remain unaffected.

When you move the Range tool over an already selected range, it will change temporarily to the Arrow tool. With this tool you can click and drag to move all the ranges along the track or to other tracks easily. If you do this, it will split the region at the range boundaries to create new regions from the ranges. However, if you use the Arrow tool over a range and click and drag downward

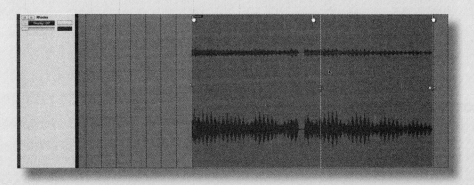

FIGURE 3.5
Studio One has perhaps the most adaptive cursor, which can used to select regions or ranges, to adjust the Volume Handles and Fade Handles, and to change the length of a region or to time-stretch a region, all by either moving to a different position within the region or by holding down modifier keys.

only slightly, it will not move the ranges but will split and create new regions from them. This can be a very quick alternative to using Edit > Split Range or Cmd[Mac]/Ctrl[PC] + Alt + X to do the same thing.

Usefully, you can use the **Cmd[Mac]/Ctrl[PC]** key to temporarily switch between the Arrow tool and the Range tool. Whichever tool you currently have selected, holding down the **Cmd[Mac]/Ctrl[PC]** key will switch to the other tool for as long as you hold down the key. This also works with the Shift modifier key to either select multiple regions (Arrow tool) or ranges (Range tool).

The *Split* tool (press 3 to select) is quite self-explanatory in that it is used to split a region to create two separate regions. This tool will work only on the region that is clicked upon with the tool, unless multiple regions are selected, in which case it will function across all selected regions. You can use the tool to select multiple regions as well by clicking on the Arrange View background and then dragging over a number of regions, but, obviously, selecting regions this way will always select regions on adjacent tracks. If you need to select regions on nonadjacent tracks, then you will need to select the regions first using the Arrow tool and then switch back to the Split tool to make the edit. A shortcut to doing this is to hold down the Cmd[Mac]/Ctrl[PC] key, which will, once again, temporarily activate the Arrow tool to enable you to make your selections, and then, on releasing the Cmd[Mac]/Ctrl[PC] key, the Split tool will take effect again. An alternative way to split regions is to click on the timeline at the position you wish to split the regions, and press Alt + X. If a region(s) is selected when you do this, then only the selected region(s) will be split. However, if no regions are selected (you can click on the Arrange View background to do this or press Cmd[Mac]/Ctrl[PC] + D), then all of the regions at the timeline position will be split.

The Eraser tool (press 4 to select) is also very obvious in its usage. When selected, any region that is clicked on will be deleted, and, if multiple regions are selected, they will all be deleted by clicking on any of them. There are a couple of caveats, though. If a region is selected and the Eraser tool is used, then only the region that is clicked on will be erased, and the selected region won't be affected. Also, if you have ranges selected, then clicking on one will erase the entire region and not just the range. If you wish to delete the range, then you will have to split it into a new region first (using the methods described above) and then the Eraser tool will be able to delete it. You can also press the Backspace key to erase regions and, interestingly, ranges. While the Eraser tool will not erase a range until it has been split to a new region, pressing the Backspace key with ranges selected will split those ranges into new regions and erase them in one step.

The Mute tool (press 6 to select) is another of the obvious tools, and it works in a very similar way to the Eraser tool described above. It also has similar caveats in how it deals with ranges. If you wish to mute a range, you have to split to a new region before the Mute tool will work. And, once again, the keyboard shortcut Shift + M (not to be confused with M on its own, which mutes the channel of the currently selected region) is actually more effective when working with ranges, as pressing it will split the range into a new region and mute it in one step.

With any cutting, pasting, or moving edits, it is often very important to be precise in location, so you will find yourself zooming in and out quite a lot. The easiest way to do this is with the keyboard, and the shortcuts for zooming horizontally are E (zoom in) and W (zoom out), while the shortcuts for zooming vertically are Shift + E (zoom in) and Shift + W (zoom out). Potentially much more useful, though, is the Edit view, which is essentially an enlarged view of a region, which is enabled by double-clicking on a region with the Arrow tool. Doing so opens up the Edit View window at the bottom of the screen. At the top of this window, you will notice all the editing tools and other controls from the main Arrange view.

If you click on the region inside this window, you can then use the horizontal zoom controls to zoom in or out independently of the Arrange View zoom settings. Any edits you make here will be immediately mirrored in the main Arrange view. It really is easiest to think of this as a separate zoomed view that you can use for detailed editing work and to use the main window for the more broad changes. Once this Edit View Window is open, clicking on a different region in the Arrange view will change the region in the Edit view to match.

Moving on from the edit tools, there is one other very interesting feature worthy of mention. While many DAWs provide the ability to split a region into a number of equal-length parts (such as eighth notes or sixteenth notes), Studio One has a trick up its sleeve in allowing you to split a region into a number of different regions based on the quantization grid selected. Setting the quantization settings to eighth notes or sixteenth notes will yield the familiar results, but if you set the quantize value to a note value with swing and then use Event > Split at Grid, you will see that the newly split regions are not equal in length but rather are divided at the positions where the "swung" notes would be. This can be exceptionally useful if you have used a particular quantization value for a song and then wish to chop up a longer region to match the quantization feel of the rest of the song. The only other way to achieve this with anything close to ease would be to use *Tab to Transient* (see chapter 9) and then use the *Split at Cursor* keyboard shortcut Alt + X, but that would be nowhere near as instantly gratifying as this *Split at Grid* command.

Cubase

Like most DAWs, Cubase has a number of very familiar tools available for basic editing tasks. There are the usual tools for selecting, splitting, joining, and a few other, more specialized ones. Equally, with the growing feature sets of modern audio software, some of these tools have multiple variants, and some have multiple uses, depending on what part of an audio region they are pointing at. Many of the tools are more relevant to later chapters, so we will look at them in due course. For this chapter we will just focus on the main tools that you will use in your editing tasks, and we will start by having a look at the default *Object Selection* tool.

The Object Selection tool is the first tool to be found in the main Cubase toolbox at the top of the main window (and also available by right-clicking anywhere in the main arrangement window) and is, in conjunction with the *Range Selection* tool, most likely to be the tool that you spend the most time using. If you prefer to use a keyboard shortcut, then you can press 1 to choose this tool, and subsequent presses of 1 will cycle through the different variations for this tool. Its main purpose, clearly, is to select objects, and these can be audio regions, MIDI notes, and a number of other things. You can use this tool by clicking on an individual region, by holding down **Shift** and clicking on multiple regions, or by clicking on the Arrange Window background and then dragging a selection rectangle over a number of regions. In addition, if you have regions already selected and you hold down **Shift**, you can then drag over regions to add them to the current selection.

The other two variations for the Object Selection tool are called *Sizing Moves Contents* and *Sizing Applies Time Stretch*. The only difference with each of the Object Selection tool variants is in the behavior of the tool when it comes to clicking and dragging on the end of a region. If we use the default option, *Normal Sizing*, and move the cursor to the bottom corners of a region, we will see the cursor change to left and right arrows, and if we click on the small square in either of the bottom corners of a region, we will be able to drag the region to change the length of it. The second option, Sizing Moves Contents, works slightly differently, in that moving the end of the region results in the audio being "anchored" to the end of the region, while moving the start is the opposite. Using this method changes which part of the audio plays (by changing the region length) and when it plays as well. (The final option works differently again, and we will look at that more at the end of chapter 12.) In addition, if you place the cursor at the boundary between two audio regions, you will see the cursor change to the left and right resize arrows but with two vertical bars in between them. Clicking and dragging using this tool will simultaneously shorten the first region and lengthen the second one. This functionality works the same, no matter which of the Object Selection tool variants is chosen.

FIGURE 3.6
Cubase has a number of different ways of using the Object Selection tool. In addition to actually making selections, it can also be used to adjust Volume Handles and Fade Handles by positioning the cursor either at the center top of the region (volume) or in the corners (fades).

Cutting, Copying, Pasting, and Moving

The Range Selection tool (selected by pressing 2) also serves to allow you to make selections, but, instead of being limited to selecting whole regions, the Range Selection tool allows you to click and drag over whichever parts of a region you are interested in. Ranges selected this way can be confined to a single track or spread across multiple tracks and can be spread across different regions as well. Once you have selected a range this way, you can then move the cursor over the selected range, at which point the cursor will change to a hand tool, and you can click and drag on the selected range to split it into separate regions and move it. If you just wish to split the selected range into new regions, you can press Shift + X or go to Edit > Range > Split.

One very interesting aspect of ranges selected in this way is that they can be moved with the arrow keys. If you selected a range of, for example, two bars over four adjacent tracks, you can then use the arrow tools to move the range up and down by one track at a time or left and right by an amount determined by the *Grid Type* control in the main toolbar. This can be very useful for splitting multiple tracks into predetermined length sections. You simply select a range equal to the length you want, split the range to new regions, and then move the range along the timeline, using the arrow keys before splitting again. As well as this, if you have a range selected across multiple tracks, you can add an additional track by pressing **Cmd[Mac]/Ctrl[PC]** and then clicking on the desired track (within the boundaries of the range), and you can delete an individual track from a range selection by doing the same but clicking on a track that is already selected.

The Split tool itself is very basic, in that it serves only one purpose. You can click on the arrangement background and drag over multiple regions to select them before using the tool itself, which can reduce the amount of time spent changing between tools. The Split tool can also be activated temporarily by holding down the Alt key with any of the Object Selection tools active. If you wish to use the song position cursor to split tracks, then you just need to click on the timeline to position the song cursor at the point where you wish to split the regions and then press Alt + X or go to Edit > Functions > Split at Cursor.

With all these (and many other) editing tasks, it can often be quite important to have a good overview of the audio that you are working on, so being able to zoom in and out quickly is very important. Cubase has the usual zoom level sliders located toward the bottom right-hand corner of the main Arrange window, and these can be clicked and dragged to vary the levels of horizontal and vertical zoom. Many people often find, however, that keyboard commands can be quicker to use, so Cubase has a number of these already set up. Horizontal zooming can be accomplished by pressing G or H (zooming in or out one level with each key press) and vertical zooming by pressing Shift + either G or H.

Using these will zoom all tracks equally in a vertical sense, but, if you wished to increase the vertical zoom on an individual track, then you should select it in the track headers (to the left of the actual Arrange window) and then press

Cmd[Mac]/Ctrl[PC] + up or down arrow to zoom only that track. If multiple tracks are selected, then the zoom on all the currently selected tracks will be changed. If you press Z, all currently selected tracks will be zoomed to the maximum vertical zoom level, and an additional press of Z will revert to the previous zoom level. This is often a much quicker method than multiple key presses if you simply wish to zoom in quickly to check something before returning to where you were.

Finally, pressing **Shift + F** will adjust the horizontal zoom level (up to the maximum possible level), so that the entire duration of the song or project is visible on screen at once. Similarly, if you have a number of regions selected, and they don't have to be contiguous or adjacent, pressing **Alt + S** will adjust both horizontal and vertical zoom levels to the maximum possible in order to fit all of the selected regions on-screen at once. Given that **Shift + F** will zoom out to show the entire song only in a horizontal sense, if you wished to zoom out to show the entire song (or as much as possible) in both horizontal and vertical senses, then you could select all regions by pressing **Cmd[Mac]/Ctrl[PC] + A**, then use **Alt + S** to adjust the zoom to the selection (all regions, in this case), and then use **Shift + Cmd[Mac]/Ctrl[PC] + A** to deselect all.

CHAPTER 4
Fades and Cross-Fades

DIFFERENCES BETWEEN FADES AND CROSS-FADES

Both fades and cross-fades are very useful in that they allow us to very quickly and easily make sure that the beginning and end of any audio regions are smooth, with no unexpected glitches. The biggest difference between them is purely a functional one. If there are no regions overlapping on the same track, then you should use regular fades. If your aim is to have one region gradually fade into another, and they are on the same track, then a cross-fade would be used. If, however, you wish to achieve the same thing, but the two regions are on different tracks, then you would still use fade-ins and fade-outs. Equally, if you wanted to have different shapes applied to the fade-out and in portions of a cross-fade and your DAW didn't give as much freedom to customize the cross-fades as you might like, then you could move the regions onto separate tracks and apply different "shapes" (as discussed in the following section) on each of the fades. The other benefit to doing it this way is that you can also apply different fade times to the out and in portions, which a "regular" cross-fade wouldn't allow you to do.

DIFFERENT FADE SHAPES

As mentioned above, there are a number of different "shapes" (or curves) that can be used for fades and cross-fades. These shapes are basically a reference to the rate at which the level change happens over the length of the fade. While many DAWs offer you the option to fine-tune or customize the fade curves, here we will be looking only at the commonly offered "preset" curves. There are four main types to consider: linear (sometimes referred to as equal gain), logarithmic (sometimes referred to as equal power), exponential, and S-curve. In order to fully understand the subtleties of difference between the different fade shapes, it might be wise to explain the decibel scale (dB) and the differences between sound levels in your DAW and perceived volumes. There is quite a complex subject, but there is a short summary of this available on the accompanying website, which should give at least an introduction to the concepts referred to below.

Different Fade Shapes

FIGURE 4.1
The curve on the left represents the change level in dB over time, while the curve on the right represents the change in perceived volume over time of a linear fade.

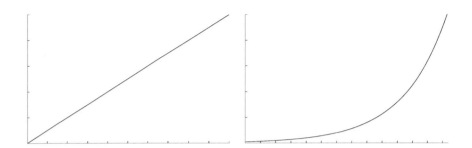

Linear

The simplest of all fade curves, linear, represents an equal rate of gain increase or decrease throughout the duration of the fade. As we have already stated, though, an equal rate of change of gain doesn't equate to an equal rate of change of the perceived volume. If you look at the diagram above, you will see the gain curves (for fade-in and fade-out) on the left and the associated perceived volume curve on the right. As you can see, this "linear" fade-in curve actually sounds like it stays quite quiet to start with, and then the rate of volume increase builds toward the end. Conversely, the fade-out would sound like there was quite a large drop in volume initially, and then it would seem more consistent toward the end of the fade.

This is quite often the default fade shape in DAWs and for shorter fade-ins and-outs, the kind we might use at the beginning or end of a region, just to ensure that there are no digital clicks at the region borders; these linear fades are perfectly acceptable. Equally, in situations where your audio has a natural ambience or reverb to it that you want to minimize, the quick initial drop in perceived volume of these linear curves could work quite well to seem to shorten the ambience.

This fade curve would do nicely in some situations where longer fades are needed. If you had a sound that you wanted to give the effect of accelerating toward you, then a linear curve might work. An exponential curve (see below) could be used if you wanted to further emphasize this acceleration effect. But for a natural-sounding, longer fade-in or -out, then a linear curve probably isn't the best choice.

The story with cross-fade is slightly different, though. Because the perceived volume drops more quickly at the beginning of the fade, you can see that, at the halfway point in time (halfway across from left to right), the perceived volume is noticeably below 50%. The effect of this on a linear cross-fade is that, at the midpoint of the fade, both sounds are below half of their maximum perceived volume, and as a result the sum of the two will be below the maximum level of either. If the two sounds are both of different levels anyway, and the cross-fade time is long enough, then this might not be a problem, but if you are using a short cross-fade for an edit that is perhaps in the middle of a single note of a performance, then you could have a perceptible dip in volume in the middle of the cross-fade. The diagram below illustrates this.

Fades and Cross-Fades CHAPTER 4

With this in mind, it is advisable to at least consider one of the other fade curves for short cross-fades between similar levels' audio regions.

Logarithmic

The perceived volume of a sound has a logarithmic relationship with its level in decibels, so it seems sensible to assume that having a fade that works on a logarithmic scale would counteract the curve of the linear fades, and, if you look at the diagram above, you will see that it does do exactly that. If you now compare these curves with the ones for linear fades, you may notice a similarity. In the case of the fade-ins, the shape of the perceived volume curve looks like we have taken the level in decibels curve and pulled the middle of the line toward the bottom right corner. Similarly with the fade-out curves, it looks like we have pulled the line toward the bottom-left corner. In the case of linear fades, it takes the straight line and introduces a curve, and in the case of the logarithmic fades, it takes an already curved line and straightens it out.

The sound of a logarithmic fade should be that of a consistent and smooth increase in (perceived) volume over the whole duration of the fade, and because of this it makes sense that this curve shape can be used for most general-purpose fading tasks. Logarithmic fades are often a good choice for long fade-outs (at the end of songs for example) because of the perceived linear nature of the fade. If you are applying fade-outs to regions with natural ambience, then logarithmic fades would seem very "neutral," in that they wouldn't tend to reduce the sound of the ambience, as a linear fade would.

They are also very useful for creating natural cross-fades, especially when you are cross-fading in the middle of a sustained note or word. When you apply a logarithmic cross-fade, the perceived volume at the midpoint of the fade is around 50% of the maximum, so, when the two regions being cross-faded are summed together, you get an output volume that seems pretty much constant.

For those of you interested in the mathematics behind how a logarithmic relationship works, I have included an explanation on the accompanying website.

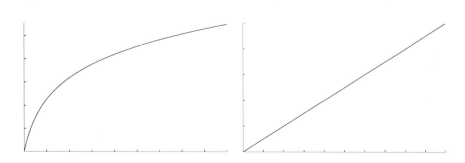

FIGURE 4.2
The curve on the left represents the change level in dB over time, while the curve on the right represents the change in perceived volume over time of a logarithmic fade.

Different Fade Shapes

FIGURE 4.3
The curve on the left represents the change level in dB over time, while the curve on the right represents the change in perceived volume over time of an exponential fade.

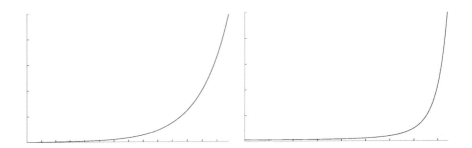

Exponential

Exponential fade curves are in many ways the exact opposite of logarithmic curves, in that the fade-ins seem to start off increasing in volume very slowly and then only shoot up really quickly at the very end of the fade, and the fade-outs seem to drop very quickly from the maximum volume and then seem to decrease only very slowly over the rest of the duration. The diagram above illustrates this and also shows that in many ways the perceived volume of an exponential fade could be seen as an exaggerated version of a linear fade. Naturally, it isn't quite as simple as that, but thinking of it in those terms can help to establish what it might be used for. Fading in with an exponential curve would give the impression of a sound accelerating rapidly toward you, which could be useful in certain situations. Equally, using an exponential fade-out at the end of regions with a natural ambience will certainly help to suppress that ambience, and that is very useful in many situations.

Given the shape of the perceived volume curve, it isn't hard to imagine that exponential curves are much less suitable for both sides of cross-fades, unless a very specific effect is required, simply because, at the center point of the cross-fade, both signals are quite low in level, so the resulting cross-fade will have a definite dip in the middle. Again, how much of a problem this is depends very much on the length and context of the cross-fade. Longer cross-fades on more ambient sounds may sound fine, with the associated dip giving a little "breathing space" in the middle of the cross-fade, but, if you are trying to use them in the middle of notes or words or for very quick transitions in general, the resulting dip may be very off-putting.

S-Curve

The S-curve is quite a difficult one to think about, because it has attributes of each of the previous three types and is further complicated by the fact that there are two different types of S-curves. Like a linear curve, at the midpoint, the level of the sound (decibel level) is at 50%, but the shape of the curve before and after this midpoint is not linear in nature. As you can see from the diagram above, you could almost view S-curves as a combination of a logarithmic and an exponential curve (although mathematically it isn't quite that). In the case of the first example (which I have called Type 1), one that I would consider a more traditional S-curve, you could view fade-ins as an exponential curve up to

FIGURE 4.4
From left to right, the curves above represent the change level in dB over time and the change in perceived volume over time of a Type 1 S-curve and the change in level in dB over time and the change in perceived volume over time of a Type 2 S-curve.

the midpoint, followed by a logarithmic curve from the midpoint to the end, and fade-outs as logarithmic up to the midpoint and then exponential from the midpoint to the end. Conversely, in the second example (Type 2), the situation is reversed, and the fade-ins are more like a logarithmic fade up to the midpoint and then an exponential fade from there to the end, and fade-outs are exponential to the midpoint and logarithmic from the midpoint to the end.

When considering applications for S-curves in cross-fading, it can be helpful to think of the Type 1 curves as a kind of halfway house between a linear cross-fade and a simple "cut" between two files. If we had two files directly adjacent to each other with no cross-fades, then the second would start playing at the exact moment the first one finished. If we are lucky, and if we have made sure our regions all start and end at zero-crossing points, the result will be a very abrupt transition from one region to the other. If the edits regions aren't trimmed nicely to zero-crossing points, then we could well get unwelcome glitches or pops. In order to avoid this, we could use any of the cross-fade methods described above, but, in all but the very shortest of cross-fade times, there would be a period when both sounds were audible simultaneously. The same applies with S-curve cross-fades. If there were no point when both sounds were playing simultaneously, then it wouldn't be a cross-fade after all. But what the S-curve (Type 1) does is to minimize the amount of time that both sounds are playing simultaneously. This allows to make edits that sound like direct "cuts" from one sound to another but which have that little extra smoothness and polish to them.

Type 2 S-curves, on the other hand, are generally more suitable for longer cross-fades, where we want the smoothness of a cross-fade and the consistency of a cross-fade in terms of overall level but the ability to have both of the cross-fading sounds audible for as long as possible. In essence what happens is that there is a relatively short period at the start of each cross-fade where the outgoing sound drops quickly toward 50% and the incoming sound rises correspondingly to around 50%. The rate of changeover then slows, and both sounds will appear to stay at almost the same level for most of the middle half of the cross-fade, and then, toward the end, there is a final quick changeover for the final few percent of the fade.

DIFFERENT FADE SHAPES: SUMMARY

To further complicate our descriptions of these basic curve shapes (not to mention custom curves), many DAWs now offer the ability to change the actual shape of the logarithmic, exponential, and S-curve cross-fades. I am not going to go into the mathematics behind these but will simply say that it is equivalent to pulling the logarithmic and exponential curves (and their corresponding

FIGURE 4.5
(From left to right, top to bottom) Decibel level change and perceived volume change for strong logarithmic fade, dB level change and perceived volume change for logarithmic fade, dB level change and perceived volume change for linear fade, dB level change and perceived volume change for exponential fade, and dB level change and perceived volume change for strong exponential fade.

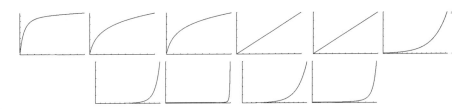

equals in the S-curves) away from the position where a linear curve would be. The diagram below illustrates this better, and there is, for consistency, a corresponding perceived volume curve for each one.

In the case of logarithmic fades, changing the shape of this curve affects how close to the start of the fade the sound will rise above 50% and how close to the end of fade-outs it will drop below 50%. Conversely, with exponential fades the change in shape will affect how close to the end of fade-ins the sound will rise above 50% and how close to the start of fade-outs it will drop below 50%. Making changes to the shape of Type 1 S-curves determines how quickly (around the midpoint of the fade) the change will happen, and, in the case of Type 2 S-curves, the change determines the amount of time that both sounds are at an approximately equal level.

USING FADES

The most basic use of fade-ins and fade-outs is as a means of making sure that individual regions start and finish without the risk of any audible glitches, pops, or other unwanted noises. If there is a good clear section of silence prior to the start of the actual audio, then I recommend a linear fade-in of around 500 ms as a nice, smooth introduction. Equally, and again assuming enough silence at the end of the region, I would recommend a similar 500 ms linear fade-out at the end of each region. If there is insufficient clean space at the beginning or end of the region, then you can always reduce this accordingly, but where possible I like to have these kinds of pre-audio fades as long (up to 500 ms or so) as possible just to make sure everything is nice and gentle. These kind of edits are really just good "housekeeping" edits and aren't always strictly necessary if you've got good basic recordings to work with, but it's always good practice to do them anyway.

Another very common use of fade-ins and -outs is as a means of changing the characteristics of a sound. In its simplest form, this is something as basic as putting a fade-in on the very beginning of a sound to soften the attack of the sound. For vocals this can be very useful in softening up the plosive sounds. These are vocal sounds made by closing the oral passage and then rapidly opening it again, and they occur at the beginning of many different vocal sounds, but the most common culprits are words beginning with "b", "d," and "p." This rapid release of air, especially if the mouth is quite close to the mic at the time, can lead to lots of low-frequency energy and a very pronounced, sharp attack to the sound. Even if we filter out the low frequencies, there might still remain an unnecessarily sharp attack. This isn't always a bad thing, as it can help with the diction and intelligibility of the voice, but, especially in situations where the vocal is otherwise quite soft and gentle, it can make that one word stand out quite a lot. Using a fade-in (probably linear or logarithmic) with a very short time (around 10 ms or so) would be a good starting point, and then the fade time should be adjusted while listening, until the attack has been softened without changing the character of the sound and the intelligibility too much.

You could also consider softening up the attack of drum and percussion sounds. If you have a particularly busy rhythm section in a song with lots of hi-hats, percussion, and even very "snappy" rhythm guitars, the whole thing can start to sound a little messy. This might be down to small timing variations between the instruments and performers. If this is the case, and if these variations make things just a little *too* loose, then you might consider moving a few things around in order to make them work better together. But there is another potential solution. Instead of trying to match up the timing of a number of elements with sharp attacks to the sounds, you could always try softening up the attack slightly on one or two of the sounds. A fairly short fade-in of something in the region of 25 ms is a good starting point, as that fade-in will give a tiny degree of ambiguity (and therefore flexibility) to the timing of that sound.

Equally, you can use fade-outs to change the character of sounds by changing their decay characteristics. A snare drum with the snares quite loose has a very distinctive "rattle" to the sound, which can be very interesting tonally but which can also be quite messy. If there isn't too much spill, then a noise gate could perhaps be used to shorten the decay. This would work, however, only if the level was fairly consistent. Any changes to the level of the snare drum going in to the noise gate would change the way the noise gate responded and would mean that some notes (quieter ones) would seem like they had been truncated more than the louder ones. That might be something that works in the context of the editing that you are doing, but, if you want something more consistent, you could always achieve a similar result using fade-outs. If you wanted this consistency, then you would want to do a little preparatory work and make sure that not only was each snare drum a separate region but also, crucially, you would also need to make sure that each one was trimmed nice and tight at the front end and also that they were all pretty much equal in length. Once you know that all of the snare drum regions are the same length, then you can simply select all of them and apply a fade-out. A linear or logarithmic curve would be a good place to start, and it is really just a case of starting with a fade time of around 100 ms and then just adjusting things from there.

Using fade-ins and fade-outs to change the characteristics of sounds in these ways is something that can be very useful in freeing up some sonic space in complex projects, but it can also be very time-consuming, as these kinds of edits need to be performed on a note-by-note basis. However, it is something that I would certainly recommend that you try for yourself and familiarize yourself with, as the effects can be quite special, and, like most editing tasks, the more often you do them, the quicker you get at doing them.

USING CROSS-FADES

The whole point of a cross-fade is to provide a smooth transition between two separate pieces of audio. But within that definition there are a few different situations when we might need to do this. One involves transitioning between two entirely unrelated or at least very different sounds, and the other is when we have two similar sounds and we want to try to make them sound like one continuous one. The process is basically the same, but we have a different set of parameters and criteria to work with in each situation.

The simpler situation is when we wish to apply a cross-fade between two very different (in both tone and pitch) files. In most cases it is simply a case of positioning the two files at the appropriate positions and then selecting an appropriate cross-fade curve and time and making minor adjustments to fine-tune things. In most situations like this, that will be all that is required, simply *because* the sounds are quite different. There may be times where there is some phase cancellation (see below), but, given the context that these kinds of fades are likely to be used (long cross-fades for background ambiences, cross-fade between different whole songs, and things of that nature), those problems aren't likely to be a major factor all that often. Unfortunately, when we are dealing with two sounds that are quite similar (in tone and pitch), we have a few more things to take into account.

In the last chapter we looked at the importance of zero-crossing edits and also the phase alignment of the two regions. We found perfectly good reasons for this phase alignment simply in the editing process as a part of the pursuit of a natural sound, but there are other implications that become more relevant when we are dealing with cross-fades. In order to fully understand this, we need to briefly look at phase cancellation.

PHASE CANCELLATION

Every sound has a waveform, and every waveform has a phase. In essence, phase is simply the position of any given point in time on the waveform cycle. This is easiest to illustrate with a simple sine wave. Starting from the "neutral" position, the wave first moves upward until it reaches a turning point, then moves downward toward another turning point, and then returns to the neutral position, and the whole process begins again. That initial up-and-down-and-up represents one full cycle and, because phase measurements are in degrees (as used to measure angles), one full cycle is considered to be 360 degrees. Phase alignment (or phase correlation, as it can be called) simply refers to the process of ensuring that the two audio files in question are at the same relative phase angle (measured in degrees) or, at the very least, at a very similar point in this cycle: on an upward trend or a downward trend. The phase of an individual sound doesn't matter in isolation. In fact, our hearing system cannot detect absolute phase. What happens, though, when we don't listen to sounds in isolation, and we start mixing them together?

Any waveform can be considered to have both "positive" and "negative" parts. These correspond to the periods when a speaker is moving toward you and away from you, respectively, and are represented by the parts of the waveform shown to be above or below the imaginary midpoint (often shown by a horizontal line in the waveform display). What is important is the relationship of the sounds that we are cross-fading to each other in this positive/negative sense. If both of the sounds are moving upward at the same time, then, when we add them together, we will get a cumulative effect. This is what we are looking for. Conversely, if one sound is moving upward while the other is moving downward, there is a chance that one will "cancel out" the other. The diagram below illustrates this with simple sine waves.

The top two regions on the left represent two sine waves that are the same frequency and in phase, and the result of cross-fading the two regions is the third region below. With the "in phase" sine waves, you can see that not only is there no change to the waveform, but also there is no change to the volume, because an equal-power cross-fade was used. On the right we have a similar setup, only the two sine waves are now 180 degrees out of phase. If we look at the result of cross-fading the out-of-phase sine waves, we can see there is something present at the beginning (when only the first wave is audible) and something present at the end (when only the second wave is audible), but, at the midpoint of the fade, when both are being mixed at equal levels, there is a period of silence owing to the phase cancellation.

Now I will admit that the times when you will be required to cross-fade two pure sine waves of exactly equal level and frequency are going to be rare, to say the least. And as soon as we change the waveform shape (by adding in harmonics or other notes or sounds), the frequency, or the level, we will not get 100% cancellation as shown above. What could happen instead is that there are *parts* of the sounds that cancel each other out, which, in reality, can actually sound even more

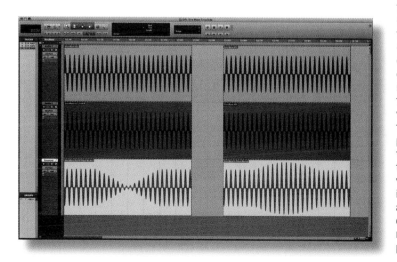

FIGURE 4.6
The top two tracks on the left are sine waves that are 180 degrees out of phase, and the third track is the result of cross-fading those two regions. Note the dip in volume at the center of the cross-fade. On the right the two top tracks are sine waves again, but this time they are in phase. When these two are cross-faded, there is no loss in volume. The slight increase in volume is a result of the cross-fade curve's not being of equal power.

obvious than a simple temporary volume drop. And even the most experienced and technically gifted editor wouldn't be able to tell exactly what the frequency content of a sound was just by looking at a waveform display. That doesn't mean that the waveform display is utterly useless in this regard, though.

PERIODIC PATTERNS

As you zoom in (horizontally) to a waveform display, you will, at a certain point, and in most (but not all) sounds, start to see a pattern emerging. It is very unlikely to be a simple pattern—like that of a sine wave—but there will almost certainly be some kind of cyclic repetition. This is what you need to look for, even if you don't know what frequencies it represents. If you are trying to cross-fade two sounds with similar tones and frequencies (as we are discussing here), then you should be able to see a similar pattern emerging in both sounds. If you have a similar pattern at the end of the first region and the beginning of the second region you are trying to cross-fade, then there is a good chance that you can get things sounding very smooth indeed.

Once you have found these points, you can start to position the regions to get the best result. As you move one of the regions (usually the second one) backward and forward along the timeline, there will start to be an overlap between the two. Making sure you have an appropriate zoom level set (to where you can see a good handful of the "cycles" that you have visually identified on either side of your cross-fade point), you can then visually line up the second region so that its start point (the start of one of these cycles) lines up with the start point of one of the cycles in the first region. Once you have completed this visual step, you should have a listen back. The chances are there will be some kind of audible issue at the edit point. It might not necessarily be a click or a pop or any other artifact of a bad edit, but there could be a perceptible change in pitch or tone (depending on how different the two regions were to start with), and this is where the cross-fades come in.

In the case of two regions with subtle pitch and tone differences, you should start off with a very short cross-fade (just a few milliseconds) and then increase

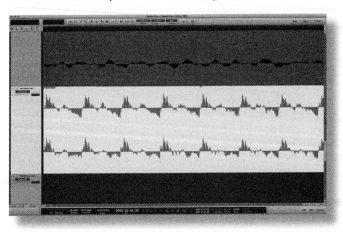

FIGURE 4.7
While it is far from a simple waveform, the vocal in the image above does have a clearly repeating pattern. Ideally you want to make any edits at a similar point in this cycle for maximum smoothness and consistency.

the length, until it sounds right. The reason for this is to minimize the risk of any strange cross-fade effects such as phase cancellation by minimizing the amount of time the cross-fade takes. Because the two regions are very similar, there is a good chance that, even with quite a short cross-fade, we wouldn't hear any major differences. If the two regions were quite different, though, we would hear an abrupt change.

The same applies to sounds with a larger tonal or pitch change, but with the difference that, unless you are specifically happy with a sudden change, you are likely to have to make the cross-fade quite a bit longer before there is any subtlety to the edit. And of course the longer the cross-fade, the more likely we are to have problems. If you find that you just can't get it to sound smooth, then you could consider changing the shape of the cross-fade curve (perhaps even making it an asymmetric one), because, if you absolutely need to get these two regions cross-faded, it might be that a small dip in volume (as a result of the curves used) in the middle of the cross-fade could help to mask a little "bumpiness" in the transition.

Even with all of the options that we have for cross-fading, there will still be times when we can't quite achieve what we want to. One example that comes to mind would be a situation where we wanted to take two sustained vocal notes with different pitches and blend them together smoothly. Using the techniques that we had for cross-fading simply wouldn't do this smoothly, as the fades are not content-aware. The fades don't know what the pitches of the sounds being faded are; they work simply on the volume/level of the sounds. If we used a cross-fade on our two vocal notes, we would end up with a period of time during which both notes could be heard simultaneously rather than gliding from one pitch to another, which we would expect from a singing voice.

In general, though, with the exception of long, subtle fades, most cross-fade use will be when you are trying to blend similar sounds (perhaps different takes of the same performance) or to smooth off a transition when you have had to remove a part of the middle of a region (used extensively in some time-stretching scenarios) and need to move the parts on either side of the gap together. In all of these situations—where we want the end result to be inaudible—we can often get really great results with a little patience and effort.

HANDS ON
Introduction

In this section we will take a look at the practical ways to create different types of fades and the different shortcuts and tools available in each of the DAWs. Like cutting and pasting, because these operations are carried out so often, there are numerous ways in which you can achieve the same result, and there is sure to be a solution to fit your individual work-flow and needs. The options presented here are just examples of some of the easier and more convenient ways in which you can get results.

Logic

At the end of the last chapter, we saw that it was possible to change the default preferences of Logic to include an option for the Marquee tool to be automatically selected under certain circumstances. Equally, it is possible to set things up so that the Fade tool is automatically selected. In order to do this you need to go to Preferences > General > Editing (Tab) and make sure that the check box next to *Pointer Tool in Arrange Provides: Fade Tool Click Zones*. Like the Marquee tool, choosing this option means that the standard cursor/tool will change to the Fade tool under certain circumstances. In this case, when the cursor is positioned in the top left or top right corners of an audio region, the Fade tool will be automatically chosen. This tool looks like a vertical bar crossed by a horizontal line, with left and right arrows at either end. Please note, however, that if you have *Flex Mode* activated on a particular region, then you will need to turn the *Flex* view off (so that the Flex markers are not showing on the region) in order for this to work correctly. If the tool is showing as it should, then all you need to do is click and drag to the right (at the start of a region) or left (at the end of a region), and a fade-in or -out (as applicable) will be created that extends as far as you have dragged.

Once the fade has been created in this way, you have a number of options. At the start (fade-out) or end (fade-in) of the fade, there will be a vertical line overlaid onto the region, and if you place the cursor at the top of the region at this point, you will see the Fade tool appear again, and you can move the start or end of the fade as desired. However, if you move the cursor to the top of the region at a point within the fade area, you will see a slightly different tool appear. This time, rather than being a vertical line with left and right arrows either side, it will look like a horizontal line, with the arrows at either end and a dot in the middle. If this tool is showing, and you click and drag left or right, you will see that the shape of the fade changes. In Logic-speak this is changing the curve of the fade, but this simply equates to a change from linear to either exponential or logarithmic curves of varying strength.

The usefulness of the Fade tool doesn't end there. Both of these variations on the Fade tool can also be used to create and edit cross-fades. If the Fade tool is positioned between two adjacent (and they must be touching or overlapping) audio regions and then clicked and dragged, a cross-fade will be created that is symmetrical either side of the crossover point. Once the initial cross-fade has been created, we can change the length of the crossover, as we did before, with the difference that, in the case of a cross-fade, the length of only one part of the cross-fade will be changed. While the initial cross-fade was created as symmetrical, these length changes will not be, and the resulting midpoint of the cross-fade will move as we adjust the length of either side. Cross-fades can be made between regions with a gap in between, but there is a slight variation in method. In order to do this, you will need to position the cursor to show the Fade tool at the end of one region and then click and drag as if to create a fade-out, but then, while still clicking, drag in the opposite direction to drag over the start of the following file. Once the Fade tool has reached the

Fades and Cross-Fades CHAPTER 4

following region, a cross-fade will show, which can then be varied in the ways described above.

We can also use the variation of the Fade tool to change the shape of the cross-fade, as we did with the fades, but things are a little more complex here. Whereas for fade-ins or -outs there is a simple change from exponential through linear and on to logarithmic, with cross-fades there is a reshaping of both sides of the fade but in an inverse manner. What this means is that if one side of the cross-fade moves toward logarithmic, then the other side will simultaneously move toward exponential. Given that an exponential curve is the inverse of a logarithmic curve, this makes a lot of sense, as doing things this way will preserve a constant power throughout the cross-fade, which is, in most cases, the best option for a smooth cross-fade.

The default behavior for dealing with overlapping regions in Logic is to simply switch playback from one region to another. The first region will play until the start of the second region, at which point that one will take over immediately, even if there is a portion of the initial region still to play. There are, however, a number of other modes for dealing with situations like this, and the most useful to us in this context is the option to automatically create cross-fades. There is a drop-down box at the top right of the *Arrange Window* called Drag, and if you choose X-Fade, any regions that are dragged (or resized) and overlap another region as a result will automatically have a cross-fade applied. The cross-fade itself will last for the duration of the overlap, and once in place the ends of the cross-fade and the shape of the curve can be adjusted in the same ways as we have already described. If you prefer to have this mode set to "Overlap" in general, you can always change it to X-Fade temporarily if you are working on something specific that would benefit and then change back afterward, as any cross-fades created while X-Fade mode was selected will remain in place if you switch back to one of the other modes. If it is something that you find yourself swapping in and out of regularly, then it is possible to set up custom key commands for each of the modes to enable a quick change.

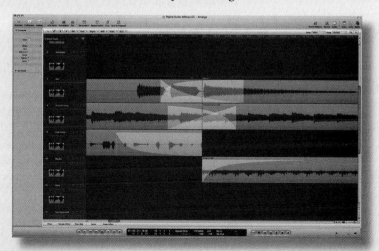

FIGURE 4.8
There are many different options for fades and cross-fades in Logic, including equal power and equal gain options and fully variable logarithmic and exponential curves.

Using a combination of these two techniques, along with some of the other options available, will mean that you spend much less time having to change tools or navigate through various menu options in the middle of an intricate editing process. Such tasks can often be repetitive and time-consuming, so any methods that we can use to speed up the process will be extremely valuable to us.

Pro Tools

The first and perhaps most obvious way to create fades is to use the *Selector* tool to highlight the section that you want to create a fade for. If you choose only the end of a region, then a fade-out will be created; choosing only the beginning will create a fade-in; and choosing a selection that overlaps two regions will create a cross-fade. Once you have the selection in place, all you need to do is press Cmd[Mac]/Ctrl[PC] + F or choose Edit > Fades > Create to open the *Fades* dialogue.

The other (some would say easier) way of creating fades is to use the Smart tool to create the fades. When the Smart tool is selected and positioned in either of the top two corners, the cursor changes to a square divided in half diagonally, which can be used to create fade-ins and outs. Clicking and dragging from either top corner toward the center of the region will create a fade-in or -out (context-dependent) of a length determined by how far you drag. If, however, you position the cursor at the bottom corners of two adjacent regions, then the cursor changes once more to a square divided into four diagonally, and this can be used to create cross-fades. As soon as you click, a cross-fade will be created that is symmetrical either side of the cross-fade point and the length of which is determined by how far you drag.

If you create fades or cross-fades using the Smart tool, then it will be created using default parameters, but you can still go in and fine-tune the settings once the fade has been created. The method used here is the same method that you would need to use if you wished to make changes to any existing fade no matter which method you used to create it. If you click anywhere within the fade area, it will select the whole fade region, and you can use Cmd[Mac]/Ctrl[PC] + F (or Edit > Fades > Create) to reopen the Fades dialogue and make changes from there. A quicker method, though, is to simply double-click on the fade region with the Grabber tool (or Smart tool at the bottom of the region), and this will also open the Fades dialogue.

Once you've arrived at the Fades dialogue, you will find that it is consistent in layout whether you are doing fade-ins, fade-outs, or cross-fades but with different options visible, depending on which type you are creating. At the very top left there are four buttons that allow you to customize the viewing and auditioning options, but more important are the actual curves themselves, as this is where all the character and feel of the fades are determined. What we have are independent sections for fade-ins and fade-outs and a central *Link* section that will create curves with a degree of symmetry.

If you are creating a fade-out, then only the left hand *Fade-out* section and a central *Slope* section will be visible. There are three options under either *Out Shape* or *In Shape*: *Standard*, *S-Curve*, and a pop-up selection box with seven different preset shapes that vary from a nearly instantaneous change through exponential, linear, and logarithmic. If you choose with Standard or S-Curve, you can click on the colored line representing the shape of the curve and drag to the left or right to fine-tune the shape of the curve. In the case of Standard, this will move from a linear fade at the center point through to exponential and logarithmic at either extreme. If you choose S-Curve, then moving either side of the central point will give you the Type 1 and Type 2 S-Curves we spoke about in the main chapter. If, however, you choose one of the preset shapes, then no adjustment is possible. Naturally, everything we have just said applies for fade-ins just as much.

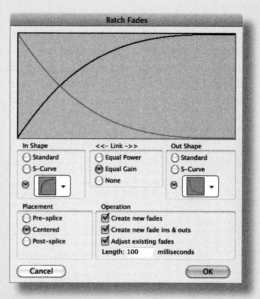

FIGURE 4.9
The Fades dialogue in Pro Tools offers a large number of choices ranging from simple equal power and equal gain cross-fade to more esoteric and customizable options.

The Slope control has a choice of *Equal Power* or *Equal Gain*. We explained in the main chapter that the decibel scale is a logarithmic one, which means that a constant interval of gain change (in steps of 3 dB, for example) does not equal a constant change in power (or perceived volume). As such, Pro Tools gives us the option to scale our fades to either scale. The effect on the curves is simply to give slightly different shapes to each of the options. An Equal Power fade should sound smoother and more consistent than an Equal Gain one, but, as always, let your ears be the judge of what works best in any given situation.

Things become a little more complex if you are working on a cross-fade, simply because you have the option to have both sides of the cross-fade linked to create a degree of symmetry if you so desire. If the Link control is set to either Equal Power or Equal Gain, then the cross-fade created will, by default, be symmetrical about the center position and have curves that are chosen to give (unsurprisingly) either an equal power or equal gain throughout the duration of the cross-fade. However, that isn't the end of the story. As with fade-ins and -outs, if either Standard or S-Curve is chosen, then clicking and dragging on the curve line will change the shape. The obvious addition here is that, as you drag on either curve and change the shape, the shape of the other corresponding curve is updated automatically as well.

If you choose one of the preset shapes from the pop-up list at the bottom, then, while you might not be able to change the shape of each curve individually, you can choose different shapes for the fade-in and -out curves. If you have the Link mode set to Equal Power, you will notice that some of the preset shapes are grayed out, simply because it wouldn't be possible to create an equal power cross-fade with those shaped curves, but, with Equal Gain chosen, you have the full choice for both fade-in and fade-out curves.

Things start to get really interesting when you set the Link mode to *None*, because not only can you choose Standard, S-Curve, or the presets separately for the fade-in and fade-out portions (with the adjustable curve shapes for both Standard and S-Curve), but also you will see black dots at each end of each curve, which you can click and drag earlier or later to further fine-tune the shape of each curve. Of course, once you start getting into the realms of this much adjustability, then it will be very difficult to preserve either a constant gain or a constant power throughout the cross-fade, but, to be honest, having this level of flexibility in the cross-fades is definitely worth risking a little unevenness, especially given that you can always choose one of the more-standard options if things go awry during your experimentations.

Studio One

As we have already seen, the Arrow tool in Studio One is very flexible and has a number of different uses. In the last chapter, we looked at the basic selection abilities of the tool, but we can also use it create simple fade-ins and fade-outs for audio regions. If you select a region with the Arrow tool, you will see two small triangles (called *Fade Flags*) in the top corners, and these can be clicked on and dragged to create a fade-in or fade-out. As you drag there will be a pop-up info box, close to the tool, that tells you the total length of the fade. The waveform overview will also update as the fade is created, to show the effects on the waveform in real time.

If you need to enter a specific duration for a fade-in or -out, you can always use the *Inspector* panel (press F4 or go to View > Inspector). This panel gives a lot of information about the track and region as a whole, including, close to the bottom of the panel, numerical values for both fade-ins and fade-outs. Double-clicking on either of these allows you to type in a value (in seconds, and up to three decimal places) to use for the fade. This can allow you to quickly enter a very specific value rather than dragging the Fade Flags to try to get an exact value. The maximum value that you can enter by typing into these boxes is 2 million seconds, and the shortest is 0.001 seconds (or 1 millisecond or ms).

Unfortunately the abundance of flexibility in fade times isn't carried over quite so extensively when it comes to fade shapes. Once you have the fade at the length you require, you can then adjust the shape of the fade by clicking and dragging on the small square at the midpoint on the fade line. Clicking on this square and dragging upward will change the fade shape from a linear one to a logarithmic one, while dragging downward will change from linear to exponential. Sadly, at the time of writing, there is no option for S curves in either fade-ins and -outs or cross-fades.

Fades and Cross-Fades   CHAPTER 4

Another useful feature is the ability to automatically add short (10 ms) fades at the beginning and end of regions by pressing **Shift + X** or by going to **Audio > Create Autofades**. This is usually explained as being especially useful as a way of avoiding any little clicks and pops if you have split regions at transients or bend markers, but it is equally useful as a precaution for just about any audio region that you have.

When it comes to cross-fades, Studio One is equally intuitive, and creating them is just a matter of selecting the adjacent regions that you wish to cross-fade and pressing X or going to **Audio > Create cross-fades**. If you select only a single region and then press X, cross-fades will automatically be created between that region and any overlapping or adjacent (touching) regions. The cross-fade time defaults to 10 ms, but you can easily adjust this by zooming in and clicking on the Fade Flag either side of the cross-fade and dragging. As you do this, you will again see a pop-up information box that displays the fade-in or fade-out time, depending on which Fade Flag you clicked on.

If the regions are touching, then, if possible, Studio One will automatically adjust the end point of the first region and the start point of the second to actually allow for some overlap. This will happen only if the regions have already been trimmed and have additional material that has been truncated. In the event that the region cannot be extended in this way, then, instead of a cross-fade, you will simply get a fade-out on the first region and a fade-in on the second one. If any one of the regions can be extended, then that one will be, while the other remains untouched. The actual area of overlapping regions is highlighted, so that you can see when the two files are being cross-faded. If the fade lines are extended outside this area (by dragging the Fade Flags beyond the highlighted area), then there will be a partial fade-out of the first region,

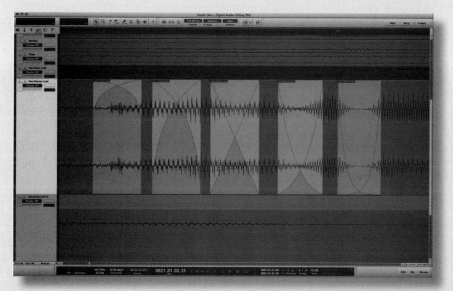

FIGURE 4.10
While there is not a huge number of options for fades and cross-fades in Studio One, it is exceptionally easy to create and edit the curves by clicking and dragging left to right to change the length, and up and down to change the shape.

followed by a period of cross-fade, and ending with a remainder of the fade-in on the second file.

If, on the other hand, the regions are overlapping, then you can extend the cross-fade time either by extending one (or both) of the regions to give a greater overlap time or by moving one of the regions to create a greater overlap. When you do either of these, the Fade Flags remain in the same relative position. What that means is if there is a fade starting 10 ms after the overlap begins, and you move the entire second region 100 ms earlier, then the cross-fade time will be automatically lengthened, so that the cross-fade still starts 10 ms after the overlap begins. This is quite different than some other DAWs, where the cross-fade time remains constant even if you move one of the overlapping regions. Whether you prefer this method or not is a matter of personal choice, as one is not clearly better than the other.

If you move the cursor over the highlighted cross-fade area, it will change to a vertical bar with left and right arrows, and clicking and dragging will move the cross-fade position. The length and shape of the cross-fade will remain the same, but its center position will move along the timeline. Also, clicking and dragging up and down on the small square on the fade line will change the shape of the cross-fade from exponential through linear to logarithmic in the same way as it did for fade-ins and -outs. The cross-fades in Studio One are symmetrical, and there is no way (at present) to have, for example, an exponential fade-out on the first region combined with a logarithmic fade-in on the second region. While this may not be as flexible as some other DAWs, these basic fade/cross-fade shapes will still be very useful in a majority of situations, even if they might not necessarily give 100% of the result that you desire in more-complex situations.

Cubase

Setting up and editing of fades and cross-fades in Cubase is not only intuitive but also incredibly powerful. In fact, the flexibility of the fades in Cubase is arguably among the best there is currently in any DAW, because there are not only options for both destructive and nondestructive approaches, but also there is an extremely well-specified fade editor to help fine-tune your fades and cross-fades in more ways than you could ever need. To begin with we will look at the destructive fade methods.

Destructive fades in Cubase are limited to simple fade-ins and fade-outs, and these are accessed by going to Audio > Process > Fade In/Fade Out. Any fades created in this way are applied over the full duration of the region no matter what its length. While this means that you will have to split a region if you want to fade only a small part of it, this particular method does have in its favor the fact that it uses fundamentally the same fade curve editor as the real-time fades and cross-fades, which gives it a great deal more scope than many other destructive-fade options in other DAWs.

The fade curve editor itself is a remarkably powerful tool while also not being overwhelming. If your fading needs are simple, then, at the bottom of the editor window, you will see a series of eight buttons that range from logarithmic curves (in three different intensities) through linear, S-curve, and then on to exponential (again three different intensities). Clicking any of these will bring up a visual representation of the curve itself in the main editor display. Here you will not only see the curve itself but also the original waveform overview and the resulting waveform after the fade. The original waveform is light in color and sits behind the new waveform, which is darker in color. There is also a *Preview* button at the bottom of the editor window to allow you to audition the fade. If you want just a simple fade-in or -out, then you just need to choose a curve and hit *Process*, and the resulting fades will be written into the original file.

If, however, you want to fine-tune things, then you can go beyond these preset curves and adapt them to make your own fades. If you choose any of the fade shapes other than linear, you will notice that there are three (five, in the case of the S-curve) dots spaced out along the length of the curve-shape line. These are nodes and allow you to make changes to the shape of that line. You can click on a node and move it, and there will be a resulting change in the curve. If you click on the line itself and not on an existing node, then a new node will be created that can, again, be moved around. If you add a node by mistake or simply want to clean things up, you can get rid of a node by clicking on it and dragging it off the top or bottom of the display.

We aren't finished quite yet, either, because there are three buttons at the top of the editor window that we can use to further change things. These three *Curve Kind* buttons allow us to adapt how the fade curve actually reacts to the position of these nodes. The default option, the middle one, is called *Damped Spline Interpolation*, which, in nontechnical terms, means that a smoothed curve is drawn between the start point and the end point and passing through all the nodes along the way. The "damped" part essentially means that the curve isn't perhaps as flowing as it could be. In fact, it is halfway between a fully flowing curve and a series of straight lies connecting the nodes. And in fact those two extremes are the other two options that we have here. On the left we have *Spline Interpolation*, which is, once again, a curved line that passes through all the nodes but that, in comparison to the Damped Spline Interpolation, is much more flowing and extreme. On the right we have *Linear Interpolation*, which just links the nodes with a series of straight lines.

Once you have chosen or created your curve, you also have the option to store it as a preset by clicking on the *Store* button and entering a name. This will save the particular curve shape you have set up in a preset list, which makes especially complex curves very simple. These curves aren't fixed in duration and simply represent the evolution of the fade from its start to its finish, however long that actually is.

Moving on now to the nondestructive and real-time fades, we find ourselves with a very simple setup that enables us to get results quickly. With the *Object Selection* tool active, simply move the cursor to either of the upper corners of a region, and you will see a small triangle. This is the fade handle, and simply clicking and dragging on this (the cursor should change to left and right horizontal arrows that position it over the fade handles) will create a simple linear fade whose duration is determined by how far you click and drag.

When it comes to editing the shape of the fade curve, you need to open up the fade editor by going to Audio > Open Fade Editor(s). Once the fade editor is open, you will see that it is largely identical to the one used for the destructive fades, with only a couple of minor differences. At the top right of the curve display, there is a *Restore* button, which will undo any changes that you have made since you opened the editor. You will also see, at the very top right of the editor window, a box where you can enter a duration (again in bars, beats, divisions, and ticks). This allows you to change the length of the fade inside the fade curve editor window rather than doing it as a separate step. Other than that, all the options work exactly the same as we have already described.

Cross-fades are handled with equal ease, and you can apply a cross-fade between two overlapping regions simply by pressing X or by going to Audio > **cross-fade**. This linear (by default) cross-fade can be edited either by double-clicking on the cross-fade area, which is marked by diagonal lines, or by going to **Audio > Open Fade Editor(s)**. The editor window for cross-fades is, by necessity, slightly different from the other fade editors but is basically composed of two separate fade editors: one for the fade-out and one for the fade-in. For each of these we have the eight preset shapes, the three Curve Kind selections, and the ability to create custom curves by adding and moving nodes. To the right of these two curve editors, there are additional controls for choosing either Equal Gain or Equal Power cross-fades and also buttons below that allow you to preview the effect of the fade-out curve only, the fade-in curve only, or the combined cross-fade. Finally there is the control that allows you

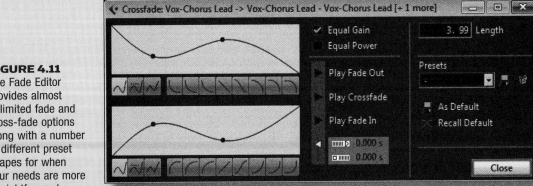

FIGURE 4.11
The Fade Editor provides almost unlimited fade and cross-fade options along with a number of different preset shapes for when your needs are more straightforward.

to adjust the length of the cross-fade from inside the editor, and also the preset list and saving options.

It should be noted that if you choose Equal Power, then you cannot actually add nodes and can move only the one on each curve in a horizontal sense and gradually change between logarithmic, linear, and exponential. If you want to have the full custom fade curve options, then you need to select Equal Gain. It is also worth taking into account the fact that, while the cross-fade editor doesn't actually show the effects on the waveform of the cross-fades that you are creating, the waveform display in the main Arrange window constantly updates to show the effect of changes; so if you select the cross-fade region and then press Alt + S to zoom in to that region with maximum zoom, and then you move the cross-fade editor window off to the side a little, you will be able to get a real-time preview of the effect of any changes in the cross-fade curves as you actually make them.

CHAPTER 5
Level Control

SHOULD THIS BE A PART OF THE EDITING PROCESS?

Some people feel that the subject of level control of any form should be left until the mixing and production stage, and there is a lot of sense in that. However, what we are talking about here, in the context of level control, is the process of correcting problems and anomalies with levels. The aim of level control in the context of audio editing is to correct any problems to make sure that there are no unwelcome surprises at the mixing stage.

Given that largely the same tools will be used for level control during editing and mixing, we could ask why we should bother making these changes during the editing process—surely it is better to keep the maximum amount of flexibility at the mixing stage. While this is true, there are certain things that will be extremely obvious, and, if we can fix *those* problems during the editing stage, then it will free up the mix engineers to focus on the more creative tasks that they have to do.

Naturally there is a fine line between things that need to be done and things that could be done, and that line can vary from person to person. There are, however, a couple of areas where I believe that it is absolutely necessary to work on level control at the editing stage, and, if that isn't done, the problems can be exaggerated later on. The two most common of these are cleaning up background noise in recordings and achieving consistency in comped tracks. The latter is especially relevant if you are cross-fading in the middle of sounds. But before we take a look at these and other situations where we might use level control during the editing process, let's first look at the different ways in which we can control the levels and address the question of whether "destructive edits" or "nondestructive edits" are preferable.

DESTRUCTIVE VERSUS NONDESTRUCTIVE EDITS

Perhaps the first thing we should do is define and differentiate between destructive edits and nondestructive changes. In essence, the distinction is simple. A destructive edit is one that results in changes to the underlying audio file.

Destructive vs. Nondestructive Edits

A nondestructive edit is essentially a change that is laid on top of a file, is calculated in real time, and does not have any effect on the underlying file. With that in mind, why would you ever choose to use destructive edits if the same thing could be achieved nondestructively?

To begin with, destructive edits on computers have a number of levels of safeguards to prevent you from doing anything irreversible (unlike any destructive edits on tape that were either permanent or a huge problem to rectify). The first and most obvious is the use of some kind of preview feature. Quite often you will be able to preview the effect of an edit before you commit to it. To do this, the software creates a copy of the file you are working on and then applies the edit to that copy. You can then compare the edited copy to the original, and once you are happy to move forward, you commit the edit, and the original is overwritten with the edited copy. This is a first line of defense against applying an edit where you have incorrectly set a parameter. What it doesn't do is allow for you to change your mind afterward for any reason.

Then we have the Undo command, which is a lifesaver if you realize that you have done something wrong only after you hit the Apply button. The Undo command has limitations, though. It will be of use to you only if you have not completed too many other steps since (the number varies by software). And given the nature of editing, it is distinctly possible that you may apply an edit, carry out several other steps, and only then think that you could perhaps have done it differently. The other downside to using Undo in this way is that, of course, if you step back through the steps you have taken to get back your original file, even once you have changed the parameters and applied the edit again, you will then have to redo all the other steps you took, and you may not remember what exactly you did.

Another possibility is for you to manually create a copy of the file and apply the destructive edits to that. By keeping an untouched "original" copy, you know that you are safe, even if there is some problem with trying to undo the changes that you have made. The most useful benefit of working this way is the ability to create different versions of your work at various stages. A typical use might be to have a completely untouched original version, to then remove any background noise or unwanted sections and apply fade-ins and fade-outs to the different regions, and then save to a new file. Then you could create a comped file and save it under a new name. From there you might adjust the timing of a few parts, apply a little pitch correction, and adjust a few levels before creating a final file. The only real downside to this approach is that you can end up with a lot of different versions of files (depending on just how cautious you are). Modern hard drives are large enough for the actual size of all these files to be not much of a worry, but that doesn't allow for the fact that, unless you are very strict with your file naming and general "housekeeping" within your project, you can easily end up very confused if you ever have to open up the project at a later date.

One useful method is to settle on a suffix scheme that you can apply when you name your different stages of edits. In the example given above, you could save the original untouched file as *Test-ORIG.wav*, the cleaned-up version as *Test-CLN.wav*, the comped version as *Test-COMP.wav*, and the final version as *Test-FIN.wav*. If this is something that you are consistent with, it can greatly help your work flow and make locating different versions very simple.

As you can see, there are a great number of ways we can stop destructive edits being especially "destructive," but in the end, we will come to a point where we are happy with the result and we have committed to it. The edits are then a part of the underlying file, and we know that if we pass that particular file over to somebody else, it will be exactly as we meant it to be on their system as well. In fact, if you are editing files for somebody else, unless you are 100% sure that they have a system that is fully compatible with yours and that they will be able to open any project files that you send them, then you will, at some point, have to bounce/render your edits anyway in order to ensure full compatibility.

Another very important consideration relates to the use of the files/regions. If we commit to a destructive edit, then, because the underlying file has been changed, any instances of (or references to) that file or region throughout the entire project will be changed in the same way. How much of a problem this could be is hard to say conclusively, as it very much depends on how you work and how much copying and pasting there has been of different regions throughout a project (and also how your DAW handles destructive edits, as some will automatically create new files for destructive edits). On the one hand, if there is a particular sound that you have used a number of times throughout a project and there is some need to change an aspect of it, if you make that edit destructively, then you know that all instances of it will be changed, and that method can save you a lot of time. Equally, though, if you don't realize that you have used the same file or region somewhere else in the project, then making a destructive edit could cause problems elsewhere that you don't even realize.

If, however, you have a file or region that is used in multiple places throughout a project and you want to apply a change to it in only one of those places, in that instance, it would be better to apply the change nondestructively to ensure that no other copies are affected. But at the same time, if you apply changes like this nondestructively, then there is always a chance that you might miss one copy of it somewhere in the project.

So there is no conclusive right or wrong approach. Each method has its advantages and disadvantages, and your choice will be largely down to the individual project and your personal preferences and work flow. I would always advise, however, that you check for multiple copies (or aliases) of any file or region before you do any destructive edits, just to make sure that you know exactly what you will be changing.

DIFFERENT TYPES OF LEVEL CONTROL

Destructive edits

The most basic of all destructive level control methods is to apply an overall gain change to a file. This is usually a simple case of adding or subtracting a certain amount of decibels consistently over the whole duration of the file or region. However, if too much of a positive gain change is applied, then the resulting file will be clipped and distorted, and this can be a big problem. In order to avoid this happening, if getting the maximum possible level is what you are trying to achieve, there is a dedicated function: Normalize.

What the Normalize function does is to first scan the file to determine the current maximum level in the file and then apply a gain change to the entire file to bring that current peak level up to the maximum possible value. That doesn't mean, however, that all or even most of the file will even be close to the maximum possible level. If that is what you want to achieve, then you will need to consider something like a limiter plug-in, and, to be honest, anything like that should probably be avoided during the editing process and left until the mixing process. In fact, I would probably advise, as a general rule, against even normalizing audio files. When a file is normalized, it means that any additional processing (during either the editing or mixing process) that applies any kind of positive gain change will result in the file clipping (briefly going beyond the loudest possible level resulting in the waveform being limited to the maximum level for the furation of the "clip" and the resulting waveform shape being changed or "distorted) at some point, so anything we can do to avoid that is welcome.

In addition to these constant gain changes, many DAWs and stand-alone wave editors also allow us to apply "gain envelopes." This simply means that the gain we apply isn't constant throughout the whole file or region. We could use this for something as simple as a gradual increase or decrease in gain (similar to fades) or perhaps a situation where gain is applied only to certain parts of the file or where something more complex is required.

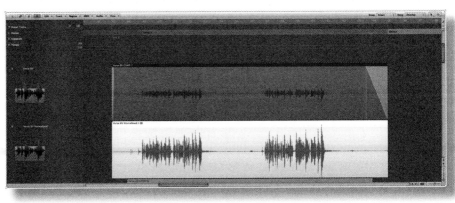

FIGURE 5.1
Normalization is one of the simplest forms of level control in audio editing. The idea is to make a recording as loud as it can be without clipping, as you can see in the example above, where the bottom track is a normalized version of the track above it.

Level Control   CHAPTER 5

Moving on, we will take a look at the many ways in which we can control levels nondestructively. They generally fall into one of three categories—static, variable, and automatic—and we will look at each in turn and look at the pros and cons and discuss possible applications.

Static nondestructive edits

The easiest way to achieve a nondestructive gain change, where applicable, is to simply add a gain "offset." This will add (or subtract) a particular amount of decibels to the level upon playback. It should be noted, though, that because this is operating on the file itself, it will change the level of the signal going in to any plug-ins on the mixer channel for that file. If you have any plug-ins that are level dependent (compressors, noise gates, amp modeling, etc.), then changing the gain in this way could change things other than just the volume of the sound. On the other hand, this may be exactly what you are looking for; perhaps one region is being compressed a little too hard, or perhaps one region is a fraction too quiet and isn't opening the noise gate. Using volume changes in this way can help to resolve those problems.

Another option, if it is available, is to use a region volume envelope, which is often drawn directly onto the region in the Arrange window. While these are generally used for more complex and variable changes, you can, if you choose, use them for constant gain changes, but you should once again remember that doing this will also change the volume of the sound *before* it reaches any plug-ins, so the same caveats apply regarding level-dependent plug-ins.

A third option is to use a gain-changing plug-in in conjunction with track/channel automation. This plug-in will have the same effect as moving the main channel fader but with the additional flexibility that it can be placed anywhere in the signal chain after the audio files itself and before the final channel fader. This has the added benefit that, when you have a gainer plug-in on a channel, it could serve as a reminder you that you have made some level changes on that track. Also, by using a gainer plug-in, you can try moving its position around to precede or follow any other plug-ins that may be on the channel, to determine where the volume change is most suitable. You would need to make any adjustments to the level on the automation for the plug-in, but that is pretty similar to making changes on a region volume envelope, so, with the exception of opening the plug-in and moving its position in the chain, the process is very similar to using the region envelopes, only it offers quite a lot more flexibility.

The final option is to actually automate the level of the main channel fader. If we have no additional processing on the channel (no plug-ins), then all four methods should give us exactly the same result. Any processing, however, would mean that one of the other options might be better. This method, like the others, does have its disadvantages. Although you wouldn't ordinarily do so during an editing process, you might find that you need to make some quick level changes during a later mixing stage, and, in that situation, you might have reverb or delay plug-ins on the channel itself. Using any of the first three methods will allow the reverb or

Different Types of Level Control

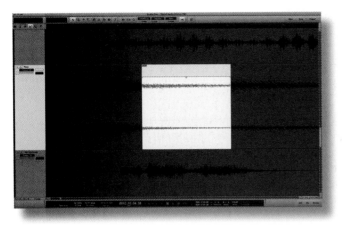

FIGURE 5.2
Volume envelopes (or Clip gain) allows for simple control of a region or part of a region within the Arrange window. This type of level control is independent of any volume automation and will take effect prior to any plug-in processing.

delay effects to tail off naturally, but if you automate the main channel fader to do a quick drop of a couple of decibels on a particular word or note, then you will also be dropping the volume of the reverb or delay tails temporarily, which can sound very unnatural. Therefore, automating the channel fader itself isn't really especially suitable for "spot" changes to volume levels, unless the channel itself has little or no plug-in processing on it, and certainly no reverb or delay effects.

Variable nondestructive edits

Each of the methods that we described above for static level changes can also be used for variable level changes, with the (obvious) exception of the simple region gain offset that we listed first. And each of them has the same benefits and drawbacks associated with them in these situations as well. In fact, the only real difference between the two situations is that of subtlety. The variable level changes that I am talking about here are not aiming to even out the volumes throughout an audio file, as that is more of a mixing task. Instead, what I am talking about is still, in essence, a corrective measure. It would be used in situations where what we are looking for isn't that different from a static change but perhaps with a little extra control.

One example might be in a comped vocal where there is one word from one take in the middle of two words from a different take. The vocalist may have sung the first take a little closer to the mic or with a little more power, so the first take is just a little louder than the second. The word we want to use from the second take starts off at a good level but then tails off slightly in volume. The beginning of the word, therefore, might sound like a good level compared to the first take, but we might feel that the end of the word just needs a little lift to bring it more into line with the word that follows. We are walking a fine line here between editing and mixing, and your own opinion may be that what I have just described is a part of the mixing process, but, even though it is a close call, I would consider something like this to be editing, purely because we are doing it as a part of the comping process, and the use of this slight volume ramp is to make the inserted region the same (dynamically) as the one we are replacing with it. In that sense, at least, we aren't making any major creative decisions. We are just making the word from the second take as close as we can (dynamically) to the same word from the first take.

Needless to say, if we are going to do that, then there will always be a temptation to just go in and poke around with things a little more. And this is where you will have to learn discipline. If in doubt, just ask how far your client wants

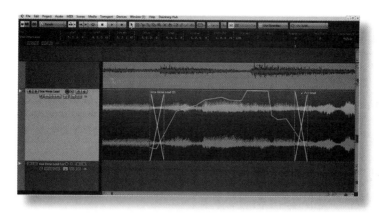

FIGURE 5.3
In some cases, volume envelopes can allow for far more elaborate control of the recording, as shown in the example above. The degree of complexity here allows for very accurate fine-tuning of the recording, before it is processed by any plug-ins, which can reduce the need for relying on compression to even out volume discrepancies.

you to go with the editing. If it is strictly an editing job, then you probably shouldn't be overly concerned with evening things out too much, as anything beyond simple, corrective level changing is more the realm of mixing and production.

Automatic nondestructive edits

The final type of level control that we can incorporate into the editing process requires the use of additional plug-ins that will take some of the guesswork out of controlling the levels. In most situations using plug-ins (dynamics plug-ins in this case) is something that would generally come further down the line, but there are a few situations when plug-ins can be used as substitutes for some of the manual techniques we have been looking at.

A simple and fairly noninvasive example is to use a compressor (on a drum sound, for example) with a fairly gentle compression ratio (no more than 4:1 or so) and a Threshold set at 2 to 3 dB above the average level, with Attack and Release setting appropriate for the sound. The overall effect of this compressor should be—nothing! If we have set this up right, then the compressor will hopefully do nothing at all but will always be just on the edge of kicking in, and if it occasionally kicks into action, then it's probably OK. If, on the other hand, it is taking action every couple of bars or even more often, then we might need to adjust the Threshold a little. This may seem pointless, but in fact we are just using it as a safety net in case anything slipped by unnoticed. You could arguably use a peak limiter instead of a compressor, but, in a situation like this, I think that a compressor is a gentler approach, and, seeing as we still have the mixing stage to consider, if we can avoid squashing the sound at any point, it gives the mix engineer the most possible freedom. It should also be noted that, for a task like this, we would want a compressor that is as "transparent" as possible and not one that adds a vintage character or anything like that. That may well be desirable at the mixing stage, but for the purposes that we have here, the cleaner, the better.

HANDS ON

Introduction

While a large amount of level control will take place at the mixing stage (and rightly so) for a number of reasons that we have outlined in this chapter, there are times when it will be a part of the editing process. With the ease of creating automation data in modern software and the proliferation of level-control plug-ins, many people will use these options, but there are other options available if you are aiming to simply raise the level of an individual region or selection by a static amount.

Logic

There are two types of nonautomation level control in Logic, and those are destructive and nondestructive. The choice of which type to use is, I find, largely dependent on what you are aiming to achieve. If you are looking for broad changes to an entire recording, such as bringing up the overall level to compensate for a quiet recording, then the destructive methods will be more appropriate. If, on the other hand, you are looking to make changes to smaller regions or selections in context, then the nondestructive method is usually the best choice, as it gives a very quick and easy way of changing the settings later if you feel they aren't quite right.

In order to apply the destructive processes to an audio file, you will first need to have the *Sample Editor* window open for the file you wish to work on. This is achieved by clicking the *Sample Editor* button at the bottom left of the Arrange window. It is possible to set up a key command to do this, but none is set by default in the Logic preferences. Once the Sample Editor window is open, there are two main options for static level control. The first of these, Normalize, can be accessed by pressing Ctrl + N. When you use a key command to start the normalizing process, you will see a dialogue box that gives you a warning that this is a destructive process and asks if you wish to proceed. If you choose the Normalize command through a menu, you don't get this warning, but that is because it is far easier to inadvertently use an incorrect key command than an incorrect menu choice, so there is this additional safeguard to protect you.

If, instead of normalizing, you want to apply a fixed gain of your choice to the region, then you should choose Ctrl + G, which brings up the Change Gain dialogue box. Here you have choices to set the gain change in absolute (dB) or relative (%) terms. Usefully this box will indicate if the desired gain change will result in clipping before you apply the processing. Once again, if this process has been initiated through a key command, you will receive the warning about the destructive process.

It is worth noting that, with both of these key commands, you will need to make sure that the focus is on the Sample Editor window rather than the Arrange window. If you select an audio file on the Arrange window and then open the Sample Editor window, the focus will automatically be set. You can tell where the focus is, because there is a lighter highlight around the edge of whichever window has the focus. If you wish to simply process the currently selected region, then you shouldn't have any problems. If, however, while the Sample Editor is open, you select a different region in the Arrange window, you will notice that the focus goes back to the Arrange window. If you now try to use your key commands, they won't work, and you will need to click back into the Sample Editor window (click on the menu bar at the top rather than the waveform display) to get the focus back before they will work.

Moving on to the nondestructive level control, there is only one real option available. which is the Gain parameter available in the Inspector panel to the left of the Arrange window. If this isn't visible, then simply click on the Inspector button (a blue circle with a white "i" in it) at the very top left of the screen. In this Inspector panel, you will see a number of parameters that all apply to the selected region(s), but the one we are interested in here is the Gain parameter. Clicking and dragging enables you to create a nondestructive gain offset between –30 dB and +30 dB in 1-dB increments. Unfortunately, as useful as this feature is, it is arguable that it could have been taken further. For many, the minimum increment of 1 dB could prove restrictive, but perhaps more of a potential problem is the fact that the waveform display inside the regions in the Arrange window doesn't change to reflect the change in gain. While any obvious clipping as a result of a gain change applied in this way would be audible, it would still be beneficial if the waveform display did change as any gain offsets were applied, as it does in some other DAWs. At the very least, it would be helpful if each individual region showed, in the region itself, if the gain had been offset and by how much, if only as a time-saver to avoid having to select each region individually and look at the Inspector panel to see what changes had been applied.

Finally, it is worth again referring back to the Hands On section at the end of chapter 3 and the use of the Marquee tool to create automation nodes, because this does give us a way, albeit a little bit of a work around, to actually get a visual display of any gain changes applied. By creating these automation nodes

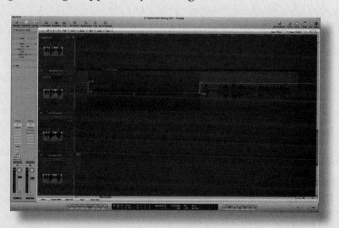

FIGURE 5.4
Other than a region gain parameter in the Inspector, Logic doesn't have any additional pre-plug-in level control options, but, if it is useful, using the Marquee tool on the channel volume automation track will automatically create four nodes at the edge of the Marquee selection to allow for quick volume changes to specific parts of a region.

for any regions or selections that you wish to change the gain of, you will have a visual reference of any gain changes when the automation is visible. The biggest caveat to all of this is that, if you apply automation to the volume, it will be *after* any plug-ins, and they may not have the desired effect. If, for example, you wanted to change the gain of a particular word on a vocal that was causing the compressor plug-in to work too hard, changing the level of the volume with automation wouldn't solve the problem. In cases like this, it can be better to load the Logic Gain plug-in into the first insert and then automate this plug-in rather than the main channel volume, as this will affect the level going into any further processing, which, in most cases, would be the purpose of level control at the editing stage.

Pro Tools

Pro Tools is one of the better-equipped DAWs when it comes to nondestructive level control. Of course there are the usual options for volume automation of the channel fader, but, if we are looking to avoid that for the reasons listed in the main chapter, then Pro Tools has the excellent Clip Gain feature that allows us to do a great deal with the level of the region before any plug-in processing takes place. The first thing that we should clear up is a terminology matter. For consistency with the main chapter, I refer to individual sections of audio as regions, but, from Pro Tools 10 onward, these are now referred to as clips. While the name difference may be misleading, the two are actually identical in the sense of what both terms mean. So when I am referring to *Clip Gain*, you can think of it as *Region Gain*.

There are two ways in which we can use the Clip Gain feature, depending on what exactly it is that we are trying to achieve. The simplest of the two, which is perfectly sufficient if we simply want to adjust the gain up or down for the whole of a region, is to use *Clip Gain Info*. We can enable this by pressing Ctrl[Mac]/Start[PC] + Shift + = or by going to View > Clip > Clip Gain Info. When this is enabled, you will see, at the bottom left of each region, an icon that looks like a small fader along with a numerical gain reader in decibels. Clicking on the small fader icon will bring up a miniature fader next to the icon, and you can drag up and down while holding down the mouse button to change the gain. As you do this, the numerical readout will change, and, as stated, the waveform overview will change to reflect this.

At the end of chapter 3, we looked at ways of zooming in to the waveform overview to get a more detailed look at it, and at the time, we stated that this zoomed-in view was not the same as a change in gain. The unfortunate thing is that, at a glance, there is no easy way to tell if a waveform that looks like it will be clipping is as a result of a zoomed-in view or excessive use of the Clip Gain. Of course the numerical readouts will show if you have applied any gain, but it is distinctly possible that you have applied a small amount of gain that hasn't resulted in clipping, but, because you are still zoomed in, it looks like you have. The best way to avoid situations like this is to always reset any vertical

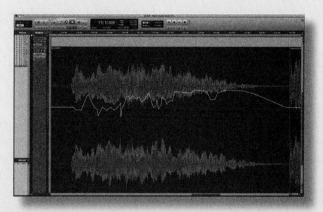

FIGURE 5.5
The Clip Gain Line in Pro Tools allows for extensive level control of audio regions. In addition to the flexibility that this gives, the waveform display also updates in real time, allowing you to get good visual feedback on the changes you are making.

waveform zooming to the minimum level once you have done what you need to do. If you follow this routine, then any subsequent gain changes will be reflected accurately in the waveform overview.

If you want to do something more complex than a simple, static gain change for the whole region, or if you want to change the gain of only a part of a region, then you can use the other method by pressing Ctrl[Mac]/Start[PC] + Shift + – (minus) or by going to View > Clip > Clip Gain Line. This will overlay a line, similar to an automation line, on top of the region. At the beginning of each region, you will notice a small circle on the line, which represents a breakpoint. Each breakpoint defines a level and a position. You can adjust the position of individual breakpoints up and down, and you can create new ones to enable more-complex changes. The specific changes that you can make to these breakpoints depend on the tool you currently have selected.

If you have the Smart tool selected, and you move the cursor close to the Clip Gain Line, you will see the cursor change. You can click and drag the line up and down using this cursor, and it will affect all breakpoints within the region equally. The region doesn't have to be selected for this to work, but, if you do have multiple regions selected, this method will apply the same gain change to all currently selected regions. As you are dragging using this tool you will notice that, at the top of the region, you get a numerical readout that shows you the absolute gain change applied (based on the original level being 0 dB) and the relative gain change that you are applying with this use of the tool. If, for example, you had previously applied a gain change of –2 dB, and you now used this tool again to add a new change of –3 dB, then the display would show –5 dB on the left, as this represents the overall change from the original region, and –3 dB on the right, as this represents the change applied at this time.

Another option with the Smart tool is to move the cursor to the top of the region, so that the Selector tool becomes active, and then click and drag a selection within a region. If you then move the cursor over the Clip Gain Line, you will be able to change the gain for only that section. If there is already a gain change happening within that selection, you will be able to move it up or down, whether it is static of changing, using this method. New breakpoints are created at the selection boundaries, and the gain change is applied only to the selection and won't change anything for the rest of the region.

If you have the Grabber tool selected, and you move the cursor close the Clip Gain Line, then you will see the cursor change to a pointing finger with a small "+" sign. Clicking on the line will create a breakpoint at that position, while clicking and dragging will both create the breakpoint and change the gain value. If there is another breakpoint after the one you are creating in the current region, then raising or lowering this new one will cause a ramp from the new breakpoint to the following one as well as from the preceding one to the new one. But if there is no breakpoint after the one you create, then there will be a ramp only from the preceding one to the new one, and then the gain change will stay constant to the end of the region. If, however, you move the cursor over an existing breakpoint, then you will see only the finger icon without the "+" sign, and if you click and drag, you can change the position (both level and time) at this point.

If you need to make only a minor (and consistent) change to a region, then you can use the key command Ctrl[Mac]/Start[PC] + Shift + up or down arrows. Any regions that are selected when you use this command will have their gain changed (nondestructively, remember) by 0.5 dB. This is the default value and can be changed if you wish by going to Setup > Preferences and then choosing the Editing tab and then changing the *Clip Gain Nudge Value* (located under the Clip heading).

Studio One

Studio One has a few different tools for level control, depending on what you want to achieve. One of the simplest things you can do is to *Normalize* the region (Audio > Normalize Audio or Alt + N). Unlike some DAWs, this isn't actually a destructive process and simply automatically calculates the required gain offset to achieve the same as normalization and applies it to the region in real time. In some ways, this could be considered a superior system to the traditional destructive version, because it means that you can normalize a region or file, perhaps from a sample library that you have, without risk of it changing the original file and that having repercussions in other projects. What it will do, of course, is add another layer of nondestructive processing that needs to be carried out in real time, but, given that normalizing is a simple gain change and therefore not too CPU-intensive, this shouldn't present a problem.

Another level control process that Studio One can take care of is the removal of silence in audio regions. Now silence, in audio terms, is very subjective. Technically silence would have a zero digital level, but, for all intents and purposes, we can define it as sound that falls below a certain threshold of audibility. In truth a very quiet, almost inaudible sound on a single track is very unlikely to have a detrimental effect on any finished product, but, when we are talking about tens or sometimes even hundreds of audio tracks, these small background noises can mount up and create something that is very much audible as a whole.

As a result, audio editors (and producers and mix engineers) are often very keen to make sure that background noise is at an absolute minimum.

Traditionally this was achieved by using a noise gate that, depending on the exact characteristics of the source material, was often very effective. However, even though there are noise-gate plug-ins available for those who like to work fully in the digital domain, there are other options that could be preferable. Studio One has a *Strip Silence* function that performs much the same task as a traditional noise gate. There are a number of different parameters for Strip Silence that can be adjusted to fine-tune the result, and to adjust these, you need to open the Strip Silence panel by going to **View > Additional Views > Strip Silence** or by clicking on the Strip Silence button in the toolbar at the top of the screen. This will open up a panel with a number of different controls, of which the most fundamental is the *Detection* control. There are four different options for the *Material* control, and of these, the first three set the *Open Threshold* and *Close Threshold* values automatically. These three options are optimized for material with different amounts of silence, ranging from the most silence (*Lots of Silence*) through to very little silence (*Noise Floor*). The fourth option, *Manual*, allows the user to set the threshold values manually.

Under the *Events* heading are controls that relate to the splitting of regions and the new regions that are created. The *Minimum Length* control allows you to define a minimum size of region that will be created. This is to avoid ending up with too many regions (that are each tiny) in particularly choppy material. The *Pre-Roll* and *Post-Roll* controls serve to extend the actual regions created by the values used here, so allow a small amount of time on either side of the silences to minimize the chances of clicks. And the last two controls, *Fade In* and *Fade Out*, work in conjunction with the Pre-Roll and Post-Roll controls to further minimize the chances of any clicks.

Once you have set the parameters how you want them, click on the Apply button, and the region will be processed according to the set values. However, there is a shortcut if you simply wish to apply to a new region using the previously selected settings. To do this you need only to go to **Audio > Strip Silence**

FIGURE 5.6
The Region Volume Envelope in Studio One, while simple, offers a very quick and convenient way of making region-by-region level adjustments that are accompanied by real-time updates of the waveform overview, so that you can see what effects the changes you are making are having.

(or right-click on a selected region and choose Strip Silence from the contextual menu), and the currently selected region(s) will be processed.

The final way—other than track automation, of course—that we can affect level changes is to make use of the Arrow tool once again to adjust the region *Volume Envelopes*. The fades that we looked at in the last chapter are one part of the overall Volume Envelope process, but there is one final use of the Arrow tool that we haven't yet looked at. If you move the Arrow tool to the top center of a selected region, you will see a small square, and the cursor will change to a finger icon. Clicking on this square allows you to apply a (nondestructive) gain offset on a region-by-region basis. As you click and drag, you will see a pop-up information box that will show two numbers. The number on the left is the overall gain offset applied, while the number on the right is the gain offset applied by this click-and-drag movement. You could, for example, create a −6-dB offset with one click and then, sometime later, come back and apply an additional −3-dB offset. In this case, the number on the left would read −9 dB, whereas the number on the right (at the time of the second change) would read −3 dB. This is very helpful to keep you informed of not only your immediate actions but also the cumulative effects of all the changes that you have made.

Cubase

Cubase has both destructive and nondestructive methods for effecting level control, and the destructive methods offer a lot more control than you would ordinarily expect in a DAW. But before we get to that, let's take a look at the more simple types.

The most basic forms of destructive level control are the *Gain* and *Normalize* functions. Both of these apply a fixed gain adjustment to the entire region and differ only in implementation. *Gain* (accessed by going to Audio > Process > Gain) allows you to create a gain offset of between −50 dB and +20 dB by either adjusting the slider, clicking on the up-and-down buttons to the right of the display (increases or decreases by 1 dB at a time), or by double-clicking on the readout and typing in a value. Beneath the slider you will see text that says "No Clip Detected," which allows you to determine whether your gain change will lead to clipping. However, this is accurate only if you click on the Preview button at the bottom of the screen. This can be very misleading if you aren't paying attention and should perhaps update automatically every time you change the gain amount. At present, though, it doesn't, so you just need to be aware, whenever creating a positive gain change, to preview it first in order to know whether it will lead to clipping.

The Normalize function (Audio > Process > Normalize) in Cubase works a little differently than some other DAWs in that it doesn't only adjust the loudest peak to maximum level. That is ordinarily what a normalize function does, but Cubase goes a step further in adapting the process, so that, instead of just applying a gain change that will make the loudest part of the region scale to full level, you can actually specify what that "full level" is. If you wanted to scale the region so that the maximum peak was at −10 dB, then you would just adjust the slider (or up and down buttons, or double-click and type in a value) to −10 dB, and then the region would be scaled accordingly. In effect, it is like a traditional normalize followed by a gain change.

If, however, you want to do something a little more creative, then you should go to Audio > Process > Envelope, where you can get far more creative. The actual envelope editor has a lot in common with the fade editor that we looked at in the last chapter. It has the same three Curve Kind settings, and it uses the same concept of nodes (that we can add, move, and remove). The biggest difference here relates to the fact that, unlike a fade, it doesn't have fixed start and end points. It always has a minimum of three nodes: one is anchored to the start of the envelope but is variable in position from maximum level to minimum level, another is similarly anchored to the end of the envelope, and the third is freely adjustable in both level and time. There are also two waveforms displayed, with the original waveform being lighter gray and the new, postenvelope waveform being a darker gray.

This additional flexibility more than makes up for lack of the preset curve shapes, which, in all honesty, aren't as relevant here, and the fact that your curve creations can be stored means that you can apply the more-complex creations that you make to a number of different regions. However, as with the destructive fade-ins and fade-outs that we looked at, there is no way of specifying a timescale for these envelopes, as they simply get applied to the entire length of the region. If you wanted to synchronize a complex volume envelope across a number of different tracks, then you would have to make sure that you used equally sized regions from each track; otherwise, the level changes wouldn't be synchronized.

Moving on to the real-time level control, we again have both simple and complex options available to us. The most-simple form of real-time level control is to use the *Volume Handle* for a region. If you have the *Object Selection* tool selected, and you click on a region, you will see, in addition to the fade handles that we spoke about in the last chapter, an additional handle, represented by a square located at the very top of the region in the center. Moving the cursor over this changes it to a pair of vertical arrows, and clicking and dragging creates a gain offset to apply to the region at playback. As you drag, you will see a pop-up box below the cursor that shows the overall gain offset applied (on the left) and the gain change applied only on this change (on the right). If you look toward the top of the screen, you will see a parameter called *Volume*, which represents the gain applied using the Volume Handle. You can click and drag on this number to adjust it, or you can double-click on it and type a value (between −60 dB and +24 dB) if that is what you prefer. Any gain change you apply using either of these methods results in a visual updating of the waveform overview to reflect the changes.

FIGURE 5.7
Cubase has both a fixed gain Volume Handle and a variable Volume Envelope that work in conjunction with each other. This makes it easy to create variable changes to even out the dynamics of a performance while still being able to click and drag to adjust the overall level of the region in comparison to surrounding regions.

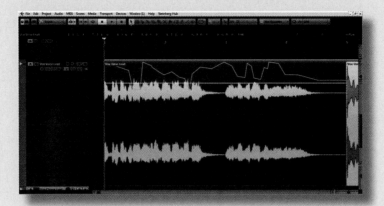

This Volume Handle, combined with the Fade Handles, gives you quite a decent amount of real-time control over the level of a region, but, in the event that you need something a little more, you can use the *Draw* tool (press 8) to create more-complex volume envelopes. Clicking anywhere on a region with this tool will create a new node for this volume envelope. If this is the first node, then a simple horizontal line will be created, but additional clicks will create additional nodes. In principle, this is very similar to the envelope editor we looked at above but with the difference that the draw tool will create only linear envelopes and not spline-based ones. Otherwise, it works exactly the same. The envelope doesn't have to start at maximum or minimum, and it doesn't have to end there, either; it can have any number of nodes making it up, and nodes can be removed by dragging them outside the region.

Volume envelopes created in this way cannot add gain. When a node is positioned at the very top of the region, it represents a gain change of 0 dB and points lower down represent negative gain changes. If you already have fades or a volume change in place, and you then create a volume envelope using the Draw tool, then both will be applied to the underlying audio region. Like the static volume change, any changes made by the volume envelope will be added to the waveform overview in real time.

CHAPTER 6
Tonal Matching

WHAT IS TONAL MATCHING?

Before we move on to our next subject, comping multiple takes into one "master" take, we need to factor in one final technique: tonal matching. What I am referring to here is the process of trying to match the frequency response of different sounds. These could be entirely different sounds but is more likely to be attempting to match the sound (in terms of the frequency response) of alternate takes of the same instrument. This is most likely to occur because any acoustic instrument (including the voice) has the potential to have different tonal characteristics between different takes.

One reason for this is that no microphone has a consistent frequency response at all distances and from all directions. In the event that the microphone moves relative to the sound source between takes, there is a good chance that the tone of the sound as picked up by the microphone will change, even if the actual tone at the source hasn't changed. Another possibility is that the different takes were recorded using different microphones or in different rooms. This isn't a very common occurrence, but it can happen if there is a need to rerecord some parts at a later date, perhaps owing to a musical change that happens at the last minute, and the original studio or equipment is not available for the rerecording session.

A final possibility with acoustic instruments and voice is that the player/singer actually performed the part slightly differently between takes. A singer could deliver the part with more "breathiness" or more power, and a violinist could move the bow to a different part of the string, creating a tonal change. Things such as fatigue on the part of a singer or even weather and atmospheric conditions can all subtly change the tone of acoustic instruments. These changes can happen over the space of hours, and, in the case of a singer who records different takes days apart, the change can sometimes be very drastic.

The process of tonal matching can be, and traditionally would have been, carried out using a normal EQ and a very good pair of ears. There really is no magic formula that can be given to make you good at tonal matching by ear; it is simply a case of setting up an EQ band with a small (a few decibels) boost and then sweeping the frequency around (within the usual frequency range that a vocal lies within) while comparing the two vocal parts, until you hear something that sounds a bit closer. It may be that you need to do this with more than one EQ band to get close to matching the sounds; you could even try using a graphic EQ for an increased number of available bands, but with a little time you can probably get quite good results. After all, we don't need a 100 percent accurate tonal match, just one that sounds within a natural range of variation for a single take. Any kind of spectrum analyzer plug-in can be useful here, because, if there are any very obvious tonal differences between the two takes, these may well show up on the analyzer, and this can give you a good indication of what frequency to use as a starting point.

If the response time (the time taken for the analyzer to react to changes in the frequency spectrum) is adjustable, then it would be very useful to slow that response down, so that what it shows is more like a constantly moving average. This is more beneficial to us, because we are trying to use the EQ to approximate an average tonal change rather than an instantaneous one. It will be easiest if you have one analyzer on each of the two takes and have them both visible while you are adjusting the EQ. Then you can start to fine-tune the frequency, the actual amount of boost (or cut, of course) and, if applicable, the bandwidth (or Q) of the EQ. The analyzers will prove useful here, as they can show you visually if the adjustments that you are making are bringing the two sounds closer together, but as always you should rely on your ears more than your eyes.

However, as with so many aspects of recording, recent technology has given us tools that can either take care of this process for us. or, at the very least, give us a very good idea of where to start. What I am referring to is EQ plug-ins, commonly referred to as matching EQs. The idea behind them is simple. You first play the sound that has the desired tone, and while this is playing the EQ will analyze the frequency spectrum and then store this as a "map" of the desired tone. Then you play through the sound that you want to change, and the analysis is carried out again. Once you have done this, a correction curve is calculated, which is a model of the differences between the two sounds and applied using something very much like a graphic EQ with a very large number of bands. Most matching EQs also offer some kind of "depth" parameter, so that you can gradually adjust between the original sound and the corrected one. There is also very often a "smoothing" parameter, which has the same effect as reducing the number of individual bands and making the overall effect a little less audible.

These kinds of matching EQs are not always 100 percent effective, but, if used carefully, they can be useful. It is also possible to use matching EQs to give you a visual representation of the EQ change needed to use as a reference point and then go in to another EQ plug-in, perhaps one with good adjustability or perhaps one that just has a good tone, and then try to re-create the EQ curve from the matching EQ manually.

Tonal Matching　CHAPTER 6

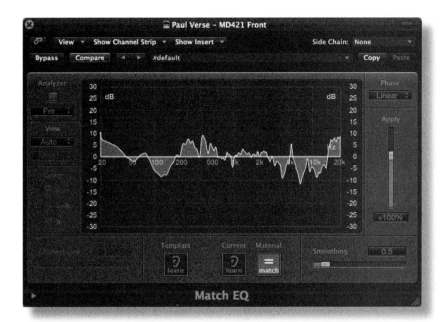

FIGURE 6.1
Matching EQ plug-ins such as the one above will analyze the spectral content of a source (or "Template") audio file and then of a destination ("Current") file, and, once the analysis is complete, will create a detailed correction curve that shows the difference in tonal balance between the two. Often this curve can be smoothed and also applied in varying degrees.

Clearly this isn't a two-minute job, and we need to consider that we could well need to do this whole process to match up several different takes or even different parts of the *same* take. For this reason, I will often leave the tonal matching until after I have completed the comping process, because, up until the comping is completed, I won't necessarily know which regions will need to be matched up against which other regions. There may be occasions when there might be an obvious single tonal mismatch that is just annoying or off-putting—as might be the case if there is a cross-fade in the middle of a word—in which case it might be worth taking a few minutes out to roughly match it up, but otherwise it makes the most sense to leave final tonal-matching decisions until after the comping choices have been made.

CHAPTER 7
Comping and Alternate Takes

THE SEARCH FOR PERFECTION

Comping, or compiling, is the process of putting together one "master" take from a number of alternate takes, using the best parts from each to create one final version. It is a very widely used technique, but possibly the most common application is vocal comping. Because of that we will focus on using all the things, we have learned so far to comp a vocal. Most of the things we cover will be common to comping all different kinds of instruments and voices, but there will be specific issues for certain types of instruments, and there is additional information on the website about these additional considerations.

But first, why do we bother comping? To answer that, we need to accept the fact that, as human beings, we can't achieve 100% perfection for 100% of the time. Even if the performers are well-rehearsed and know what they are doing and are technically competent and artistically expressive, there are still physical limits placed on us. First, even for the most experienced, a studio environment, especially when the "tape is rolling," increases the pressure to get things right. Some people thrive in this environment, while others fall apart. Furthermore, when you are expected to give it everything you have, you can't be expected to give it for too long. Fatigue sets in eventually, and for instrumentalists this can lead to less accuracy, less expression, less emotion, and there is nothing that can be done about it other than resting and coming back later.

Vocalists in particular aren't governed only by their physical abilities to "play" their instrument but are also very much at the mercy of the fact that their instrument is a part of their own body and can be affected by a number of things that are all beyond their control. Trying to push a vocalist too hard for too long will only be a downward slope. Also, most vocalists will, at some point in their career, have pushed too hard and could well have lost their voice, so they won't be in a hurry to do that again. When you couple their understandable resistance to damaging their voice with the increasing pressure as the day wears on, it is completely understandable if the quality of the performance starts to falter eventually.

So we can't reasonably *expect* a performance to be 100% perfect from start to finish. That's not to say that it couldn't happen, but the chances are incredibly small. Additionally, even if a singer were to be 100% pitch-perfect, with immaculate timing, expansive dynamics, wonderful delivery, and sublime tone, we still have to realize that each performance is just an interpretation. You could have two takes that were both perfect, but, at the same time, different in some respect (emotion, tone, etc.), and you could clearly prefer one over the other just because of the feeling it gave you. Sometimes, there is a lot more to a perfect take than just timing and pitching.

This gives rise to an interesting question of what would be preferable to us: 50% perfect 100% of the time or 100% perfect 50% of the time? The answer to this is not only very personal but also it depends largely on circumstance. For example, if we are considering a studio recording environment, where we have the ability to perform all kinds of magic on our recordings, we could make do with either of these scenarios. If the singer was 50% perfect 100% of the time, then we have tools to help get that 50% up quite a lot higher. We can adjust timing, pitching, tone (in terms of matching tone between takes), and dynamics. In fact, the only two things we really have no control over are the *basic* tone of voice and the emotion/delivery of the performance. Equally, though, if a singer is 100% perfect but only for 50% of the time, then we have the option to comp things together, to assemble a "best of" take and get that 50% much higher. In other words, we can deal with both scenarios.

Having said that, many would prefer, in a studio environment at least, someone who was closer to the 100% perfect but for a lower percentage of the time. This is purely a practical consideration. We have to consider that comping and perhaps moving a few parts around will have less potential risk of artifacts and other unwanted side effects than if we have to start time-stretching, correcting pitch, and matching up tone between takes. In addition, it is a far easier process. So for those reasons alone, it can be easier to work with somebody who is closer to perfect but for less of the time. Of course there is always a balance and a tipping point. It would be much easier to work with somebody who was 75% perfect for 50% of the time than somebody who was 100% perfect but only 5% of the time.

In summary, comping as a skill in general, and in particular vocal comping, is perhaps one of the greatest assets that you can have as an editor. In a high percentage of modern music, the vocalist is the absolute key feature, and the vocals will be riding high above everything else: loud and proud, front and center. So if you can perfect your vocal edits and comping, and in the end have it sound like the vocals were all recorded in one perfect take, then your skills will undoubtedly be in demand.

PUTTING IT ALL TOGETHER

The idea of comping shouldn't be new to us. Even if you didn't know about it before, we have referred to it several times in the earlier chapters, so it should be starting to become familiar now. We have looked at all the types of skills you would need in order to do successful comps, and now we have arrived at the point where we can put it all together into a very useful and often-needed practical use. What we will do now is to walk through the process. We will touch on technical issues but won't dwell on them overly, as most of the deeper discussion has already been done. Instead we will focus more on the process itself: the decision-making process, the things you need to consider in terms of the choices you make, and tips that can help make the whole process a little more intuitive.

The first thing you need to do is make sure that all the takes are lined up and synchronized. If you have done the recording yourself, then they will most likely already be lined up, but if you have received the files from elsewhere, you may have to do this on your own. With DAW software as widespread as it is today, it is fair to assume that most recordings will be made direct to DAW. If that is indeed the case, then the various takes will probably have been lined up in the original project. That doesn't mean, however, that all the files you receive will necessarily line up when you import them. If you perform a basic import of a file into a DAW, then the import process is essentially a "dumb" one. The software doesn't know the start position of the original file. If you are lucky, the files will all have been trimmed to the same start point, so all you would have to do is import them and line up the start of all the regions, and it will all play back in sync. What can sometimes happen, though, is that the original recording engineer will have cut the start point of the audio region in the Arrange window, so that everything lines up nicely, and will then send the files over to you. Sadly, cutting the front of the region in the Arrange window simply changes the start point of the playback and doesn't actually trim the file itself. This is possible and very easy to do from within your DAW, but it is a step that is often forgotten. So if, for whatever reason, your various takes don't line up nicely, then you have to move them by hand until they do.

Once everything is lined up, the next step is a simple one: *listen*. I know it might sound ridiculous, but a quick listen through each of the takes in isolation, one at a time, just to check that there is nothing unexpected, is a very good idea at this stage, even if just to familiarize yourself with the song and the sound of the vocalist before you really get involved in anything more complex.

MAKING THE GRADE

At this point it is useful to start to grade the performances. It can be very useful to use a color code here to help you identify things later. You could do this by splitting each take into sections (a bar at a time, a line at a time, potentially even a word at a time, but this is probably overkill at this stage) and then listen to each take, one at a time, with the music playing, and color each region according to your initial opinion. You could use green for regions you thought were good candidates for the final comp, yellow for ones that you thought might be usable, red for ones that you really weren't sure of, and black (or muted) for ones that were totally unsuitable or had mistakes. This initial scan-through and grading process can be very helpful, as it reduces the number of options you have to consider later.

Once this is done for all tracks, it's time to begin the comping in earnest. The concept is simple enough: you listen to each take of a particular line (or word, if it comes to that), and you choose which one you want to include in your final comp. This doesn't necessarily mean the line that is best in isolation, because there has to be a flow through the final comped vocal. Because of this, it is useful to not just loop around one line as you are auditioning the different takes but to loop around not only the line you are concerned with but also the one before and the one after. This contextual listening will help you get a better picture of the flow.

As for actually putting the comp together, there are two main schools of thought. The first is that you listen through all the options for a particular line, and then, once you have chosen which one you want, you mute all the others and leave your choice un-muted. The other option is that you create a new channel/track in your Arrange window and move the line from whichever take you have chosen on to this "master track." Both have exactly the same result, and the choice is really down to how you prefer to work with the master track, perhaps having the advantage of keeping things a little easier to interpret.

If you have, so far, separated only each take-out into lines, then you might, at this point, need to further split the lines into words in order to put together the best take. Once you have done this, or if you already did it earlier, you might still find a need to split a take midword (as we discussed in chapters 4 and 5). If this is necessary, I would make a quick note and come back to it once the rest of comp had been completed. Getting a good cross-fade in the middle of a word can take a bit of work, so it is usually more productive to leave that until the end to avoid interrupting the flow of the comping process and also to avoid your getting that one word stuck in your head too much at this stage. The same is true of any level matching or tonal matching. It is probably best to leave both these things until the end of the comping process to avoid getting caught up in the details and losing the feel of the vocal performance.

If you have gone through all the options for a particular line, and even the best one needs a cross-fade or level matching or tonal matching, and if the difference

FIGURE 7.1
When you have multiple takes, it is often useful to either name them or color them based on how good they were. The final comp will most likely include sections from a number of different takes, but an initial grading of takes is often helpful in speeding up the process.

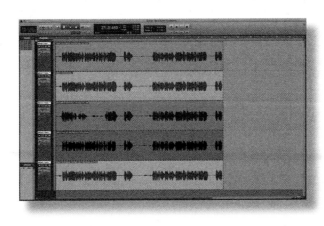

FIGURE 7.2
The traditional method of DAW comping involved cutting out sections from a number of different tracks and manually moving them to a main "comp track." While the basic principles are still the same, all the DAWs featured in this book have developed methods to make this process as automated and intuitive as possible.

is enough for it to be really obvious and off-putting, then it might be wise to at least roughly try to remedy the problem. Doing a quick cross-fade, even if it's a little bumpy, or a rough guess at a level match or a quick EQ job, might be enough to make the edit stand out a little less while you finish up the rest of the comping. The whole process can be quite time-consuming and more than a little difficult, as the differences can be very subtle at times, and there is no clear right and wrong: it is often a purely artistic decision (once obviously poor options have been removed). Once you reach the end, it is advisable to take a break, as you will be moving into a different mind-set for the next stage. Having been working with broad strokes, you now have to start filling in the detail, and a break and resting of your ears would be a good choice.

TONAL MATCHING

The obvious thing to do next would be to take care of any midword cross-fades and level matching, but to do this you may need to take care of the tonal matching first. It is important to remember that you need to tone-match any given region to the regions on either side of it. It is hoped that the regions on either side will be fairly consistent with each other, so that whatever EQ changes you make will blend smoothly on *both* sides. If this isn't the case, then you will need to make additional changes to one of the other regions as well. If you find yourself in a situation where you have too much tonal correction to do, or where you have a number of different takes and each one is tonally different from all the others, then you probably have bigger problems than a simple vocal comp. If there are too many different tones to the takes, then you can either change your choice of take for each section and try to use more parts from each take and (consequently) less different takes, or, ideally, but not especially practical, you can see if it is possible to get some additional recording done to try to get the job done in fewer takes.

When you have matched up the regions from the different takes, you have a couple of options. If the matched region is on a different track, you can't apply a cross-fade as such. Of course, you can fade-out the region on one track and do a corresponding fade-in on the other track, and, if the parameters were comparable, there shouldn't be any difference in sound. It isn't quite as convenient, though, so there is always the option to bounce the tone-matched region to a new file and then drop that new file back on to the main "master take" track and apply the cross-fades there. Which option you choose comes down to the question of convenience vs. flexibility. Keeping the tone-matched region on its own track will allow you to make further adjustments to the EQ of that region individually, but the transition between regions might take a little longer to get right. Try both, and see which works better for you.

At this point, we have reached a very important milestone in the creation of our vocal comp. By now we should have something that is sounding pretty good and quite fluid and natural. There might still be some more we can do to get it as close to perfect as we can, but we will have definitely come a long way. Now you have a choice to make. You can either push ahead and move on to the next stage of polishing, or, if you have the time and the inclination, you can go through the whole process again. The reason for doing it all again is to create another version of the comp using the best of the remaining options to serve as a "double track" vocal to thicken up the lead vocal.

The next thing two things we need to work on are the pitching and timing of the vocal. Chapters 11–13 will deal with the technicalities of this, but I will just make reference here to a few things to consider, and you can always jump ahead to those chapters if you want to read more.

PITCH ADJUSTMENTS

Vocal tuning is a very difficult subject. Technically, it is relatively straightforward, but artistically it is a little more complicated. And it is further complicated if we decide to incorporate it into the audio editing process. I mentioned right at the beginning of this book that there are many areas where editing and mixing or production overlap, and this is definitely one of the big ones. If the editing is for a project of your own, then working on pitch correction during the editing stage certainly won't step on any toes, and if you have a backing track already that the vocals were recorded to, then you have a frame of reference to use when tuning. If, on the other hand, your editing work will be passed on to somebody else for mixing and production, then, unless it has been specifically requested, I would avoid doing any kind of pitch correction at this stage—with one small exception.

The only time I would consider pitch correction to definitely be an editing task is to match up the pitches of cross-faded sections. Any need for this would have become apparent during the process of doing any final cross-fades between regions. If the difference is subtle, and if the cross-fade it long enough, then you will probably be OK, as it will sound like a gentle and natural bit of pitch drift, which can be cleaned up during production and mixing if needed. If the difference is large enough, it will need fixing, or you might even need to choose a different take. But let's proceed on the assumption that you had a very good reason for choosing that particular take at that point and need to deal with the pitch difference.

The first thing we need to establish is whether the note is out of tune overall or whether it just drifts out over its duration. If the note is out of tune overall, then it is a fairly straightforward process to correct it. Most DAWs have some sort of included pitch-shifting plug-in, so we can quickly create a new track, move the out-of-tune region onto that track, load up a pitch-shifter plug-in, and then adjust the "fine tune" control until the note is and tune and we are done—right? Actually, I would say no. As good as pitch-shifting plug-ins are, any that work in real time simply won't be of the best possible quality. And audio editing is, after all, about getting the best possible version of the files we are editing! So what we can do is use this pitch-shifting plug-in to figure out how much we need to shift by (in cents), and then, once we have that figure, we can create a new file from this region and return it to its original place on the master-take track and then go into the wave editor in our DAW (which should have a pitch-shifting option). Make sure that only the region we want is selected, and then process this region with the pitch-shifting option. Because this pitch-shifting isn't working in real time, we should get a much better result.

Notes that drift in to or out of tune will require a little more effort, and, potentially, some external software. It may not be the best idea to even consider using any kind of pitch-correction plug-in for a task like this, because, as we have already said, at this stage, we aren't trying to make the tuning more accurate. All we are trying to do is match it up to another sound, so, unless that other sound happened to be perfectly tuned, automatic pitch-correction won't work. What we need is the ability to just change the part that is out of tune with the "reference" part, irrespective of whether that part is perfectly in tune. The two most commonly used tools for this are Antares Auto-Tune and Celemony Melodyne. Both are available as plug-ins, and both are capable of doing exactly what we are looking for (and much more).

Both of them work in a fundamentally similar fashion, in that you have to record the audio in to the plug-in, which then analyzes it before you can perform any corrections. Because the audio has been pre-analyzed, we have a much higher quality than a typical pitch-correction system, and yet we have the ability to make changes to the audio as it is playing. It really is the best of both worlds. Once the audio has been analyzed, we will see an overview of what we are dealing with. Both plug-ins use a similar system—which has some of the attributes of a typical piano roll editor in that each note in the audio is presented on a grid, whereby its vertical position tells us its pitch (to the nearest note) and its horizontal position its time—but it is different in that, instead of the rectangular blocks shown in a piano roll editor, here we have the actual waveform shape shown to us. On top of this, there is usually a line overlaid that will show the exact pitch and any fluctuations from the nearest "perfect" note. This line is what we can use to determine what we need to fix, without worrying about its actual tuning. From here, we can use either of them to manually adjust the pitch of both parts of the comp to bring them into line with each other (although not necessarily perfectly in tune) and, with a bit of luck, end up with something that sounds very natural.

Once any level difference, tonal differences, and tuning differences have been dealt with and any cross-fades applied, it is time to move on to any timing adjustments that are needed. Strictly speaking, when talking about timing adjustments, we are referring to the timing of the start of words, but, given that there will be times when a move of the start point will leave a wrong or uncomfortable-sounding gap, we will also discuss about time-stretching, as the two often go hand in hand during comping.

TIMING CHANGES

With luck, any timing changes that you have to make should be pretty subtle, and as such should be done by ear instead of looking at the position of the waveform relative to the beat grid on the Arrange window. Aligning the start of the audio to the grid is very useful to get things roughly into position if they have to be moved a lot, but the final careful positioning should always be done by ear. Each region should be moved individually, except if you have cross-faded regions, in which case all the regions that have cross-fades should be selected simultaneously and moved by the same amount in order to preserve the cross-fades. There will be times, however, when moving things around isn't enough. This may lead to overlapping regions, or pauses/gaps that are too long, or a number of other reasons why we might want to time-stretch instead of (or in addition to) moving things around.

Before we start, though, let's remember one important thing. In the case of any "lead" instrument, and voice especially, time-stretching should be avoided wherever possible. Although the technology has improved greatly over the years, there are often still audible artifacts to time-stretching that we should try to avoid if possible. So if there is a gap at the end of the region, you should ask yourself if it really *is* too long or if you are wanting to time-stretch it just to keep the end of the region where it was before. If there genuinely is a need, then time-stretching might be an option. Chapter 12 looks at this in detail, so please refer to that for more information. For now, suffice it to say that, as long as the stretches aren't too long, and the factors listed in chapter 12 are taken into account, the results should still sound good.

If, on the other hand, we are dealing with a region that now overlaps with another, then we may have another option. As long as the overlap isn't too long, we can consider using a cross-fade to just gently taper off the end of one word and allow it to flow into the next. This is particularly suitable if the end of one word only slightly overlaps into a following word. But if there is a breath sound prior to the next word, then having the preceding word cross-fade into a breath doesn't sound natural at all. Instead we could try shortening the now too-long region to bring the end of it back and allow enough of a gap before the breath sound. We can then apply a fade-out to the end of the region to make sure that the preceding word tapers off naturally before the breath. The choice of whether to use the time-stretching method, the fade, or the cross-fade methods depends largely on the situation, and it is difficult to know which will be more suitable. It is perhaps easiest to try the fade or cross-fade methods first, as those are

Comping and Alternate Takes   CHAPTER 7

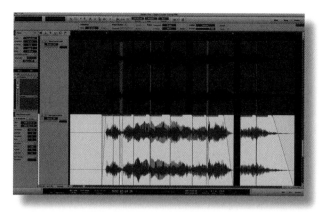

FIGURE 7.3
Sometimes when you are putting a comp together, you may find that you have gaps or overlaps to deal with. The top track above shows an example of this, which was then cleaned up by adding fades and cross-fades and time-stretching a few parts. The result is a much smoother and more flowing final comp.

probably quicker and don't have the artifacts associated with time-stretching, but if they don't work, there is always the time-stretching method to fall back on.

There is one final point of note to do with time-stretching that I would like to make, which relates to the choice of exactly what we should time-stretch. Any time-stretching should take into account the preservation of any attack portion of the sound (for transients and subtle pitch variations or bends) along with factoring in the possibility of delayed vibrato, which could sound unnatural if stretched or compressed too much. Again, there is a greater discussion of this in chapter 12, but it is worth bringing to your attention here.

As you can see, the process of comping isn't an inherently difficult one (from a technical perspective, at least) but it can be quite a long-winded one. If only there was a way to streamline even a part of the process to help make it more intuitive and a little less "clunky"…

COMPING MADE EASY

In late 2007, Apple introduced Logic Pro 8, and with it they introduced a new feature called Quick Swipe Comping. The idea was a fundamentally simple one: instead of having to manually cut each take into separate parts/lines/words, you could simply choose which take you wanted to use by selecting it with a "swipe" (a click and drag). Only one take could be selected for any given part, so swiping over a line on one take would automatically replace whichever take had been used before. Cross-fades would be applied at the edit points, and you could set up a default cross-fade time and shape as part of your DAW preferences, but this would be applied globally to all edits in the comp. Once you were happy with the result, you could either bounce it to a new file, create a new track with all of the comped regions and still leave the comp track intact, or just remove the comp track, leaving only the comped regions.

Perhaps you might see this as only evolutionary, but for many, it was revolutionary and changed the way we thought about comping. What was previously seen as a hugely time-consuming (but often needed) task had just become orders of magnitude easier. It wasn't perfect in either concept or execution (it was quite "buggy" in the earliest versions), and there were many areas that could, and later would, be improved. Yet in spite of this, it really was a very welcome addition to the Logic toolbox. And because of this, other DAW manufacturers were quick to follow with their own tools to improve the comping work flow.

Pro Tools added a similar (although some might feel slightly less user-friendly and immediate) system to Pro Tools 8 in 2008, while Steinberg's approach didn't really develop fully until Cubase 6 was released in 2011 (even though they had been developing their approach since Cubase 4). MOTU added their own version in Digital Performer 6 in 2008, and PreSonus greatly improved their comping tools in Studio One version 2 in 2011. With this flurry of activity to introduce similar features into other DAWs, Apple quietly improved, both in terms of features and reliability, their own version in Logic Pro 9, and, over the last few years, many people have grown accustomed to this new paradigm. It is true that it doesn't allow us to do anything inherently new, but for people who work in any industry day after day, work-flow improvements that allow them to complete their work more quickly (as long as the quality doesn't suffer) can often be just as welcome and important as completely new features.

While this can be a great way of quickly putting together a comp and keeping everything nicely organized, as well as giving you the option to easily store and switch between alternate comps (depending on which implementation you are using), it isn't all perfect. You still need to carry out the rest of the steps afterward, and doing it this way may not automatically take into account any timing differences between the takes. It may be that the tone and delivery on take three sounds great, but it was a little too late. Some DAWs will allow you to nudge the timing of individual parts of the comp, while others won't. But what none of them will do is automatically align the timing of different takes.

So that's pretty much it for general (including vocal) comping. We have covered all the common comping techniques, traditional and new, and, and, although the examples we have considered have been focused on vocal comps, they are still applicable to comping any instruments. As stated at the beginning of the chapter, some instruments' groups have other factors to consider, so please take a look at the website for more information about these.

FIGURE 7.4
Swipe comping, as shown here in Logic, enables you to carry out the comping process without having to cut regions and move them manually onto another track. It also allows you to create a number of different alternative comps that can be switched easily, enabling you to try different ideas with a minimum of fuss.

HANDS ON

Introduction

Comping is one of the most-often carried-out editing tasks in the world of music recording and production. Its use is not as extensive in music for picture and film scenarios, but in any case, it is a prerequisite skill for any editor to have, as it can be seen in many ways as the sum of all the corrective-editing processes. It was traditionally quite a time-consuming task, but the most recent versions of most major DAWs have introduced tools to make the editing process not only quicker but also much more intuitive and flexible. These new tools could quite easily reduce the time taken to carry out a typical comping job by 50% or more. As such, a good look at how we use them is almost essential.

Logic

The biggest tool in Logic (and many of the other modern DAWs) for comping is its so-called *Quick Swipe Comping*, which is a very quick and easy way of carrying out comping. Instead of having to manually cut regions from each of the takes and mute them as you decide that they won't make the final comp, and instead of having no simple way of comparing alternate comps, this process gives a huge amount of flexibility and makes the entire process seem far less daunting.

Crucial to the whole idea of Quick Swipe Comping is the existence of *take folders*. These are a simple way of grouping multiple takes of a given part, which, aside from the obvious benefits that we will see shortly, is also a very good way of keeping your Arrange window organized and tidy. If you are recording audio into Logic and have set up a loop of a particular section (a verse, for example) to allow the performer to repeat the section a number of times, then a take folder will be automatically be created for all takes recorded continuously. If recording is stopped and then restarted for exactly the same loop range, then new takes will be automatically added to the take folder, while, if a totally new range is selected, then a new take folder will be created. However, if a recording range is selected that overlaps an existing take folder, then the new takes will be added to an existing folder and the loop range of that folder extended.

On the other hand, if you are given a number of separate takes from a third party that aren't already grouped into folders, all is not lost. In order to be able to carry out Swipe Comping, you will need to get the individual takes into a take folder, and fortunately this is a very easy process. All you need to do is select all the regions that you wish to put into a take folder (either by clicking and dragging or by holding down Shift while clicking on the individual regions) and then right-clicking on any of the selected regions and choosing *Pack Take Folder* from the Folder contextual menu. Alternatively, you can go to the menus at the top of the Arrange window and choose Region > Folder > Pack Take Folder, or, if you prefer key commands, the same thing can be accomplished by pressing Ctrl + Cmd + F. If the take folder was created during recording, then by default all the takes will be visible, while creating one from separate regions will have the individual takes hidden. To change from one view to the other, you simply need to click on the disclosure triangle located at the top left of the take folder.

Once you have your take folder, you can start the actual comping process. The first thing you need to do is make sure that Quick Swipe Comping Mode is enabled. This is done simply by clicking on the Quick Swipe Comping Mode button at the top right of the take folder. This button looks like a rectangle with three small horizontal bars inside. If the rectangle looks transparent, then Quick Swipe Comping Mode is disabled, and if the rectangle background looks solid, then Quick Swipe Comping Mode is enabled. With Quick Swipe Comping Mode enabled, you can begin to make decisions about which parts of which take to use.

The take that is active (being used) will be colored, while the unused takes (or parts of take) will be grayed out. To change which take is being used, you simply move the cursor over one of the inactive takes (the cursor will change to two vertical bars) and then click and drag (or swipe) over the section of the inactive take that you would like to use. At this point, the swiped section (take region) will become colored, and the corresponding part of the previously active take will be grayed out. Fundamentally that is all there is to it. It really is quick.

Of course there is a little more to it than that, as we could well need to fine-tune the sections that we have chosen, and there are two main ways to do this. The first method is to move the cursor over either the left or right edge of a currently active take region, at which point the cursor changes and adds left and right arrows on either side of the two vertical bars. This tool allows us to move either end of a take region. Doing this not only changes the start or end position of the currently active take region but also correspondingly lengthens or shortens the adjacent part of the preceding or following take region. The other option that we have is to move the entire take region. If we position the cursor in the middle of a take region, it will change to just left and right arrows, and clicking and dragging with this tool will keep the overall length of the take region the same but will move it earlier or later. In combination, these two tools allow you to quickly fine-tune the initial swipe-comping that you have done.

FIGURE 7.5
A take folder in Logic expanded to show the different takes and the sections of each take that go into making up the final comp (on the very top track).

Comping and Alternate Takes — CHAPTER 7

If for any reason one of your takes is not aligned with the others properly, then you can move it within the take folder, but to do so you need to turn off Quick Swipe Comping Mode, as described above. Once it is turned off, clicking and dragging on an individual region follows a more normal behavior, where it simply moves the region either earlier or later. You should note that any take regions on the region you are moving will stay in the same place relative to the region and not to the take folder, which could mean that a movement of this kind could mess up any comping you have done so far. As such, it is probably wisest to make sure all takes are aligned using this method before embarking on the comping process. If you find that only a single part of one take is out of time, then you can split that take and move only the required part to bring it into line.

Once you are happy with your comp, you have a number of choices. You can either leave things as they are and move on to the next task, or, if you want a little more control over things, you can export the comp to a new audio track. If you click on the small arrow at the top right-hand corner of the take folder, it will bring up the Take Folder menu, and from this menu, you can choose *Export Active Comp to New Track*. This will create a new audio track and copy all active take regions into their correct positions on the new track and apply cross-fade between the regions. This then allows you to further work on cross-fades and other editing tasks on the individual regions if required. Once this is done, you can mute the audio track containing the take folder, and you are ready to move on.

Alternatively, if you are completely finished with the comping, then you can choose *Flatten* from the Take Folder menu, which carries out the same process, only it will move the take regions into position on the same track as the take folder and delete the folder, leaving you with just the comp. The only real difference here is a work-flow one. If you are 100% sure that you do not wish to make any more changes, then flattening is the better option, to minimize clutter. However, unless you are 100% sure, I would always choose the Export option, as it allows you to go back in and try something different if you need to.

And speaking of trying something different, the Take Folders allow you to have a number of options you can quickly choose between by being able to store different versions of comps within the folder. Once you are happy with your first comp, you can export it, and then, in the Take Folder menu, choose *Duplicate Comp*. This will create an exact copy of the current comp, which, in the menu, will show up as Comp 2. If you now make changes to your comp, perhaps trying different lines from different takes, those changes will be stored within Comp 2. If you open the Take Folder menu and switch back to Comp 1, you will see that your original comp remains untouched. To make matters easier, you can use *Rename Comp* from the Take Folder menu if you would rather have a reference to each version by name (Master, Double Track, Alternate Lead, etc.). Once you are done, you can again export this new comp to a new track.

These alternate comps make it very easy to create alternate takes, and the fact that you can export each one once you have finished it while still keeping it as a comp within the take folder means that very little has to be committed to that can't easily be changed later. Whether or not this is a good thing depends on how you like to work, but, whatever your views, having options during this often-critical process can only make life easier. Once you are done with all the comping for this take folder you can, of course, delete the whole take folder, but, unless you have a very compelling reason to, I would always suggest muting it and leaving it in your arrangement—just in case.

Pro Tools

Comping in Pro Tools is made much easier by the use of *playlists* on tracks. The most simple example of a playlist in Pro Tools is a single region on a single track that starts at a certain point and plays for a certain amount of time. If you add in different audio files or regions, or you split regions and move them around, then the playlist will become more complicated, but it is still counted as a single playlist. Where this functionality becomes much more useful to us for comping is the fact that each track can have multiple playlists stored. Only one can be active at any time for any given track, but multiple alternatives can be stored. This idea could be used to store alternative arrangements of the same regions in different playlists, perhaps to allow different song arrangement ideas to be tried, or it could be used to store different versions of a particular recording—in other words, different takes.

Now the reality is that if you had five alternate playlists with a different take on each, it isn't a massive step away from having five different tracks with a single take on each, but the real benefits come from the work-flow improvements. If you had five different tracks, then you would have to remember to make sure that only one was playing at any given time, and you would also need to duplicate any plug-ins that were being used across all five tracks, and this could potentially place a significant extra load on your computer. And then, if you wanted to used those five tracks to create a comped part, you would probably need to create a sixth track to put the comp together on. You would have to make sure that, as you were moving or copying each section of the comp onto the "master" track, it was copied in the same position (in the time sense) as the take that it came fro and, not insignificantly, it would create a massive amount of clutter if you had multiple parts to comp.

Using playlists, on the other hand, keeps things organized and keeps clutter to a minimum by allowing you to easily hide and show the alternate playlists for a given track. It also allows you to quickly switch between alternate playlists if you want to try different things out but be able to get back to where you started quickly, and it also provides a very quick and simple way of copying sections from individual playlists onto another playlist, so that you can put your comp together with as little fuss as possible and as few worries about timing consistency as possible.

Comping and Alternate Takes

If you are going to be recording into Pro Tools yourself, then there is a very simple option that you can set that could greatly speed up the early stages of the comping process. If you go to Setup > Preferences and then go to the *Operation* tab, you will see, under the *Record* heading, an option to *Automatically Create New Playlists When Loop Recording*. Turning this option on will mean that any recording done in loop mode will group each pass of the loop (or take) onto the same audio track and will create a playlist for each take. When you do this, the most recent take will be the uppermost; and the earliest take, the lowermost. Recording tracks in this way removes one step of the comping process. Of course there will be times when you simply record straight through a song from start to finish and then, perhaps after making some adjustments, start recording again. In this case, or if you have been given a collection of individual files from a client, you need to do a little bit of work before you are ready to actually start comping.

The simplest way to do this is to create a new audio track by pressing Cmd[Mac]/Ctrl[PC] + Shift + N or choosing Track > New to bring up the *New Tracks* dialogue. Here you can choose the number of tracks you wish to create along with various options about the type of track you wish to create. Choose *Mono* or *Stereo* as appropriate, make sure *Audio Track* is the selected option, and hit *Create*. The audio track will be created and you can rename it to something meaningful ("Vocal Comp A," for example) by double-clicking on the track name and entering the new name in the dialogue that appears. This track will now function as a container for all the different takes that we have and will also be the eventual location of our comped track.

The next thing to do is to click on the *Track View Selector* (located on the track header underneath the *Solo* and *Mute* buttons) and change the default *Waveform* to Playlists. As you do this, you will notice that a smaller sub-track appears underneath our newly created track, and this is where we can start to put together our alternate takes. If you already have all the alternate takes in the main arrangement (already synchronized to each other), all you have to do to make them available for our comping process is to drag and drop them onto the smaller track. This will move the region and create a playlist for that region, the track will become larger, and another small sub-track will appear beneath that one. Once again you can double-click on the track name to rename it ("Lead Vocal Take 1," for example), and this is always a good idea, just so that you know which track represents which take at a glance. You can then drag in as many takes as you need, one by one, into the smaller subtracks, and new playlists will be created for each one. Once this is done, we will be at the same stage that we would have been had we recorded the audio into Pro Tools in loop mode, so we can now move on to the actual comping process.

The actual process of comping itself is very easy. You can listen to any individual take (playlist) by clicking on the Solo button on the individual take track and hitting Play. The rest of the tracks in the song will play along with only the soloed take. This allows you to preview each take in context of the song. If no individual take is soloed, then you will hear what is on the main comp track. If you have not yet selected anything to be placed on this track, then you won't hear anything. Once you have auditioned each track and are ready to start putting the comp together, you have a couple of ways of doing it. If you want to move a whole region from any particular take onto the comp track, then you just need to select that region (either with the Grabber tool or the Grabber part of the Smart tool) and then either press Ctrl[Mac]/Start[PC] + Alt + V or press the up arrow located next to the Solo button for that particular take. Doing either of these will move this region to the main comp track/playlist. Alternatively, if you wish to move only a part of a region, you can select it using either the Selector tool or the Selector part of the Smart tool, and then once again either pressing Ctrl[Mac]/Start[PC] + Alt + V or pressing the up arrow located next to the Solo button to move the selection to the main comp track/playlist.

You should note that if you choose a region or selection that overlaps a region or selection that is already on the main comp track, then this new selection that you move will overwrite the existing area of the main comp track. The latest file to be moved to the main comp track is always the one that takes priority. Once the regions or selections have been copied to the main comp track, you can use the Trimmer tool to fine-tune the start or end points of each section and then apply cross-fade to smooth things off if required.

If you wish to create an alternate comp using the same takes, then you can start either by duplicating the existing comp or by creating a new, empty one. To duplicate the existing comp/playlist, you should click on the down arrow to the right of the track name and choose *Duplicate* from the pop-up and then name the new comp/playlist. If you wish to create a new comp/playlist, choose *New* instead. Whichever of these options you choose, the newly created comp/playlist will appear on the main track, and the previous comp/playlist will be moved down to join your individual takes. Now you can repeat the process and create as many alternate comps as you require. You can switch between which comp is active by clicking on the down arrow next to the main track name and choosing one of the other comp tracks from the list. Choosing any other comp moves the previous active one down into the area with the original takes.

FIGURE 7.6
While the naming might be a little different and the mechanics slightly different, the Playlist functionality in Pro Tools brings quick and efficient comping and the ability to have multiple versions of comps stored for quick access.

Once you have completed all your comping, you can either leave the track as it is and hide the alternate playlists by changing the *Track View Selector* back to Waveform or, if you prefer or have multiple alternate comps, you can move each comp to its own new track by creating new audio tracks and simply dragging and dropping the regions from each comp track onto their own new track.

Studio One

Comping in Studio One works with the concept of *Layers*. Each track in the Arrange view can be composed of multiple layers, and the best parts of each of these layers can be pieced together to make the final comp. But before we get to the comping process itself, let's take a look at how to get everything set up on the layers that we need.

If you are recording your audio directly into Studio One, then the first thing that you need to do is enable the *Record Takes To Layers* option either by going to Options > Record Takes To Layers, or, if you need to adjust other recording settings, you can open the *Record Panel* by pressing Alt + Shift + R or by going to View > Record Panel and click on the *Record Takes To Layers* checkbox at the far right of the *Record Panel*. Doing any of these means that each take of recorded audio will be placed on its own layer, underneath the main track. This is true whether you record in loop mode or whether you record noncontinuous takes.

To view all of the different layers for a track, you need to right-click on the track header and choose *Expand Layers*. This will show all of the separate layers and will give you the (very advisable) option to rename each individual layer. To rename an individual layer, either double-click on the layer name or right-click on the layer name and choose *Rename Layer*. This ability is especially useful, as, when creating the comp, the name of each layer will be transferred to each part of the comp, so you will be able to see (even without the layers expanded) which part of the comp came from which take. In addition to naming the layers, you can also right-click on the region on the layer itself, and, by clicking on the colored box next to the track name on the pop-up, you can choose a different color for each take/layer. Once again this can be helpful when creating your comp, so that you have a visual representation of how the comp was created. You can also click on a layer header and drag up or down to reorder the layers. This can be useful if you wish to keep the most recent takes near the top, or if you wish to organize them according to any particular criteria that you have. It is also very important when it comes to multi-track comping, which we will read more about in the next chapter. Once this is done, you are ready to start creating your comp.

If, on the other hand, you are working on audio files that have already been recorded as discrete tracks, then you have to approach things slightly differently. The first thing you need to do is create an empty track for your comp by going to Track > Add Audio Track and choosing the *Mono* or *Stereo* option as required. Alternatively, if you needed to create multiple tracks for multiple comps at once, you could go to Track > Add Tracks (or press T), and you will be presented with the *Add Tracks* dialogue, which gives you the option of creating multiple tracks along with a few options for type and format. Once the new track is created, you then need to start adding your audio files/regions into layers.

You then right click on the newly created track header and choose Layers > Add Layer to create a new empty layer. However, once you have done this, it won't look like anything has happened. This is because, although the layer has been created, the track itself hasn't been set to show the layers. To do this, right-click on the track header once again and choose *Expand Layers*. Now you will see the newly created layer located underneath the main track. You should then repeat the process to add as many layers as you will need for your comp. Once all the empty layers have been created, you can start moving your regions into them to prepare for comping.

Populating the layers is as simple as dragging and dropping (or holding down Alt while dragging to copy instead of moving) into the desired layers. However, one important word of warning here is that, once your regions are in the layers, they cannot be moved back or forward along the timeline. As such you will need to make sure all the takes are synchronized *before* you move them into the layers. Because of this it could be sensible to copy any regions into the layers rather than moving them, so that you have a copy on a track that can be moved around if need be and then placed into another layer. Once you have copied (or moved) your regions into the layers, you can rename the layers or change the color of the regions as described above, and you are then ready to move on to the actual comping process.

When you have a track composed of a number of layers and you move the cursor over any one of the layers, it will change to a cross-hairs cursor with a vertical line that runs the whole height of the Arrange view. Clicking and dragging with this tool will highlight a range on a particular layer and, once the mouse button is released, will automatically add that range to the comp track. If there is an existing range from another take already in place, the newly selected range will overwrite the existing one for its duration. If there is any overlap between comped ranges, then cross-fades will automatically be created at the boundaries of all overlapping ranges.

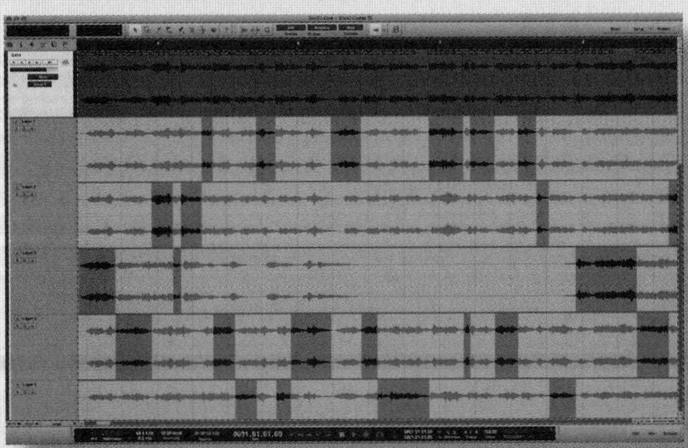

FIGURE 7.7
The Layers in Studio One serve as the basis for the advanced comping features and can make the process of putting a complex comp together a matter of minutes rather than hours.

If you move the cursor to the boundary of a highlighted range on a layer, then the cursor will change to a vertical bar with left and right arrows and a small "x" to the right. Using this tool, you can click and drag on the boundary to change the start or end position of this range. Doing so will automatically lengthen or shorten the adjacent range on the other layer to prevent any gaps forming. If you are resizing a range that doesn't (yet) have an adjacent range from another layer, then only the layer you are resizing will be affected.

If you move the cursor up to the main comp track and position it in the middle of the cross-fade between adjacent ranges, then you will be able to click and drag the cross-fade along the timeline. Doing this will lengthen and shorten (as appropriate) the ranges on the two layers that make up the cross-fade. This is effectively the same as resizing either of the ranges on layers directly, as described above. However, if you move the cursor the either edge of the cross-fade, you will be able to lengthen or shorten the cross-fade itself. Dragging the left edge of the cross-fade will lengthen or shorten it without affecting the point at which the second range starts, while clicking and dragging the right edge will not only lengthen or shorten the cross-fade itself but also will change the starting position of the second range to be in line with the end of the newly adjusted cross-fade. Finally, if you move the cursor to the small square on the middle of the cross-fade lines, you will be able to click and drag up or down to change the shape of the cross-fade.

If the main comp track already has a range selected for a particular time period, then you can quickly swap that time period for an alternate take by double-clicking on one of the other layers directly below the range. For example, if you had chosen Take 2 for the period between Bar 1 Beat 2 and Bar 2 Beat 2, then you could swap to Take 3 by double clicking on the layer for Take 3 anywhere between Bar 1 Beat 2 and Bar 2 Beat 2. This system is inherently easier than trying to click and drag exactly the same range on a different track and means that you can very easily switch between alternate takes for a given part of the comp. If there is no layer currently selected for any given part of the main track, then this double-clicking technique will not work, and you will have to click and drag to create a comp range from any of the layers before being able to switch to others.

Once your comp is complete, you can hide the layers (right-click on the track header and remove the Expand Layers option) and then leave things as they are, or you can merge all the separate parts into a new region by highlighting them all and pressing G (or Event > Merge Events), and then you can easily copy or move that around. Note that this is still an editable collection of separate regions that is merely grouped visually on-screen. If you wish to create a totally new file, then press Cmd[Mac]/Ctrl[PC] + B (or Event > Bounce Selection), and the comp will be bounced to a new file that will be placed on the comp track in the right position.

Alternatively, you might wish to create an alternate comp from the same takes. The easiest way to do this is to right-click on the main track and choose Layers

> Duplicate Layer. This will take your comp and create a copy of it on a new layer under the main track. You could then name this ("Main Comp," for example) and then create an alternate comp on the main track again. You can repeat this process as many times as you like to create multiple alternate comps. Of course, you can only copy what is on the main track playing, so you need to then move each of these comps on to a new track. To do this, you can use the up arrow button located on each layer header. Pressing this button will move that layer up to be on the main track. Whatever was previously on the main track will then be moved to a layer track. Using this method you could create multiple comps that would be moved down onto layers each time they were completed, and then, once you were fully finished, you could click on the up arrow next to each comp layer in turn, move that to the main track, merge the comp into a new region, and then move that region off onto a new track.

Cubase

As with all of the DAWs that we are covering in this book, Cubase has some great features to help make the comping process easier. And like all of the other DAWs, it relies on the use of subtracks within a given track. The Cubase term for these is "lanes," and, once we have our audio in these lanes, the comping process can begin. The actual way that we get audio into the lanes depends on if we are recording audio directly into Cubase or if we are working on tracks that we have been given.

If we are recording audio directly in, then we have to make sure that the recording mode is set to either *Keep History* or *Cycle History + Replace*. To do this you need to make sure that the *Transport Panel* is open (press F2 or go to Transport > Transport Panel to open it if it isn't already open) and then go to the top left-hand corner where, by default, you will see Keep History selected. If you wish to change this, then simply click here, and a pop-up selector will appear for you to select the *Audio Record Mode*. The Keep History option will create new lanes for new takes whether or not cycle recording is enabled, while Cycle History + Replace mode will create new lanes for recordings made in cycle mode but will replace existing recordings if cycle mode is not enabled.

Once your recording is completed, you will see a number of lanes created below the main track, with the most recent takes at the top and the earliest at the bottom. If you need to hide or show these lanes, then you will find the *Show Lanes* button in the track header, which is located directly to the right of the *Lock* button in the track header. If your vertical-zoom level is too low, then you may not see the button, in which case you might need to zoom out a level or two before you can see them. The exact position will also depend on width of the track header. If the header is wider than the Show Lanes button (along with the *Time Base* and Lock buttons which are always grouped together) may be on the right of the header, but, if the header is narrower, it may be located below other buttons. If you have recorded your audio into Cubase, then you are ready to start comping.

Comping and Alternate Takes

If you are assembling a collection of separate tracks for comping, then the process is slightly different. You first have to create a new, empty audio track by going to Project > Add Track > Audio and then clicking on the Show Lanes button. This will show that a single lane has been created ready for you to start copying your takes into. You start to assemble your takes by dragging (or press Alt and dragging to copy) the first take onto the main track using the *Object Selection Tool*. Doing this will move the take into the first lane while simultaneously placing it in the main track. In addition, a second lane will be created below the first. If you were to now drag your second take onto the main track, it would simply replace the first one. Equally, if you tried to drag it directly on to the second lane, it would instead place it on the first lane and replace the first take. In order to successfully add a second take, you will first need to move the first take down on to the second lane (just drag it to do this), and this will create a third lane and free up the first lane. You can then drag the second take to the main track, and it will be placed in the first lane. From there you simply repeat the process of moving the take down to the last empty lane and dragging your next take into the main track. Once you have added your final take and dragged it down to the empty lane at the bottom of the list, you will be left with an empty first lane. You can get rid of this by right-clicking on the track header and choosing *Clean Up Lanes*, which will get rid of this empty lane. You are now at the same position that you would be if you had recorded the audio in directly, so you can move on to the next step.

With all the lanes visible, you should now select the *Comp Tool*. Strangely, there is no default key command for selecting this tool, but you can create one of your choosing if you wish. With this tool selected, you simply click and drag over a selection of any of the takes to make it a part of the main comp that is visible on the main track. When a selection has been chosen on any take, all takes will be split at that point, so that, as the comp progresses, each take (and the main comp track) is divided into separate regions. This will become useful later on in the process. You then work your way through the entire track, choosing your favorite takes for each section, which are moved to the main comp. If you need to listen to a particular, take you can use the Solo button on the lane header, at which point that whole take will be played irrespective of what is selected in the comp track.

With the Comp tool still selected, you will notice that each region still has the fade and volume handles in place. Usefully these handles allow you to carry out simple fade and gain adjustments while carrying out the comp. Often this level of fine-tuning is left until after the main comp is completed, but this kind of in situ control makes it easier to figure out how the final comp will sound without having to commit to anything during the comping process that you might wish to change later.

If you move the cursor to the boundary between two take regions, you will see it change to left and right arrows with two vertical bars in between. This cursor allows you to click and drag to resize both regions at once. If you are shortening

FIGURE 7.8
Instead of Take Folders, Playlists, or Layers, Cubase has Lanes, but the functionality is very familiar to anybody who has used this type of "Swipe Comping" before.

one region, then the adjacent region will be lengthened, so that there are no gaps in the comp. On the other hand, if you actually want to shorten a particular region without automatically resizing the adjacent region, then you can move the cursor to either of the dark squares in the bottom corners of the region, and the cursor will change to left and right arrows. Clicking and dragging now will resize only the current region (and will only shorten it, not lengthen it) without changing the length of any other regions. It should be noted that doing this changes the length of this region for all takes simultaneously.

If you wish to change to an alternate take for any particular section, then, with the Comp tool still selected, you need only to click on the alternate take for that region, and it will be automatically moved to the comp take. This is a very simple and quick way of switching between different takes for a particular section without having to click and drag over a selection each time. This can also be achieved if you have the Object Selection tool enabled, although the process is slightly different. Simply clicking on a region from an alternate take will select that take but won't move it to the comp track. If you wish to move it as well as select it, you need to click on the small square in the middle of the bottom of the region. The Object Selection tool can similarly be used to adjust fade handles and the volume handle for regions within the comp, but it can also be used to move or even copy regions within a take or even between lanes. The most obvious use would be to manually adjust the timing of a particular word that wasn't quite lining up in a particular take, but it could be used for whatever purpose you have in mind.

Once your comp is complete, you can leave it in place and hide the lanes, or you can commit it to a new file. The latter option is good practice for simplicity, but is also essential if you want to create multiple alternate comps. To create a new file, you need to make sure the entire comp is selected in the main comp track and then go to Audio > Bounce Selection. You will then be asked if you wish to *Replace Events*. Choosing *Replace* will get rid of the current comp and replace it with one new, bounced region. If you choose *No*, then the bounced file will still be created but will be added only to the audio pool and not placed in the arrangement. If you wish to create multiple comps from the same takes, then you can use this method to bounce the comp into the pool and then create a different comp and repeat the process. Each comp can then be dragged to a new audio track.

Alternatively, when you are ready to bounce your comp, you can right-click on the track header and choose *Duplicate Tracks*. This will create a new audio track that includes all the lanes and the comp that you have carried out. You can then select all regions in the comp on this duplicate and bounce the selection, choosing to replace the events, and then hide the lanes. This newly copied track will now contain your first comp, and you can then return to your original comp track and work on your alternative comp versions. It isn't necessarily the most intuitive way of doing things, but, equally, it isn't hugely difficult either.

CHAPTER 8
Multi-Track Comping

IDEAL-WORLD MULTI-TRACK COMPING

The very first complication that we have to deal with when comping a multi-track recording is the issue of making edits across multiple tracks at once. The actual process isn't any different: make sure takes are all aligned, remove any silence if necessary, cut the takes into relevant-sized sections, listen through and choose the best take for each section, and then assemble a "master take" from the best parts. What makes it much more difficult is the fact that you have to do this across multiple tracks at once. Throughout this chapter, I will make use of examples of a multi-track drum kit recording, because, in truth, this is the most likely situation you will come across and the one that is potentially the most difficult to deal with. Other instruments that are commonly recorded on multiple tracks (and by this I don't mean a simple stereo recording) include electric and bass guitars and pianos, and there is additional information about the issues that you could face when comping them, and how to resolve them, on the accompanying website.

If we are using our traditional "manual" comping method, then, although time-consuming, the process is a relatively easy one. You simply have to make sure that all the tracks you want to edit are selected and then cut/copy/paste/move as needed. But when you have multiple tracks (with drums this could be anything from four up to sixteen or more) and *each* track has multiple takes, it can get very hard to keep track of what is what. Color-coding tracks can be very useful here, so that all the kick drum tracks are, for example, red, all of the snare drum tracks are yellow, the high toms are green, etc. If we do this, though, we won't then be able to color-code the different takes according to quality. This isn't necessarily a major problem, even if you use some kind of color-coding system for vocal takes and the like, because the way that you comp drums is generally a little different.

Most people, when comping drums, don't listen through each section of each take and try to pick out the most suitable one. Instead it is much more common to listen through the various takes and decide which one is closest and then

114 Ideal-World Multi-Track Comping

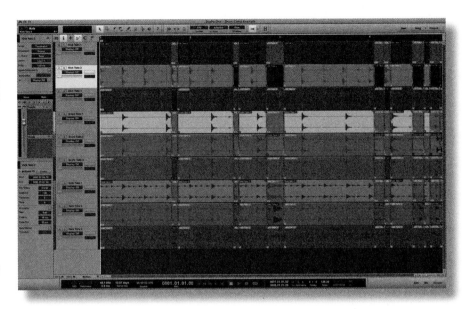

FIGURE 8.1
Drum comping tends to be a little different in that generally there will be one take which features more heavily than the others, with the alternate takes used when necessary to cover any minor mistakes or timing errors. Equally, it might be that alternate takes are used as the basis for different sections of the song.

only replace the parts that need replacing. If you were to listen through and try to comp a multi-track drum recording drum-by-drum, hit-by-hit, not only would it require an almost unimaginable amount of work, but, because of the nature of the sounds, it is unlikely that the improvement in the take would warrant the time spent. Any time that is going to be spent would be far better split between a relatively simple comping process and making sure that the groove was sitting right. Of course, that is just one person's opinion, and there is nothing to prevent you going in to microscopic detail if that is what you prefer.

When you have decided on a method to make everything clear to you, so that you know what tracks and takes belong to which drum, it is time to start actually making the edits. All the same decisions and criteria apply as with normal comping, but you just need to remember that you are dealing with multiple tracks. So when, for example, choosing which take to use for the first beat of the bar (a very simplistic example, of course) you need to remember that you have (let's say) eight different tracks to make that edit on. Traditional thinking, with good reason, is that you choose the same take for all the tracks at the same time. That's to say, if you want the kick drum from take four for the first beat, then you also use take four for all the other tracks as well.

Now, in our ideal scenario that we are considering, and if we are using our "manual" comping method, that wouldn't be a necessity, because each drum on each track would be "clean," so we wouldn't have to worry about which take we used. This would mean that we had complete freedom to use, for example, Take 1 for the snare drum, Take 2 for the kick drum, Take 4 for the hi-hats, and Take 6 for the overheads. This would give us the most freedom, but sadly it is almost impossible to achieve this in a real-world situation.

On the other hand, if we are using "swipe" comping, then by default, the same take will be selected for each track. If you choose the kick drum for take one, then all the other tracks will use take one as well. The reasons for this will become apparent when we step outside of our ideal-world scenario, but it's a factor if you do happen to have a pretty clean drum recording. If you want the most flexibility to pick and choose which takes to use for each drum, then the manual method, although more time-consuming and potentially more prone to confusion and errors, is far more flexible.

REAL-WORLD MULTI-TRACK COMPING: SPILL ISSUES

Before we look at what the next stages of multi-track comping, we need to move beyond the theoretical and into the practical and start to think about a more realistic drum-kit recording. No microphone, no matter how good, will pick up only sound from directly in front of it. Many, called *hyper-cardioid* microphones, are very directional and will pick up a vast majority of their sound from the direction that they are pointing, but they still pick up some sound from other directions. As well as the potential pickup of other sounds in the direct field of the microphone, there is also the possibility that the mic will pick up reflected sound as well. All rooms, with the exception of anechoic chambers, have a natural reverb. Sometimes this will be very controlled and dampened, and other times this will be a bigger, more "open" sound. In any case, this sound will fill the room and will to some extent be picked up by any microphone in that room.

This spill can actually help to make a recording sound cohesive, as our ears don't hear the sound of a drum kit as a number of separate instruments but rather as a whole sound. Of course we can pick out individual sounds if we focus on them, but the overall effect is of a combined drum kit. In addition, we don't naturally hear sounds directly into our ears. The room that we are in will contribute to the sound we hear (unless we are listening on headphones, of course) so we are used to hearing sounds with an "ambience" to them. As a result any very "direct" recordings can sound unnatural.

There is, though, another way that we can get a sense of this natural ambience, and that is through the use of "room" mics. These mics (sometimes a single mic but usually a stereo pair) are positioned at some distance from the drum kit, and they are used to pick up the sound of the overall kit as it would be heard by our ears in that position. The benefit of using room mics over relying on spill is that the room-mic signal is inherently more controllable and can be mixed in and out independently of the main drum mics.

So with this potential advantage of spill or room mics, why would it cause us problems? You may remember that we discussed phase cancellation in chapter 4, and we found that if the same signal is overlaid on top of itself, there is an increase in volume, but if that same signal is delayed slightly (making it out of sync) and then overlaid, it can in some instances cancel itself out and lead to silence. Of course, these represent the two extremes. In the first example, the signals are 100% "in phase," and in the second, they are 100% "out of phase." If the situation lies

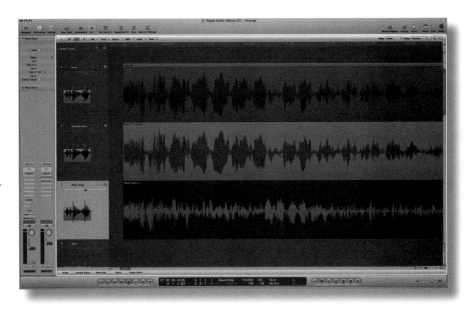

FIGURE 8.2
The top two tracks represent the same sound but delayed relative to each other by a few milliseconds. The result of this is shown in the bottom track. It is clear that the resulting sound is very different from the source sound.

between these two extremes, then the effect is, instead of complete cancellation, a change in the shape of the resulting waveform.

As you can see from the top row of illustrations above, when the identical waveforms are delayed relative to each other by differing amounts, there are gradual changes to the resulting waveform, and a changing waveform means a changing sound. The second row of illustrations gives perhaps a more familiar view of this phenomenon. Below each of the top-row illustrations is a spectral analysis of the resulting sound similar to what you might see on an EQ plug-in. In the first (100% in phase) example, you can see a single frequency peak, which represents the frequency of the waveforms as they stand; but, as the time difference changes, you can see additional frequencies starting to appear and then gradually they all disappear when we reach 100% out of phase.

This effect is only really pronounced when the two versions of the same sound are very close together. With a constant, unchanging waveform, as used in the illustrations above, there would be a gradual cycle. The two eaves would start off 100% in phase, and then, as the time difference between them increased, they would slowly move to 100% out of phase, then back to 100% in phase, and so on. The more practical effect of an increasing time difference would be that we would cycle through this phase cancellation up to a point, and then, because the sound changes over time and isn't static, there would come a point where we wouldn't hear the two sounds as a combined, phase-cancelled sound but rather as a distinct echo. The exact point at which this transition occurs depends greatly on the sound. Shorter, more percussive sounds will move to the distinct echo at a shorter time difference than longer, more consistent and sustained sounds. But no matter what the sound, anything up to around 20 ms would probably not be perceived as a distinct echo.

If we select the same take for each track, then we shouldn't have any problems with phase cancellation at this stage, because all the different tracks are synchronized. What I mean by that is that the exact same sounds are being picked up by all the mics at the same time. Actually that isn't *strictly* true, because if we have a close mic on a snare drum, it will pick up the sound immediately, but any sound from the snare would be picked up by the low tom-tom mic around two to three milliseconds later, owing to the fact that the sound has had to travel a certain distance (around one meter or so in this case) before reaching the other mic. This in itself can cause some problems if there is enough pickup on other mics, but it is at this stage at least a constant factor that doesn't change from take to take.

If, however, we wanted to use only the snare drum from a different take (take two) while all other tracks used take one, perhaps because there was an out-of-time note that we wanted to correct, then we have a very good chance that the timing will be slightly different. This in itself won't cause a problem. What *will* cause a problem is the fact that this new "direct" snare sound won't now be in sync with the spill from the snare picked up on all the other mics from take one. And, depending on how much of a difference there was, this could either be perceived as a distinct echo, or there could be phase-cancellation issues. Now obviously the more spill there is, the larger a factor that phase cancellation could be.

We can avoid this if we are prepared to make sure that, whatever edits and comping we might choose to do, for any given part of the comp, we always use the same take for all the tracks. It is a bit of a compromise, because that doesn't mean we will necessarily get the best part for each track, but it certainly minimizes the risk of phase cancellation owing to spill—at least for now.

REAL-WORLD MULTI-TRACK COMPING: TIMING ISSUES

Let's say we have only two takes of a drummer playing a groove: the first is perfect tonally (the right dynamics and tonal variations), but the timing is a little loose, and the second is inconsistent in dynamics and tone while being perfectly in time. We have to decide what is more important to us here, because whatever decision we make will be a compromise. If we choose the take with inconsistent dynamics and tone, then we have a lot of work to do evening out the levels, and even then the playing may just be a little too "soft," and therefore the snare drum might not have enough "bite" to it. But if we choose the take with the right tone and dynamics, then we have to start moving things around. Based purely on this, I would probably choose to go with the take that had the right sound but needed the timing cleaning up. That's an easy enough process: we select all the tracks and move the timing around. By selecting all the tracks, we keep everything in sync and avoid potential phase cancellation.

But what if the timing of the hi-hats, the kick drum, and everything else is fine, and we have a problem only with the snare-drum timing? At this point, we would need to start moving the different tracks around *relative to each other* and that is exactly where the phase-cancellation problems could start. It may occur to you to just substitute out the particular snare hits that are out of time with ones that are in time from another take. Even if we assume that the tone and dynamics were OK on this alternate take, we are still back to square one, because we have the fact that the spill on the other mics won't match up exactly with this new snare drum. It seems like we could have problems whichever option we choose, so is there anything that can be done?

HOW CAN WE FIX IT?

All of these situations have one thing in common and that is the fact that these potential phase-cancellation problems are caused by the spill in the recordings. In our earlier look at an ideal-world scenario, we didn't have to worry about all this. So in order to move our real-world recording closer to our ideal-world scenario, all we have to do is get rid of the spill.

The biggest problem when it comes to spill is the sound that any given mic is picking up when the drum that it is aimed at isn't making a sound. Fortunately, this is also the easiest area to address. Noise gates are (almost) perfect for this, because they offer a simple and automatic way to cut out any unwanted background noise when a specific set of circumstances exists. Any sounds above the Threshold level will mean that the noise gate opens (allows sound through) and any sounds below that level will mean that the noise gate is closed (no sound passes through).

On the surface this would seem like a good fit for what we are trying to achieve. After all, if the correct mic has been used, then the level of any spill should be quite low compared to the sound of drum it is intended to pick up. Therefore, if we set the Threshold at a high-enough level, it should mean that all of the background noise is stopped. In theory that works great, but, as we all know, drum sounds don't just start and stop instantly. The attack portion is generally pretty snappy, but the decay is a bit more prolonged. And, toward the end of that natural decay, the actual sound we want to record will probably drop below the Threshold level and be cut off along with all the background noise. This would give us an unnaturally truncated sound. If the spill was minimal enough, we might be able to set the Threshold level low enough that we didn't, in the context of the whole drum kit, notice the truncated ending, but that is a big "if."

To help with situations exactly like this, there are usually Attack and Release parameters that help to smooth off the edges of the noise gate's effect and mean that it doesn't open and close instantly, but rather the attack and decay parts can be made to be more gradual. In our example, above where the end of the drum sound gets truncated once it drops below the Threshold level, we could adjust the Release control so that there is still a natural decay back to silence once the Threshold level had been crossed. This would mean that, for that final period of the decay, the background spill would still be audible, but it would be falling away with the level of the sound we wanted, so it wouldn't be as noticeable. Presuming that the Release time had been set correctly—not too long that it lets through too much spill but not too short that it cuts the sound off unnaturally—it could substantially reduce the effect of the spill. In addition, because there would be some form of decay rather than an instant cut to the sound, we could actually set the Threshold level a little higher without fear that too much of the sound we wanted was actually going to be gated.

FREQUENCY-DEPENDENT GATING

A further possibility to improve things exists in the form of high- and low-pass filters on the sidechain input to a noise gate. What this means is that a copy of the input signal is taken, and then high- and low-pass filters are applied to narrow the frequency range of the sound. It is then *this* sound that is passed to the level-detection circuit to determine if the Threshold level has been exceeded. It should be noted, though, that it is the unfiltered sound that is actually passed through (or not) the gate itself, and the filter settings don't affect the actual audio being processed. In practical terms, this gives us a way to further refine the gating process.

If we have a noise gate on the low tom-tom channel, then we can use a low-pass filter on the sidechain input. We can set it to quite a low frequency, and this will have the effect of filtering out all the high-frequency spill (perhaps from cymbals or snare drums) in the level-detecting circuit. We could further add a high pass filter to try and reduce the spill from the kick drum. Many noise gates that have a sidechain filtering circuit like this have some way of listening to the effect that the filters are having and, using this we can adjust the filters to narrow the frequency range as much as possible and try to isolate the drum that we are working on as much as we can. It is unlikely that you will able to completely remove the other sounds, but that isn't really the aim. By getting rid of as much of the spill as possible, you can then set up a more effective Threshold level. In our specific example, by filtering all of the high frequencies out of the sidechain input, we can remove the possibility that an especially loud crash cymbal hit could open the gate even when the low tom-tom wasn't playing, because the sound of the cymbal will never even reach the level detection circuit.

FIGURE 8.3
A noise gate can be very useful in reducing spill and is especially useful if there are high-cut and low-cut filters in the sidechain, as these can really help focus on the sound you wish to keep.

Expand Your Options

The biggest problem with a noise gate is that, in spite of all of these options to help us, it remains an essentially static tool. It won't take into account the natural dynamics of a track. You might have really gentle drumming in a verse and much more aggressive drumming in a chorus. If you set up the Threshold level to be appropriate for the verse, then pretty much everything, spill included, will be above that Threshold level in the chorus, and it won't be effective. Alternatively, if you set up the Threshold level for the chorus, then even the sound that you want to keep might fall below the Threshold in the verse. We can get around this through automating the Threshold level, and this is, in fact, what a lot of people do.

Another alternative, and a much more controllable one, is to do what the noise gate is doing but with a manual process—that's to say, cut the spill out from in between drum hits, and then apply fade-outs to the end of each one to simulate the Release control on the noise gate. Clearly this isn't a viable option if you have to repeat the process thousands of times, but it could be a good option if you have just a few hits on a particular track that the noise gate isn't dealing with especially well. Alternatively, depending on the style of the music and on the preferences of the producer/client, it may be possible to do this kind of manual spill removal if there are going to be large parts of the drum track copied and pasted throughout the song. If you realistically have to do this only for an intro, one verse, one chorus, and a middle eight, and if the drum track is relatively simple and doesn't have 16 separate tracks to deal with, then it might just be worth your while.

EXPAND YOUR OPTIONS

Something else that might be useful is an expander. An expander can in some ways be thought of as an anticompressor. While a compressor will take any signals that are above a given level and proportionately reduce them, an expander will take any signals that are *below* a given level and reduce them.

FIGURE 8.4
An expander can be a useful alternative to a noise gate, because it can be a little less aggressive and a little more forgiving if you have a very wide range of dynamics in your audio.

So it works in a similar way to a noise gate (sounds below a certain level are affected) but works in a gentler way with a proportionate decrease rather than a full removal. Because of this, expanders are inherently more subtle and are probably better suited to situations where the spill is not too bad and just needs pulling down in level a little; alternatively, it can sometimes be useful to use an expander before a noise gate, as the expander will drop the level of the spill to a slightly lower level in a gentle fashion, and consequently the Threshold of the noise gate can actually be set a little lower than it would have been, meaning there is less chance of cutting off quieter notes completely.

There is one tool that may actually help us clean things up a little better. It is manufactured by Zynaptiq and is called Unveil. Its original intent was to de-reverberate drum and other recordings, but it has a useful additional purpose in that it offers a way to control other kinds of "ambient" noise. Strictly speaking, spill is an ambient noise in the sense that it is a noise that is outside of that which you wanted to record. There have been other similar tools in the past of the "transient designer" type, but this one offers way more in terms of the control that is available to the user. In many ways, it is like an expander but one with a degree more "intelligence" and awareness of what constitutes the part of the signal that you want to keep and what is the part you want to reduce or eliminate.

Like all the other methods here, we would use Unveil to try to reduce the spill. Whether this more-intelligent method would be right or whether one of the other, simpler, methods would be right depends on the circumstances. There is very rarely one right tool for a job, and what works well for one occasion may not be the best for another. I would encourage you to try all these techniques, or as many as possible, so you can not only learn which works best for you generally but also be equipped for situations where your go-to solution doesn't quite fit the bill.

FIGURE 8.5
Based on psycho-acoustic research and the very latest technologies, Unveil is a plug-in designed to remove reverb and ambience from recordings, but, used creatively, it can also help to diminish spill in a very natural way. Used in conjunction with some manual editing and either a noise gate or expander, this could help you get some nice, clean sounds to work with.

The truth is, though, that it is almost impossible to fully eradicate the spill from multi-mic drum recordings. Even if we can get rid of the spill that comes after the actual sound we want using the techniques described here, there could still be sound from the other drums in the background in the portion of the sound we have kept. There is the potential to use spectral editing software to remove it, and this is one of the many applications discussed in chapter 16. But even if we could do it successfully and consistently, we are really pushing hard into the area of diminishing returns here. Future developments in software

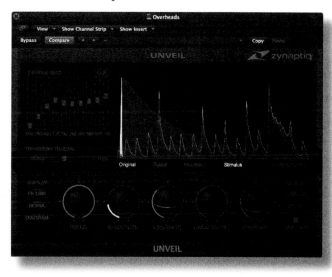

may automate this process to some extent, and at this point, it might become a realistic proposition. Currently, though, the idea of using spectral editing to remove mic spill from multiple drum tracks over the course of an entire song is just not a good use of time. Perhaps if there is one particular point in the song where a unique set of circumstances has meant an unacceptable level of spill in an otherwise acceptable recording, then yes, in that situation I could see a use for it.

Other than that, this is really one of those situations where we might need to curb our enthusiasm and insatiable drive for absolute perfection (you *do* have that—right?), realize that our time is better spent in other ways, and ultimately make sure it is not *the best in can be* but rather *the best it can reasonably be given the amount of time we have available*. I don't think that is too much of a compromise or anything to be ashamed about.

HANDS ON

Introduction

During this chapter, we spoke about some of the potential issues with comping a multi-track recording. The main problems that we discussed were issues of spill and timing, and we proposed a number of ways of dealing with those issues. If we could resolve those issues, then, in a perfect world, we could comp each individual track separately without having to worry too much. But that is rarely possible, and as a result the simplest way to deal with multi-track comping (using our example of the multi-mic recording of a drum kit) is simply to comp each track in the same places. By doing so, we won't have to worry about phase cancellation from spill, because what we are effectively doing is the same thing as comping a stereo mix of the entire drum kit on one track. Of course, by comping multiple tracks at once we retain all the flexibility of keeping those tracks separate but gain all the advantages of treating it like a single track.

Logic

The process of comping multiple tracks in Logic is no different at all in terms of the process. Obviously you have to be more careful with your selections, as what you choose for one particular track (for example, a snare drum) will be mirrored across all tracks. This may mean that you need to make some compromises between what is the best take region to use for one track and what will be the best take to use across all tracks, but that is inherent in the process, and there is nothing we can do about that. The only real difference in technical terms is actually getting your comping to have an effect across multiple tracks and take folders simultaneously. Fortunately this isn't overly complex and, subject to a few considerations, can be achieved in a matter of minutes.

The first two things that we need to look at are relevant only if you are taking separate tracks and then manually creating the take folders yourself. If you are recording multi-mic recordings directly into Logic, then these two steps will have happened by default. The first thing we need to do is make sure that all takes for all tracks start at exactly the same position and are exactly the same length. Assuming that all your tracks are synchronized (if not, then you should probably do this first anyway) but have different start and end points, the easiest way to do this is to select all the tracks and then, using the *Scissors* tool, simply cut them all to be the same length. If you use the Scissors tool at a point that is within all the regions at both the beginning and the end, then the start and end points (and hence the length) will all be identical. Any regions either side of these main regions that are created by this trimming should be deleted.

The next thing that you need to do is make sure that all the takes are in the same order for each track. For example, if you had a simple four-mic setup (kick, snare, hi-hats, and overhead) and you had three takes for each track (Take 1, Take 2, and Take 3), you would need to make sure that all of the Kick tracks were grouped together in the Arrange window in order (Kick Take 1, Kick Take 2, then Kick Take 3) and then repeat the process for the other tracks. Once this is done, you would pack all the Kick takes into a take folder and do the same for the others. The empty audio tracks can now be deleted in order to keep things tidy. You should now be left with one take folder for each track with three takes (in ascending order) inside each take folder. This is the same point you would be at automatically if you had recorded the tracks into Logic yourself, so we can now move forward from this point.

It can be very helpful if you color the different takes and take folders in different colors to help you identify what you are looking at quickly and easily. It is easiest to carry out *Quick Swipe Comping* if you are zoomed in quite a lot, and as a result it is unlikely you will be able to see many tracks on-screen at once. This will mean that you might need to scroll up and down to check the effects that your comping on one track is having on the others. The quickest way to bring up the *Color* selector box is to press Alt + C. If you select regions in the Arrange window and then click on one of the colors in this selector, the regions will change to that color. It is best to do this at this stage because grouping the tracks, which is the next step, will mean that any changes you try to make to colors will affect all tracks at once, meaning that you can't color tracks individually. Another alternative is to color the equally numbered takes (Kick 1, Snare 1, etc.) the same so that you can see, in your final comp, which parts of the comp came from which take. Each of these two options has its advantages, so you can decide which works best for your given situation.

As stated above, the next stage is grouping your different tracks, and to do this you need to open the *Mixer* window at the bottom of the screen and select all the channels for the take folders that you wish to work on. Click on the *Group* box (located directly below the track name and above the automation state button) and a pop-up menu will appear with a list of groups. Choose one of these groups, and the *Group Settings* window should appear with a list of all groups at the top and the group number that you just selected highlighted. If you double-click under the *Name* column on the group number, you will be able to type in a name to help you quickly identify your (in this case) drum group. Once you have done this, you need to make sure that the *Editing (Selection)* check-box is ticked (if you don't see this option, then you may need to click on the disclosure triangle to the left of *Settings*). When this is active, the *Phase-Locked Audio* check-box should be ticked by default. You can now close down this window, return to the Arrange window, and start comping. If you need to reopen the Group Settings window at any point, you can click on the channel Group box again and choose *Open Group Settings* from the top of the list.

The biggest consideration that you have when carrying out comping on multiple tracks at once using this method is remembering that all tracks/take folders will be affected at once, and what might be the best choice for any given track might not be the best for all the others. It is always possible to ungroup the take folders and make comping adjustments on a single track at a time, but, as we discussed in the main chapter, this can cause a few problems, so, if you are adopting this multiple-tracks-at-once approach, then you may need to try to find the best compromise.

You can choose any take folder to actually carry out the Quick Swipe Comping, and you can change from one take folder to another at any time. As mentioned above, the use of colors for different tracks or takes can help greatly here to visualize exactly what is going on at a glance. Otherwise, the actual process of the comping is exactly the same as for a single track at a time. All of the duplicating, renaming, moving, exporting, and flattening possibilities that we looked at for a single track are equally valid here. The only thing to remember is that all functions in the Take Folder menu apply to all grouped tracks at once. One very useful benefit of this is the fact that, once you are done with your comping,

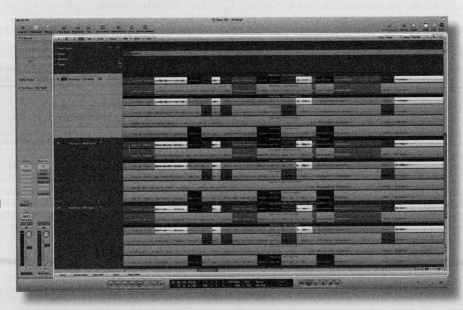

FIGURE 8.6
Multi-track comping is easy in Logic once you have everything set up correctly. The hard work is really in making sure everything is grouped correctly, because, once that is done, the rest is (for the most part) no more complex than comping a single track.

exporting the comp to a new track will do so for all grouped tracks at once, which can be a safeguard against forgetting an individual track.

As a footnote, and in advance of later chapters, this track-grouping method using the Editing (Selection) option also allows for using *Flex Mode* across multiple tracks at once, thereby keeping everything nicely synchronized, even if you wish to change the timing. While the obvious use of this is for using Flex Mode to re-groove drum recordings, it can also be useful for any groups of tracks that share a similar timing, such as lead vocals and backing vocals or harmonies.

Pro Tools

Comping multiple tracks at once in Pro Tools follows the exact procedure as comping a single track does, but, in order for your comping steps to be synchronized across a number of tracks, you need to group them. Before you do this, though, you will need to follow the steps in the previous chapter to get to the point where you have each track created with the various takes in playlists below the main comp (and empty) comp track. Using the familiar example of a multi-track recording of a drum kit, you would need have a number of different drums, each on their own track, and a number of different takes for each drum. To get the multi-track comping to work properly, you need to make sure that the takes (first take, second take, etc.) are arranged in the same order for each individual drum. It doesn't matter what order the takes are in, as long as the order is consistent across all tracks. It also doesn't matter what order the tracks are in in the Edit window, as you can drag these to change the order if required in exactly the same way as you can drag to change the order of the playlists under each track.

Once you have the playlist tracks ready for each track that you wish to edit simultaneously, you should select all the tracks by holding down Shift and clicking on the track names in the track header area, then press Cmd[Mac]/Ctrl[PC] + G or Track > Group to bring up the *Create Group* dialogue. The first thing I recommend that you do is change the default name of the group to make it easier to figure out which group is which if you use other groups later on. Next you should select an option for the *Type* of group you wish to create. An *Edit* group will group the tracks together only in the *Edit* window and will mean that any cutting and pasting, moving, time-stretching, *Clip Gain* adjustments, and, importantly, comping (and so on) will take place for all grouped tracks simultaneously. A *Mix* group, on the other hand, will group tracks together only in the *Mix* window and will link things such as channel volumes, solo and mute, and send mutes and levels. Solo, mute, and send mutes and levels are, in fact, selectable when you create or edit the group, and you can select or deselect them individually by using the check-boxes to the right of the Create Group dialogue. You can also give the group a unique *ID* consisting of a number (1–4) and a letter (a–z). For what we are doing here, you can just leave this as whatever the default value is.

Hands On

The main body of the dialogue is made up of two lists: *Available* and *Currently in Group*. These two lists are where you can define which tracks will be included in this group. By selecting all the tracks that you wish to group together before opening this dialogue, you should see the names of all the tracks that you selected already in the Currently in Group list. If you wish to add an additional track (now or later), you simply select it from the Available list and click on *Add >>* to move it to your group. Equally, if you wish to remove a track, then you select it from the Currently in Group list and click *<< Remove* to remove it. Another way of doing this is to double-click on a track in the Available list to add it to your group or double-click on a track in the Currently in Group list to remove it. Once you have you all the tracks you need added, click *OK*, and the group will be named for you.

Each group that you create will be visible in the Groups list located at the bottom left corner of the Edit window. If you wish to modify the settings for any of your groups, you simply press Cmd[Mac]/Ctrl[PC] + Ctrl[Mac]/Start[PC] + G or right-click on any of the group names in the Group list and choose *Modify*, and the *Modify Groups* dialogue will appear. Here you can choose which group you wish to modify, using the ID control at the top right to navigate to the correct group. Once selected, the parameters and track lists for that particular group will be shown, and any changes can be made and then saved. This can be useful if you add in additional tracks later that you wish to include in a group or if you make a mistake and add a track into a group that doesn't need to be there.

Another very useful feature of the Groups list is the ability to enable or disable groups simply by clicking on the group name in the list. Clicking on a group in the list will toggle between active (dark background) and inactive (light background) states. While these groups are essential to enabling us to carry out our multi-track comping, there may be times when you wish to make a change

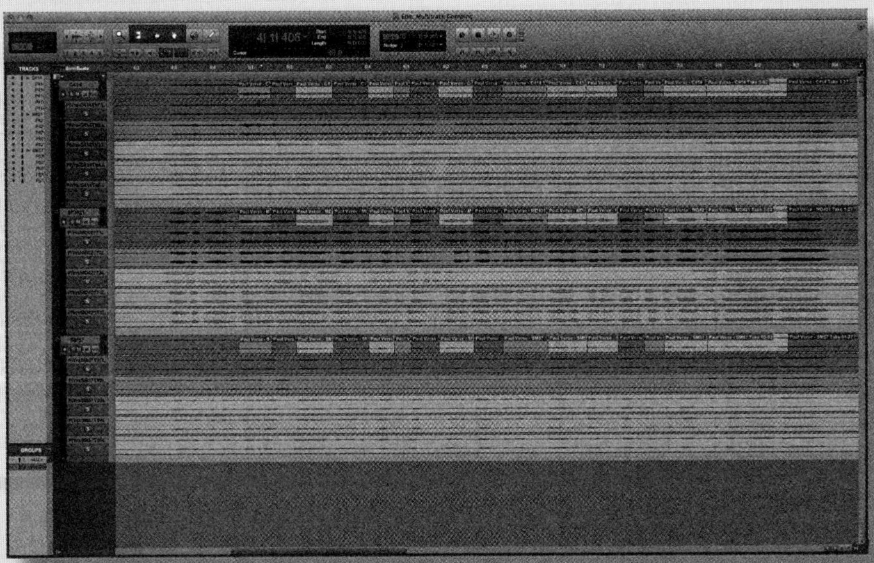

FIGURE 8.7
Pro Tools, like all of the other DAWs we are looking at, extends its single track comping with playlists into a multi-track format by way of setting up groups that will carry out edits on all grouped tracks simultaneously. This means that there will be coherence across your multi-track edits, with no chance of missing something by mistake.

to just one of the tracks within your group before continuing with the comping process. Using this method makes it very easy to have all of the benefits of these Edit Groups but also allows us to make changes to an individual track within the group if we need to, without any difficulties.

Once all this is set up, you can start the actual comping process in exactly the same way as we spoke about at the end of the last chapter. Your comp can be put together using the same methods and techniques that are used for comping a single track, but, because of the Edit Group that you created, the process is mirrored across all tracks in the group. This should allow you to create your multi-track comps quickly and easily; however, for the purposes of fine-tuning the start and end points of regions within each comp, and for fine-tuning any cross-fades, I would recommend that you make the Edit Group inactive, to allow you to make these detailed adjustments using parameters that are suited to each individual track rather than applying them en masse, across all tracks at once. Grouping the tracks when deciding which parts of which take to use is a huge time-saver, but it should be seen as a universal panacea that can be applied to everything with equally good results, and you should definitely do the final polishing on a track-by-track basis.

If you wish to create completely new tracks for your final comps, perhaps because you want to create an alternate version, you can do this easily by creating the appropriate number of new audio tracks (Cmd[Mac]/Ctrl[PC] + Shift + N or Track > New…) and then changing the *Track View Selector* back to Waveform. This will remove the various takes/playlists from view and leave you with just the three main comp tracks. Now you can click on any part of any one of the tracks and then press Cmd[Mac]/Ctrl[PC] + A or Edit > Select All, and all of the comps will be selected, and you can now drag them to the newly created tracks. Once you are done, you can change the Track View Selector on the comp tracks back to Playlists and continue with your comping.

Studio One

Multi-track comping in Studio One, as with most modern DAWs, is really just an extension of the principles and techniques used for comping a single track, with the added fact that a number of tracks are grouped together for the comping process. This phase-locked grouping means that any decisions made on which take to use for one of the tracks will be carried over to all of the grouped tracks. Using the kick drum from Take 2 of a grouped multi-track drum recording will mean that the snare drum, toms, overheads, and whatever other tracks were recorded will also use Take 2. This is by far the easiest way to deal with multiple tracks, as it removes the possibility of timing and phase issues that can occur when using different takes from different tracks at the same time. It isn't always the way to get the absolute best result in terms of the performance, but it certainly makes getting a coherent-sounding comp much easier. It is always possible to use this form of multi-track comping as a basis and get fairly close to the perfect comp, and then later you can always carry out more-specific work on an individual track if required.

If you are working on tracks that have been sent to you and that you haven't recorded into Studio One yourself, then you will have to carry out a similar process to that which we looked at in the last chapter, where you create an empty track and then empty layers before moving regions onto these layers. However, given that we will be grouping the tracks, there are a few shortcuts we can take. Start by creating the empty comp tracks that you will need, and you can do this by pressing T or going to Track > Add Tracks. This brings up the *Add Tracks* dialogue, and we can choose how many tracks we need and whether they are mono or stereo. We can also help make our life easier by using the *Name* box at the top. If we type something in to this box—for example, "Drum Kit Comp"—the tracks created when we hit *OK* will be numbered: Drum Kit Comp 1, Drum Kit Comp 2, and so on.

Once the you have created the tracks, you should then make sure they are all adjacent in the Arrange view (they should be), and then you select them all by clicking on the first and **Shift**–clicking on the last to select all. When they are all selected, you can press **Cmd[Mac]/Ctrl[PC] + G** or go to Track > Group Tracks, and this will create an Edit Group, which will allow you to comp the tracks as if they are one. By grouping the comp tracks in this way, you can simplify the process of creating the layers, as this, as well as many other aspects of the comping process, will be mirrored on all grouped tracks. Right-click on the track header of any one of the newly created tracks, choose **Layers > Add Layer**, and then right-click on the track header and choose Expand Layers, and you will see that a layer has been created for all your comp tracks. You will note that all layers have the name Layer 1. If you then repeat the process to create additional layers, you will see that they are all created for all tracks at once and are all named the same across the grouped tracks.

You should continue to add layers, until you have enough for all your different takes that you have for each track, and then you can start copying or moving your regions onto the tracks. At this point, it is very important that you make sure that the takes are copied onto the layers in the same order for each track. That's to say that if you copy the first take for the kick drum (for example) onto Layer 1, then you have to make sure that you copy the take for the snare drum, toms, overheads, and so on, onto Layer 1 on their respective tracks as well. This is crucial, because the comping process for grouped tracks works on the basis of moving parts from equal layers up to the main comp track(s) at the time of comping. If the takes are in different orders on each track, then there will be no consistency in the comping. When all your takes are on the correct layers on all tracks, you can start to work on your multi-track comp, using exactly the same process and techniques as you did for a single track.

If you are actually recording the audio in to Studio One, then the process becomes easier to a certain degree. All the takes should automatically be on the correct layers, owing to the fact that they would each have been recorded at the

Multi-Track Comping CHAPTER 8

same time, and this means you can skip straight to grouping the tracks (using the same method as described above), and then you will able to get started on the comping.

When you are carrying out the actual comping, you should remember that you can click and drag the comping tool over any take/layer to actually do the comp. This can be very useful if you are dealing with a complex multi-track recording, as you can switch your focus to whichever of the tracks requires the most detailed work at any given time. One thing to remember, though, is that it isn't only the choice of layer that is mirrored across all grouped tracks. Volume envelopes/gain offsets, cross-fade times, cross-fade shapes, even the layer names themselves are all linked, so any changes you make to one are made across all tracks at once.

Having said that all changes on a single track will be mirrored across all tracks, there could well be times when you want to adjust a cross-fade on one track without adjusting the others. Technically this is possible by simply pressing Shift + Cmd[Mac]/Ctrl[PC] + G or going to Track > Dissolve Group and then making the change on whichever track you want. However—and this is a big caveat—while you can group the tracks again after the edit that you make, that will also mean that the changes are linked again, and, if you happen make a change on one of the other tracks, it could undo the edit that you made while ungrouped. If at all possible, it is better to wait until the comp is complete, and you can then move or copy the final comps onto new, ungrouped tracks, where you will have the freedom to make fine adjustments to levels, cross-fade, and the like.

When you have completed the comping process, you can duplicate the main comp tracks if you wish to create alternate comps, or you can merge or bounce the comps as you would for a single comp track. Pressing G to merge the regions will create a single merged region for each track, and pressing Cmd[Mac]/

FIGURE 8.8
Phase-locked groups mean that multi-track comping is a breeze in Studio One. Make your edits across a number of tracks simply by grouping the tracks before you start.

Ctrl[PC] + B will bounce a completely new region for each track at the correct position. If you do now wish to move these comps onto other audio tracks, then you either need to dissolve the group and move then individually onto other tracks, or, of you wish to keep the group intact and move them en masse, you will need to create an equal number of new empty tracks adjacent to each other (but not necessarily adjacent to the comp tracks), and then you will need to remove the Expand Layers option on the comp tracks, so that the layers are no longer visible. Once you have done this, you can select all the regions on each of the grouped comp tracks and move (or copy) them onto the empty tracks.

Cubase

If you have mastered single-track comping in Cubase, then multi-track comping requires only a few extra steps. The only real difference is that you have to group together a collection of individual tracks, at which point any comping on one is mirrored on all the others. While some DAWs simply require to you set a parameter or two in order to group tracks, Cubase does require an additional step before you can actually group the tracks together.

The first thing you need to do is set up your collection of individual tracks with multiple lanes. If you have been recording your audio into Cubase, then this will already have been taken care of, but if you are collating tracks that have come from outside of Cubase (or this project, at least), then you just need to follow the procedures in the last chapter for each of the tracks. You also need to make sure that, within each track, the takes are in the same lanes (Take 1 in Lane 1, Take 2 in Lane 2, etc.), as the comping will be based only on lane numbers and won't take into account the naming of takes or anything like that. It can be helpful to color each of the tracks to help you visually identify which take is being used for which part of the comp. Coloring each take is as simple as selecting the region for each take and then clicking on the *Select Colors* button to the right of the main toolbar. It makes sense to keep all the colors consistent between tracks by using, for example, Color 1 for Take 1, Color 2 for Take 2, and so on. This isn't strictly necessary, but, for the amount of time that it takes, it is something worth considering from a work-flow perspective.

When all your individual comp tracks are in place with all the lanes prepared, you then need to move the comp tracks into a folder. This is done by selecting all the tracks (Cmd[Mac]/Ctrl[PC]–clicking on each of the track headers that you want to select) and then right-clicking on the track header of any one of them and choosing *Move Selected Tracks to New Folder*. This will create a new folder (which you can rename) and move the tracks into it. If you have an existing folder and you want to add another track to it, all you need to do is click on the track header and drag the track into the folder header. As you move the track into the folder header, a green arrow will appear on the folder header, and this indicates that you can release the mouse button. The track will then be added to the folder below all the other tracks. If for any reason you wish to reorder the tracks within the folder, then you simply need to click and drag them to the new position that

Multi-Track Comping CHAPTER 8

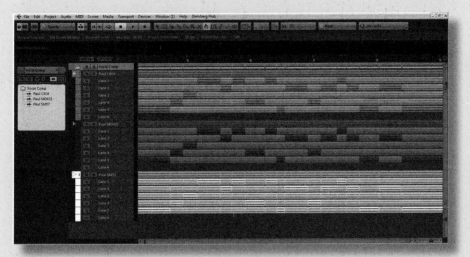

FIGURE 8.9
The combination of the Lanes in Cubase and grouped tracks give the option to carry out perfectly aligned multi-track comping and make it seem simple. These methods allow you to get the job done in a fraction of the time it would have taken using more traditional methods.

you want them to be in. As you click and drag the track, a horizontal green line will show the new position of the track once you release the mouse button.

Once the folder is created, you can engage Group Editing by clicking on the Group Editing button on the folder header (which looks like an "="). This mode is what makes simple multi-track comping possible. Any comping decisions that you make on any one of the tracks will be carried through to the others. However, that comes at a price, in some respects. Not only are all of the comping choices grouped, but also any adjustment of *Fade Handles*, the *Volume Handle*, or any repositioning of individual takes or regions are grouped. As a result it is probably best to correct any parts of your takes that are out of sync before you enable the Group Editing. You can, if you like, work on the fades or adjust the Volume Handle prior to grouping, but this isn't so crucial, as you can temporarily deactivate Group Editing and make a level adjustment or add a fade before activating again.

In theory, you can also move regions while the Group Editing is deactivated, but if you do, when you reactivate it, you may well be presented with a warning stating that the tracks in the folder are not completely in sync and that the group editing could fail. To be honest, the chances of having to move an individual take or region are quite minimal if you are working with multi-mic recordings, because all the takes should be inherently synchronized. If you have grouped different instruments (perhaps lead vocals and backing vocals or harmonies), then you may need to do a little work on the individual timings before you get heavily involved in comping.

When all this is done, you can start working on the actual comping process. This is no different in execution to working on a single track, but you do have to take a few other things into consideration. There is always the possibility that you might want to fine-tune the specific position of the transition from one take to another differently on different tracks. In the main chapter, we have looked at the reasons why it can be difficult using different takes on different tracks, so, for the most part, you will want to use the same takes on each track. But there can be occasions where positioning the transition accurately on one track means compromising the position on another. In a situation like this, you should aim for the best compromise while working on the comp, and then, if necessary, deactivate Group Editing and work on individual transitions on individual tracks (perhaps including cross-fades) before moving on to the next stage.

If you are intending on making only a single comp, then you can leave everything as it is once your comp is completed. On the other hand, if you want to create a number of alternate comps, then you have the same options as you do for a single track. You need to bounce the comp for each take, and fortunately this can be done in one step by selecting all the regions on all the comp tracks at once (it is usually easier to hide the lanes for all the tracks before doing this, so you can just click and drag of the comp ranges) and then going to Audio > Bounce Selection. Doing this will bounce each track individually and then give you the option to replace the selections. As with a single track, if you choose to replace these selections, then your newly bounced file will replace the comp you have created (and all the takes in the lanes). This is obviously not what we want if we want to work on an alternative comp, so you should choose not to replace the selection. If you choose this option, then each of the bounced comps will be added to the audio pool for you to bring into your arrangement manually.

The alternative method that we discussed in the last chapter works even more effectively when comping multiple tracks at once. If you click on the folder header and choose *Duplicate Tracks*, the entire folder will be copied. That means that all comp tracks, comps, and individual lanes and takes will be copied. At this point, you can hide all the lanes, select all the comp regions, go to Audio > Bounce Selection, and choose to replace the selections, and you will be left with a folder containing each of the bounced comps on its own track. If you choose to show the lanes again on these bounced tracks, you will see a number of empty lanes where the takes have been replaced. You can right-click on the track header for each track and choose *Clean Up Lanes* if you want to remove these empty lanes.

In either case you can now mute the new folder and continue working on an alternate comp with the original tracks and lanes and then repeat the bouncing process as many times as you need to, until you are done with your comping. Given that each set of comp tracks resides in a folder, you can name these folders as you choose, and, given that each of them will be a variation on a theme, you can then select all the folders (Main Comp, Alternate Comp 1, Alternate Comp 2, etc.) and further group them into another folder. You don't have to do this, of course, but it can help to keep things organized in your arrangement if things start to get very busy.

CHAPTER 9
Transient Detection

WHY USE TRANSIENT DETECTION?

Throughout the previous chapters, there have been numerous references to "transients" and the "attack" of sounds. These are particularly relevant when it comes to fine-tuning the timing of different regions or notes. So far we have always suggested that this is done manually, as the waveform display can be misleading in this regard. But what happens if the recording that we have is such that it is hard to pick out where the transients or start points of different sounds or notes are? What if, for example, we are dealing with a fully mixed stereo drum-kit recording where the waveform display doesn't allow us to clearly pick out all the transients?

The ability to pick out transients (or just the start of a particular sound or note if it doesn't have a sharp attack) is essential to us if we are to be able to make sure that everything is in time with our edits. Therefore, if we are unable to do this quickly and easily with more-complex source material, it can slow down our work or even prevent us from doing what we are able to do. Naturally, in even the most difficult situation, we can give it our best guess, but, if we aren't sure, then the process can be made longer, because we have to guess, then adjust, then perhaps guess again and adjust again. Add to that the fact that we base our guess on what we see on-screen and what we hear, both of which can be misleading in more-complex scenarios, and we have a lot of room for error.

Fortunately there are a number of ways of analyzing audio files through software that can give us further insight into aspects of the sound that we cannot clearly quantify through listening alone. Foremost among these is the use of spectral analysis and its ability to identify changes and patterns of different frequency levels and volumes over time. This extra information can allow us to separate seemingly similar sounds, and, in doing so, be able to be much more accurate in our estimates of where a sound might start.

For example, in the case of our mixed drum-kit recording, there is a clear separation between a kick drum and a crash cymbal on the basis of frequency content alone—but now let's consider the comparison of a crash cymbal to a hi-hat.

Why Use Transient Detection?

FIGURE 9.1
Transient detection is only half the story, because we need to look at other factors. In the image above, you can see that the hi-hats are very short in duration, while a crash cymbal lasts a lot longer. Separated, it is easy to pick out the hi-hat transients, but, if this were a mixed-overheads recording, the crash would most likely obscure the transients in the hats, making automatic detection difficult.

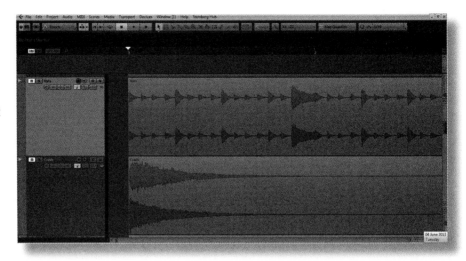

Both are cymbals, and both will have similar frequency content. If we imagine a situation where we have a crash cymbal hit immediately followed by some hi-hat hits, then, based on frequency alone, it would be somewhat difficult to separate the two sounds and decide where the hi-hats were placed. However, if we consider the time domain as well, then we would see that the crash cymbal is a much longer sound, and changes to the levels of its component frequencies will happen over a more protracted period of time. A hi-hat, by comparison, has a relatively short duration, so the sound energy created by the hi-hat would be focused into a much shorter period of time.

Listening to the combination of these two sounds may make it difficult to pinpoint the exact timing, as they could be quite similar in level. Looking at a conventional waveform display could also be difficult, as they are complex waveforms anyway, and it wouldn't be immediately obvious what, on the waveform display, constituted a part of the crash cymbal and what constituted a part of the hi-hat. We could also consider "scrubbing" through the audio to help us find the particular point in time where the hi-hat occurred.

"Scrubbing" is a technique that was an essential part of tape editing. It basically involves manually moving the tape back and forward past the tape head, until you hear (probably at a much lower pitch, owing to the slower speed that you are moving the tape) the sound that you want to edit. In the case of an obvious sound like a kick drum, this is very easy to identify, but, with our crash cymbal and hi-hat scenario, it might be a little more difficult. Scrubbing still has its place in a DAW and can be even more useful, because you do also have the waveform display to give an additional visual clue to try to help you identify the correct point.

The more complex the audio, the more difficult it will be to identify the correct edit points. The methods described so far could, depending on the complexity, allow you to find what you are looking for through a combination of good ears, good eyes, good luck, and, most important, patience. This is all well and good if you are just looking for the odd few transients or edit points in this way, but if you had a large number to find, this process would be exponentially more difficult. In a situation like this, we would benefit hugely from some kind of automated system. Fortunately, such systems do exist, and they come in a number of different flavors from the simple to the complex.

SIMPLE LEVEL-BASED DETECTION

The simplest way to detect a transient is by level, or, more specifically, a change in level. Actually, to be more specific still, it is most effective to look at the rate of change of level. The reason for this is that absolute level does not determine whether a transient is occurring. If you imagine a Hammond organ, which has a very constant level to its sound, you can see why absolute level wouldn't work. Even in the case of something like a snare drum, if we were defining transients by level alone, then the very beginning of a quiet hit might register as a transient; but, equally, 50% of the duration of a louder hit could be above our transient detection level, so that would clearly not work.

If we take things a step further and try to detect transients by a change in level, there are further problems. A violin played in a crescendo style would increase in level throughout its duration but would have a transient only at the very beginning of each note. Equally, the human voice is capable of building in volume over the course of a note. In fact, any sound that is generated through constant interaction with the sound-generating mechanism is capable of this. The only sounds that aren't are those that generate sound through a single energizing event. The obvious example is any drum or percussion sound, but things like guitar and piano also fall into this category of instruments that have a fixed volume envelope.

One definition of transients (in sound) is as follows:

> Attack transients are the initial phase of an independent sound source and occur together with note onsets. They show fast changes of the sound characteristics. A transient is a short burst of energy caused by a sudden change of state of the sound production system. An onset refers to the beginning of a sound. All musical sounds have an onset but do not necessarily include an initial transient.

This definition points us toward a solution for detecting transients. When we factor in the actual rate at which this change occurs and limit our transient detection to sudden and quick changes in level, then we are more likely to get an accurate result. In many cases a simple level-based system like this may be all that is necessary to detect the transients accurately, but there is always room for improvement, and the simplest way in which we can make these systems more "intelligent" is by making them frequency-aware.

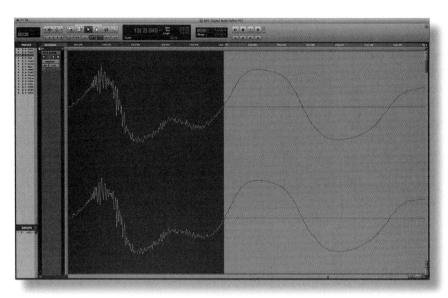

FIGURE 9.2
This close-up view of the start of a kick drum shows the burst of high-frequency energy right at the very beginning. This is caused by the impact of the beater hitting the drumhead. This initial high-frequency energy dissipates very quickly, and this is one of the hallmarks of a transient.

ADVANCED LEVEL-BASED DETECTION

Many audio transients, in addition to being short in duration, also have proportionately more high-frequency content than the remainder of the sound. Therefore, if we add in some frequency awareness to our rate of change of level awareness, we can use that to further increase accuracy. The actual manner in which this is achieved could vary, though. The analysis process could simply apply a variable high-pass filter to the sound source, so that we could progressively focus more and more on high frequencies. This may be a little too simplistic, though, so perhaps a better option would be to split the source sound into at least two (and possibly more) frequency bands and perform the analysis on each but with a control that allows us to shift the bias in favor of either one or the other, dependent on the type of sound we were analyzing.

With this more adaptable multi-band system in place, we could consider further improvements. Perhaps we could add some kind of Sensitivity control, so that transients only above a certain level were detected. This would mean that we could choose between trying to pick out every little transient there was or focus on only the more obvious ones. This could prove very useful to give more fine control over the end result.

Perhaps, if our system were not expected to work in real time but could pre-analyze the audio file, we could also specify an additional Decay parameter, which would mean that only sounds that met all our criteria would be considered transients. Only sounds that increased in level quickly enough, had a level that was above the Sensitivity threshold, and also reduced in level (particularly in the high-frequency content) at a relatively quick rate would be considered to be transients. This would definitely be a big step forward and would certainly raise the accuracy of the system. Once we had our "markers" in place, we could then make use of them in any number of ways.

There is one additional thing that this transient detection system could do for us, though. In addition to telling us the timing of these transients, if we could have them give us an indication of the volume of the sound at the point that it triggered the transient detection; then that would give us the possibility to do something very useful: drum replacement.

The two most obvious characteristics of any drum hit are timing and volume. As we have already seen, there are tonal variations depending on where the drum is hit, but for the most part these will be secondary to timing and volume in importance to the feel of the performance overall. Our transient detection system, if we can get it to not only be level-aware but also to quantify that level as well as the timing, gives us the two fundamental things we need to replicate the drum groove. We will look at this in more detail in chapter 11, but for now let's consider one last option for transient detection that takes the principles we have applied here even further.

SPECTRAL DETECTION TECHNIQUES

We have spoken about transients as being abrupt changes in the level and frequency content of a sound over a short period of time. We have suggested ways in which this could be tracked and implemented to allow a transient detection process. And, as we have said, this would work in a number of situations, but, for something far more complex, we can move over to some very advanced mathematical processing.

IRCAM (Institut de Recherche et Coordination Acoustique/Musique) is a research center dedicated to research into all aspects of sound and acoustics. It is located in Paris and has over the years carried out some truly amazing research that has led to advances in computer software and audio technology. One of their more recent avenues of research has been in transient detection (as a part of a larger research project). They have published their papers and research, which are available to view, should you be interested. They are not light reading by any stretch of the imagination, though. They contain some very advanced calculations, which make understanding their transient detection method almost impossible to understand, unless you have a degree in mathematics!

In simple—and by no means fully explanatory—terms, the sound is divided into a number of "frames," where each frame is a snapshot of the spectral makeup (frequency balance) and changes to the spectral makeup of the sound at a particular point in time. These frames are very short in length (around 35–45 ms in this IRCAM system) and take note of any changes that happen in any particular frequency band during that time period. The comparison of a number of these frames gives a value for "transient peak probability," and each frame has its own value. Then, once all these calculations have been performed, a transient is considered to have happened if, in any of the frequency bands, the transient peak probability for the given frame is larger than the transient peak probability in the preceding frames.

Of course that is a huge oversimplification of the process, and if any of you are interested in reading the whole article, it can be found at http://articles.ircam.fr/textes/Roebel03b/index.pdf. I guess, in the simplest possible terms, this would be one situation where you don't necessarily have to know *how* something works fully in order to understand that it just *does* work. These techniques go way beyond a simple two-band level change detection circuit, and, as much as I would love to, I couldn't even begin to understand the full extent of the calculations behind it. There are an increasing number of situations like this in audio editing and production, where the methods used behind the scenes to generate results are simply too complex for the average person to understand—which is a shame in some ways, because I am a firm believer in always trying to understand the *how* and *why* as much as the *what*. In my experience that not only improves my use of the tools available but also allows me to think of other, less-obvious ways in which they could be used.

POSSIBLE APPLICATIONS AND LIMITATIONS

Each of the methods described can be useful, to varying degrees in different situations, in helping us to identify transients, but, if we look back to the definition that we had for transients, there is a very distinct difference between a transient and a note onset. Some sounds, both acoustic and synthetic, simply don't have clearly defined transients, so identifying the note onset by use of a transient becomes much more hit and miss. Most modern transient detection processes will still do a fairly good job of identifying note onsets as well, especially if there is a period of silence prior to the note onset. Even though the criteria that we apply to a transient (rapid change of level, etc.) might not be applicable in this situation, the system could still detect a change of level (from silence to something), and, if the sensitivity was set accordingly, these note onsets could be detected as well.

Another problem lies in the detection of individual notes within a legato passage. Even a sound such as that of a violin, which has a clearly identifiable transient, could prove problematic if the passage was played in a certain way. In the case of a violin, the transient occurs when the bow first scrapes across the string and causes it to vibrate. Once this initial event has taken place, there is a continued transference of energy from bow to string, right up until the point when either the bow is removed or the bowing direction changes, and another transient is initiated. Owing to the nature of the violin, though, it is possible that the pitch of the note can change midway through the bowing action simply by players moving the position of their fingers on the fingerboard. This would result in a change in pitch and note but no discernible change in volume or sound energy. In this case, all of these transient detection techniques would be useless, and the marker would have to be placed by hand. To be fair, though, these systems are designed to detect transient so the fact that they can sometimes detect note onsets as well is a bonus.

Changes in the note are trackable, though, and many pitch-manipulation software tools will "split" an audio waveform and show each note as a distinct block on the screen. Sadly there is currently no way of using this information to split the underlying audio region into separate parts. As in many other areas of audio editing and music production, the technology that we need often exists, but there are times when we would ideally want to have certain facilities from one application and certain facilities from another combined into a single process. If we could incorporate pitch-tracking into our transient detection tools and use changes in pitch as an additional indicator, then we might be able to get a lot more done in a single step. For now, though, even though they have limitations, these transient detection tools can massively speed up the process of finding edit points.

Aside from the obvious help that these systems will give us when it comes to finding the correct points to do our edits, there are other uses for them to help us get a little more creative with our editing. The most obvious of these is to allow us to take longer files and make sure that they are in time very easily. By using the transient markers as timing references, and assuming of course that they have been checked and possibly reduced in number, it is more than possible to apply quantizing to audio files to either tighten up the timing of the files as they stand, or radically change the groove and feel. They would also allow us to quickly and easily cut up a sound into meaningful parts (individual hits of a drum beat, chords of a guitar, notes of a bass line, or words of a vocal performance) that we could then move or, more creatively, rearrange.

Transient detection is also a fundamental part of the some of the better time-stretching tools available. We have already seen that transients are very short in duration, and they are often crucial in establishing the character of a sound. String and brass instruments often have very similar waveforms in their steady, sustaining state but have very different transients that help us to identify which is which. If we start time-stretching sounds, then we can substantially alter these transients, and, in doing so, the character of the sound. More recent time-stretching tools detect the transients and then, once they have been identified, they will exclude these from the overall stretching process, so as to preserve their duration and integrity. Doing this greatly improves the perceived quality of the result.

Another consideration with transients, which we haven't really spoken about as yet, is that they often contain high proportions of *nonperiodic* components. These are sounds which aren't cyclic in nature and aren't necessarily related to the pitch of the note (such as the actual breath sound with a wind instrument). Once the wind instrument has been "energized," this breath noise would blend into the sound, or, in some cases, all but disappear as a sound in its own right. If the transient can be identified, then it is possible that these nonperiodic components could be separated and not treated during pitch-shifting operations as well, thus preserving the character of the instrument further.

In terms of actual usage, much of the underlying theory is hidden away from us. There are a number of tools available to help us with transient detection. Almost all DAWs have some form of automatic transient detection built in to identify and show us the position of transients. Often these tools will also automatically slice the file into separate regions at these transient markers. There are also tools that can take this information and use it to trigger other sounds (more information in chapter 11), and more-advanced pitch shifting and time-stretching tools simply make use of transient detection without even letting us know. So it is a very useful ability of modern software with a number of different uses from the relatively simple to the more exotic, and it is something that, while we can live without it, and while we could duplicate many of the processes manually, certainly makes our life that little bit easier by having it automated to a certain degree.

HANDS ON

Introduction

Transient detection is useful in its own right, as we have seen, but for many it is most useful as a preliminary step toward another process. In some cases the transient detection is built in to another process, and, as such, we don't have discrete control over it, but many DAWs offer the option of transient detection as a separate process that can help us in our other editing tasks by providing us with visual cues and markers to help us quickly locate the parts of the audio files that we want to edit.

Logic

The best way to approach transient detection in Logic is to use the *Sample Editor* window. Some might feel that using *Flex Mode* (which creates the transients for you) is better, but, in reality, once the transients are detected and the transients markers created in the Sample Editor window, the same markers are used for the Flex mode analysis and processing anyway. So if you want quick-and-easy, use Flex Mode to create the markers, but if you want more control, it is best to use the Sample Editor method first and then apply Flex Mode.

To get started, choose an audio region and open the Sample Editor window. At the top of the window, next to the menus, you will see the *Transient Editing Mode* button (a vertical bar with left and right arrows at the base), and, after clicking this for the first time, there will be a short delay while the file is analyzed. Once the analysis is completed, you will see the transient markers that Logic has placed in the audio region, marked as vertical lines overlaid on the waveform display. Each of these lines represents a single transient, and in some cases this might be all you need to do if there are relatively few transients and each is clearly recognizable. But if you need to do a little more work, there are a few options available.

Possibly your first port of call should be the "+" and "−" buttons to the right of the Transient Editing Mode button. These two buttons will increase or decrease the total number of transient markers created and could be thought of as a kind of "sensitivity" control. The initial analysis is often very close, but it might be that there are just a few too many markers placed for your purposes. Using these buttons, you can decrease the number to a level that is more appropriate for your needs. But using this method doesn't have any real direct control over which markers are deleted or where new markers are added. It might be that you want a marker in the middle of a single word, and adding more markers everywhere doesn't solve the problem. Equally it might be that the markers are generally right, but there is a section of the region that you want to remove the markers from without affecting the others. In this case you need to manually edit the markers.

If you move the cursor over the top of the transient markers, you will see it change to the *Transient Editing* tool. Clicking and dragging an individual marker with this tool allows you to move the position of that marker without changing any of the others. By zooming in you can get a great amount of detail and be very precise about the exact position, but it is always worth remembering that what *looks* like the right place for a transient marker to be isn't always the right place for the purpose you have in mind, as we have discovered in this chapter.

If you wish to add or delete markers individually rather than using the "+/−" control, then this too is very easy. If you wish to add a marker, simply hold down the Cmd key; you will see the cursor change to the *Pencil* tool, and clicking with this tool will create a new marker. You don't have to worry about being overly precise as to the location, because you can always zoom in and reposition it as described above. Perhaps even easier is the process of removing unwanted markers. This simply involves positioning the cursor on an existing transient (at which point the cursor will change to the Transient Editing tool) and then double-clicking to remove that transient. And that's really all there is to transient detection and transient markers in terms of actually creating them. But there are a couple of other things that are worth mentioning to help you get the most out of them.

One technique that I find incredibly useful, particular if we intend to use Flex Mode on this particular region, is to create additional markers to separate out the parts of the region we need from the "silences." It may seem unnecessary, given that we will want to move only the actual sounds and not the silences in

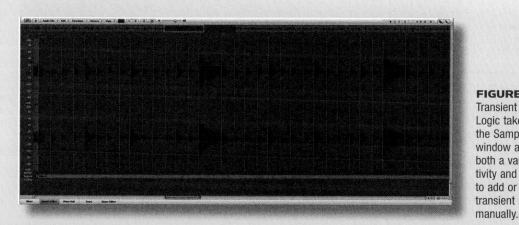

FIGURE 9.3
Transient detection in Logic takes place in the Sample Editor window and offers both a variable sensitivity and the ability to add or delete transient markers manually.

between, but in fact this is exactly why it is extremely useful to separate them in terms of the markers. Flex Mode works by allowing us to move the position of the markers and then automatically time-stretching or compressing the part of the region between the markers. In some cases, perhaps with quite a busy drum loop, this is perfectly fine, as we want to keep the fluidity of the overall audio file. But in cases where there are noticeable gaps between notes (a vocal being the main example that comes to mind), it would be more useful if we could reposition a marker to change the timing of a word without changing the length of the word preceding it.

By default, if we have a marker at the beginning of each word and wish to move a word back in time, it will automatically compress the preceding word. If the preceding word is very close, and there is little or no gap between it and the following word, then this behavior is ideal. But if there is quite a noticeable gap, then it seems unnecessary to compress the entire word just to allow us to move the following word closer. In a case like this, placing a transient marker in the silence between words will allow us to shift the timing of the following word back without compressing anything unduly, as only the silence would be compressed. Conversely, though, if we wish to move the timing of a word to the right, then we wouldn't really want a marker in the silence, because that would mean that the word gets compressed more than necessary. Therefore, I would advise against taking the time to put markers in to separate all the silences out, as it will not only take time to do so but also ultimately could mean that some of the silence markers would need to be removed anyway. It is far easier to just place them as needed in any situation where there is a silence and the timing of the following note/word needs to be moved earlier.

Another very useful benefit to having transient markers in place within a region is that it allows you to quickly select individual parts of a region (between transient markers) in the Sample Editor window without having to split the region in the Arrange window or without having to click and drag over the waveform display in the Sample Editor window. If you have transient markers in place, then double-clicking between two markers on the waveform display has the effect of selecting only the portion of the region between those markers and allowing you to then process that part selectively using the various Sample Editor tools.

Pro Tools

Pro Tools, like many DAWs, uses transient markers for a lot of different purposes. The three main uses of the transients are in *Elastic Audio Warping*, *Beat Detective*, and for the *Tab to Transient* function. While these three uses are quite diverse, they all share a basic need to have transient markers in place. Beat Detective does have its own transient detection process built in, and we will look at that more at the end of the next chapter, so, for now at least, we will focus on the Tab to Transient feature and how we can adjust the amount and position of the transient markers that it uses.

Transient Detection — CHAPTER 9

With its default settings, the Tab to Transient feature works quite well and will pick out all the transients in all but the most difficult of situations, but there may be times when it doesn't work quite as well, such as for long, flowing legato pieces where note changes aren't always accompanied by a distinct transient. On the other hand, it may work *too* well, and you might find yourself having to tab through a large number of transients when fewer would be more convenient and still give you the flexibility that you need. While there isn't an immediately obvious "Transient Sensitivity" control, there is a way to achieve exactly that, and it is to be found as a part of the *Elastic Audio* properties of a track.

The first thing that you need to do is select all the regions that you wish to adjust the sensitivity for. If it is a general increase or decrease in sensitivity, then it makes sense to use **Cmd[Mac]/Ctrl[PC] + A** or **Edit > Select All** to select all regions on the particular track, but if you have a more-specific purpose in mind, then a single region may be better. Once this is done, you need to enable Elastic Audio for the track. This is done simply by clicking on the *Elastic Audio Plug-in Selector* and choosing one of the different modes. Given that we are dealing with just the transient markers at this stage, it doesn't really matter which mode is selected, but we will look at these different modes in a later chapter. The location of this varies, depending on which level of vertical zoom you have selected. If you have track height of micro or mini, then it isn't visible at all, but if you have a track height of small, then it is located directly to the right of the *Track View Selector* and *Automation Mode Selector* buttons. If you have a track height of medium or higher, then it will be located underneath the *Automation Mode Selector*.

Once Elastic Audio is activated for your track, you then need to switch to *Analysis* in the Track View Selector. You will see the background color of the region change and a series of vertical lines overlaid on top of the waveform display. These lines are your default transient markers and are the markers used by in the default setting for Tab to Transient. There are a number of ways we can adjust these markers, but we will start by looking at a simple sensitivity adjustment. You need to open the *Elastic Properties* dialogue by pressing **Alt + Num Keypad 5** or by going to **Clip > Elastic Properties**. Annoyingly, the keyboard command will work only if you actually have a separate number keypad or a laptop that has a function whereby some of the normal alpha keys can be temporarily converted into a numeric keypad. If you don't, then you will either need to use the menu option described above, or you can right-click on the region itself and choose Elastic Properties from the pop-up menu.

There are a few different options available in this dialogue, but the only one we need to look at now is the *Event Sensitivity* control. By default it is set at 100%, which means that you can only reduce the number of transients from the default number using this method. You can click on the numerical readout and drag up or down to adjust the sensitivity, or you can click once and type a value in (in whole percentage, from 0 to 100). As you adjust the sensitivity, you will see the number of transient markers in the region(s) changing.

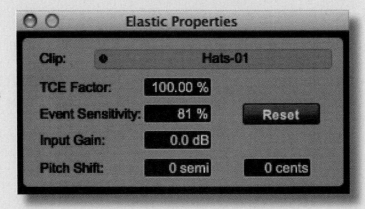

FIGURE 9.4
The Elastic Properties dialogue offers an Event Sensitivity parameter, which can be used to determine the density and number of the transient markers placed by Pro Tools.

The unfortunate thing about this, and many other sensitivity-type controls in DAWs, is that the results aren't always predictable. If you had a sampled drum loop that was a one bar loop repeating four times, there is no guarantee that reducing the sensitivity would remove the same transients in each bar, even though the region was exactly the same in each bar. That's not to say that this sensitivity control is pointless, though, because there are many times when it will work exactly as needed and thin out the transient markers that you wanted it to.

If you need to be more selective about which markers are removed, or if you actually need or want to add additional markers, or if you want to fine-tune the position of some of the markers because they aren't quite in the right place, then you can do all of this while still in the *Analysis* view. Make sure that the Grabber tool is selected, and then move the cursor over one of the transient markers; you will see it change to a vertical bar with left and right arrows on either side of its midpoint. This tool allows us to reposition the transient markers manually. While this is a great ability to have and one that is useful in many situations, there can be a temptation to zoom right in and find the exact moment that the sound starts. Technically this is the start of the sound and therefore the transient, but we also need to remember (especially if we are going to use these transient markers to quantize our audio) that the point that is technically the start of a sound isn't always the point that we would like to "anchor," so it certainly isn't worth obsessing over getting all the transient markers into sample-accurate positions, unless you feel that something isn't sitting quite right with the default settings.

As well as moving the existing markers, we can both add additional ones and remove existing ones. Adding new markers is achieved either by doubling-clicking at the desired position, or, alternatively, by holding down Ctrl[Mac]/Start[PC] while clicking. Removing markers simply involves positioning the cursor over an existing marker (the cursor will change to the one we use for moving markers) and then either double-clicking again or holding down Alt while clicking. Finally, you can, with the Pencil tool selected, just click on the region to create new markers.

When you are happy with the positions and amount of markers, you can change the Track View Selector back to Waveform, and you can, if you wish, remove the elastic audio processing by opening the Elastic Audio Plug-in Selector and choosing *None—Disable Elastic Audio*. The changes you have made either through changing the sensitivity or by manually adjusting the markers will now be carried through into the other transient-related features. If you now try using Tab to Transient, you will see that the changes you made are reflected here.

Studio One

Studio One doesn't have a dedicated transient detection process. Instead it is a part of the *Audio Bend* processing engine and is closely tied to the *Bend Markers* used for various audio-warping functions. This makes a lot of sense in some ways, because modern DAW software makes much more use of transient markers in this context than in any other. In addition, the fact that the transient detection process in Studio One is tied to the Bend Markers doesn't mean that that is the only way they can be used. It is more of a streamlining process when it comes to work flow.

To start the transient detection process, you need to open the *Bend Panel*. You can do this either by pressing on the *Audio Bend* button on the toolbar at the top of the main Studio One window (to the left of the *Quantize* settings) or by going to View > Additional Views > Audio Bend. When opened this panel will drop down below the main toolbar, and it includes a number of different parameters that are used to set up and customize the Audio Bend capabilities of Studio One. The controls are divided into four groups: *Detection, Bend Marker, Track*, and *Action*. While all four are important to the actual Audio Bend capabilities, only the first two are crucial to the actual transient detection process, as the last two relate to the processing of the audio and what is actually done with the transient markers. We will, of course, look at those in due course, but for now we will focus just on the first two.

The first thing that you need to know is that, while transient detection is carried out on a region-by-region basis, the settings are chosen on a track-by-track basis. Once you have chosen your settings and carried out an initial analysis, all the regions on the current track will change to a darker-region background to show that the Audio Bend processing is active on that track, but only regions that have specifically been selected will show the transient markers.

Moving on to the controls themselves, you will see that there is only a single control under *Detection*, and that is a *Mode* setting, which can be set to either *Standard* (default) or *Sensitive*. It is recommended to start with the Standard setting, as this will work perfectly well in most situations. If you find that this setting isn't accurately picking out the transients that you need—for example, if you have an overhead recording of a drum kit that has a lot of "splashy" cymbal noise in addition to the kick and snare transients that you want—you can switch to Sensitive, and this will, one hopes, do a better job of find all the transients. Once you have chosen the mode, you wish just click on the Analyze button, and, after a short processing delay, you will see the transient markers appear on whichever regions you had selected.

Having only two modes may seem restrictive compared to some other DAWs, but that isn't the end of the story, because you can still adjust the effective sensitivity of the process by moving on to the Bend Marker section. Here we have a *Threshold* control, which, when you click and drag on the horizontal bar, acts like a sensitivity control for the transient detection process. In truth, it doesn't adjust the sensitivity of the detection but does increase or decrease the amount of transients that are marked. However, this process, as with most sensitivity-type controls for transient detection in DAWs, is essentially a "dumb" process. What I mean by that is it works purely on a threshold basis. As the threshold is adjusted, the number of transients is increased or decreased, as you might expect, but there is no real control over which transients are removed. A transient may be removed at a certain position on one bar, but a virtually identical one may remain in the following bar. If you then adjust the Threshold control further to remove the transient marker in the second bar, it may remove an additional one from the first bar that you didn't want removed. So, at least until such times as the transient detection/adjustment process is made a little more "intelligent," you may need to supplement the automatic process with a little manual work.

In order to either add or delete new markers or manipulate the ones that are already there, you will need to select the *Bend* tool by pressing the *Bend Tool* button in the main toolbar or by pressing 7. With this tool activated, you can work on the Bend Markers (and therefore transients) and work around some of the obstacles that the Threshold control may put in your way. As you move the Bend tool over the regions you are working on, you will see the cursor change to a pencil with a vertical bar and a left arrow at the foot of the bar. In addition there will be a vertical line that runs the entire height of the Arrange view. This is to allow you to line up transient markers on one track with transients on a different track visually. Clicking anywhere on the currently selected

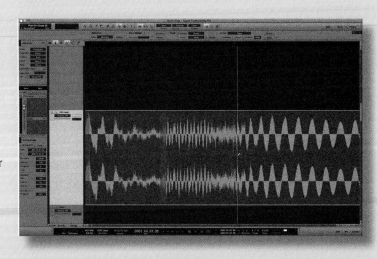

FIGURE 9.5
Studio One offers a variable Threshold for the placement of its Bend Markers but also allows you to place or remove additional markers manually.

track will create a Bend Marker at that point, and the newly created Bend Marker will be yellow in color rather than the usual light blue.

If you now move the cursor over any existing Bend Marker, you will see it change the same vertical bar with an arrow at the foot, but this time with a little waveform shape to the right. With this tool in place, you can move the underlying audio. We will look at this in later chapters, as this is a part of the *Elastic Audio/Audio Bend* processing and not strictly related to the detection of transients. However, staying with this tool for a moment, if you press the Alt key while the cursor is located over an existing transient, this tool will now change to a vertical bar with the arrow at the foot but two additional left and right arrows at the top. This is the tool we need for moving an existing transient marker (without moving the actual audio itself). Most material with clear transients (drums and percussion being the obvious examples) should have the transients automatically placed in the correct places, but for some material, especially instruments and sounds that have much softer attacks to them, it can be hard to detect a clear transient. It may well be that a little fine adjustment of the transients here and there can substantially improve things in cases like this.

You may also wish to remove a particular transient manually if an adjustment of the Threshold control isn't working quite how you would like it. To do this simply position the cursor over the transient you wish to remove and right-click and choose *Delete* from the pop-up menu. If you prefer you can position the cursor over a transient, click on it to select it (it will turn yellow), and then press the Backspace key to delete the marker.

Cubase

Transient detection in Cubase is based around *Hitpoints*. These are markers placed by the software to indicate important points in the audio, which not only includes transients but also melodic changes. This extended functionality comes in handy for the pitch component of the *Elastic Audio* processing that Cubase does, but at this point we will concentrate on the ability of the Hitpoints to serve as transient markers.

In order to start the transient detection process, you have two choices. The first choice is to select your region(s) in the Arrange window and then go to Audio > Hitpoints > Calculate Hitpoints. The regions will then be analyzed and Hitpoints calculated by Cubase and then displayed as a series of dashed vertical lines on top of the waveform overview. Once this initial analysis is completed, you have a number of options in the same menu. You can choose to create *Audio Slices* (Audio > Hitpoints > Create Audio Slice from Hitpoints), which we will make significant use of when we come to the later chapters; you can create *Markers* (Audio > Hitpoints > Create Markers from Hitpoints); or you can split the region in to smaller regions by going to Audio > Hitpoints > Divide Audio Events at Hitpoints. Finally there is the option, in this menu, to remove all the Hitpoints, should you wish to do so. While this method is great for many purposes and is especially useful, as it can be done without leaving the Arrange window, there will undoubtedly be times when you need a little more control or need to change the sensitivity, adjust the position of or remove markers, or edit them in some other way, and for that you will need to go into the Sample Editor window.

Double-clicking on an audio region will open the Sample Editor window, and to the left of the window you will notice a series of buttons that are used to carry out various editing tasks on the currently selected region(s). The one we are most interested in here is Hitpoints (fourth from the top), and pressing it will reveal a number of other controls and buttons below. If you have already analyzed the region, perhaps using the Arrange window method, then you will already see a series of vertical gray lines overlaid on the waveform overview. These represent the current Hitpoints. If, on the other hand, no analysis has been carried out as yet, then you will need to press the *Edit Hitpoints* button, and the initial analysis will be carried out and the Hitpoints displayed. Because we are now in the Sample Editor window, we can actually manipulate these Hitpoints and carry out a little fine-tuning if we need to.

The first thing that we can do is to adjust the number and position of the Hitpoints by using the *Threshold* slider. This is, effectively, a way of controlling the sensitivity of the whole transient detection process, but, unlike some other DAWs, adjusting this slider shows a visual representation of where the threshold is on the waveform overview. This is actually extremely useful in getting the correct setting if you are working with quite a complex region, as you can, to a certain extent, figure out where transients are likely to be detected and adjust the threshold down to that point. If you want a little more control—and this is especially useful with drum and percussion parts—you can use the *Beats* control below the Threshold slider to restrict the placement of transient markers to only on quarter notes, eighth notes, sixteenth notes, or thirty-second notes. This can be a very intuitive way of placing transient markers only at key points rather than necessarily trying to find every single transient when you need only the ones that represent a single beat (quarter note), for example. You can, of course, not use this Beats control and go ahead and remove the Hitpoints manually.

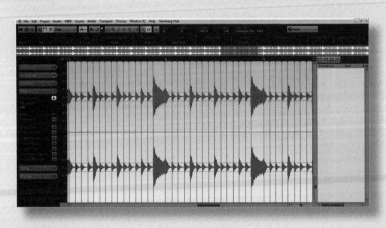

FIGURE 9.6
While the variable Threshold offered by Cubase isn't particularly unusual, the fact that the Threshold level is depicted in the waveform overview does make it much easier to predict where the Hitpoints will be placed.

Transient Detection — CHAPTER 9

Click on *Edit Hitpoints* again if necessary to activate this mode (the icon to the side of the control will illuminate, and the Hitpoint lines will become thicker), and you will be able to delete a Hitpoint by clicking on the Hitpoint that you wish to remove and moving the cursor to the top of the Hitpoint line. As you do this, the cursor will change to an "x," and you will see a pop-up message saying "Disable Hitpoint." Clicking now will remove the vertical line but keep the small triangle at the top of where the line was. This removes the *Hitpoint* but still allows you to add it again by clicking on the small triangle. In addition to removing or replacing the Hitpoints in this way, you can also change their position by moving the cursor over an existing Hitpoint, at which point it will change to left and right arrows, and you can click and drag to reposition the Hitpoint.

When dealing with something as important as the placement of the Hitpoints, it makes a lot of sense to carry out any repositioning while zoomed in quite a lot. As a result you will probably make a lot of use of the keyboard commands for zooming in (press H) and zooming out (press G), but you can also use key commands to step through the Hitpoints by using Alt + N to locate the next Hitpoint and Alt + B to locate the previous Hitpoint. This is basically the same concept as the Tab to Transient feature in some other DAWs but with different key commands and different terminology. Needless to say, it makes it very easy to keep quite a high zoom level and navigate back and forward through the various Hitpoints in your region to make sure that each of them is accurately lined to where it should be.

You then have a number of different things that you can do with these Hitpoints in this window. The first option, *Create Slices*, will (as we have already mentioned) be of great use to us when it comes to creating Recycle-like options, and the second option, *Create Groove*, allows us to use the Hitpoints that we have detected to create a quantization map to be applied to other audio or MIDI regions. *Create Markers* will, quite unsurprisingly, create a marker for each Hitpoint, which can have a number of different uses, while *Create Regions* will separate the file/region into smaller regions at the Hitpoint locations, and these Regions will show up individually in the *Audio Pool*. The main Arrange window will still show just one "block" if you choose Create Regions, because, in Cubase terminology, the blocks that you see in the Arrange window are called *Events*. Therefore, if what you want to achieve is to split the region you are working on into smaller blocks (what we have referred to generically in the book as "regions") then you need to choose *Create Events* from the Sample Editor window, and this will have the desired effect. The last two options are *Create Warp Markers* (which we will look at more at the end of chapter 12) and *Create MIDI Notes*. This last option will prove invaluable when it comes to both Recycling and drum replacements (chapters 10 and 11, respectively), and we will look at it in greater detail then.

Once you have carried out any changes that you want to make to the Hitpoints themselves and have finalized any additional tasks that use the Hitpoints, you can close the Sample Editor window, and, when you return to the main Arrange window, you will see (if you zoom in sufficiently) that the changes that you made are now shown by the dotted vertical lines on top of the waveform overview. You can, of course, go back in to the Sample Editor window, and you can also use the Audio > Hitpoints menu choices to create slices, markers, and events to work on these edited Hitpoints.

SECTION 2
Creative Editing

CHAPTER 10
Beat-Mapping and Recycling

INTRODUCTION

In this section of the book, we are going to start to look at editing tools and processes that can not only serve a corrective purpose but also can be put to more creative uses if the need or opportunity arises. And as much as some of these techniques cross the line between corrective and creative uses, equally some of them cross the lines between editing, production, and sound design. So, first up, and following on from our last chapter on transient detection, it makes sense to start off with a pair of related but very different ways of using those transients that the software kindly worked so hard to figure out for us.

Beat-mapping is the process of analyzing an audio file—often longer files or even full songs—ad using the transient markers created during the analysis as a way to subtly (or not so subtly) alter the timing of the piece while keeping it fluid and recognizable. Recycling is now a generic term for a process of cutting up an audio file—normally one that is quite short in length—into "slices" and then creating a sampler patch and associated MIDI file based on the timing of the slices. The name originates from the *Recycle* software released in 1994 by Propellerhead Software, but it has become such a well-used process that now, even though other tools, software, and methods have been developed to do the same process, the whole idea of loop-slicing has now become synonymous with the term "recycling."

Although both processes rely on transient markers, that is where the similarity ends. Each of the two has its strengths, its weaknesses, and its common uses.

BEAT-MAPPING

As mentioned briefly above, beat-mapping is set up more as a corrective tool. It will analyze an audio file and create transient markers throughout the whole length of the file. These transient markers can then be moved and the resulting groove and feel of the track changed. Crucially, though, as each transient marker is moved, the audio between it and the following marker is time-stretched, so that there are no gaps or overlaps between the markers. The whole thing is very fluid and consistent. This makes it ideally suited for quickly and easily changing the groove of a long audio file by choosing how the timing should be changed (usually quantizing) and then letting the software do the rest.

Another very useful consequence of the way that this system works is that, once the transient markers are in place, the whole track will then be easily mappable to a standard "bars and beats" format instead of being purely based on absolute time. Instead of starting at 1:02 (minutes and seconds), the chorus can now be thought of as starting at bar 33. This shift from minutes and seconds to bars means that, even without actually changing the groove or quantizing, the file now has the ability to be tempo-independent, as any changes in tempo will mean that, instead of the software having to calculate that an event at 1:02 now has to occur at 1:05, and doing so for every "event" in the file, the software knows that what happens at bar 33 *stays* at bar 33.

Changing the tempo of an audio recording has been possible for a very long time, so being able to speed up or slow down an audio file isn't new, but the ability to do so without changing the pitch only comes with time-stretching. Equally, time-stretching itself isn't new but traditionally would require a file to be processed, and, once the destination tempo had been chosen, the only way to change it again would be to reprocess the file. Furthermore, any gradual changes to tempo would be impossible with regular time-stretching, so this method of actually having the tempo completely free to be set at whatever tempo you choose or even to change midway through a file represents not only a huge workflow improvement for static tempo changes but also a whole new range of possibilities for changing the tempo as the song progresses.

One of the other things that beat-mapping allows us to do from a corrective perspective is to make tracks that were not recorded to a click track—and consequently have tempo variations throughout their length—conform to a consistent tempo.

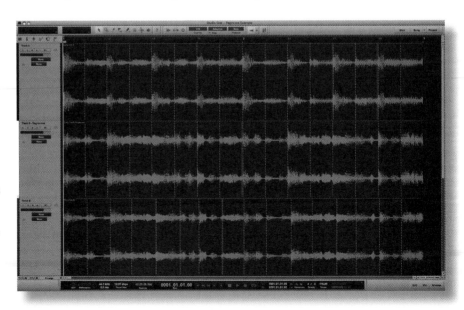

FIGURE 10.1
This is an illustration of the power of beat-mapping. The top two tracks show two different songs that would play back nicely in sync. The bottom track shows the original timing and positioning of the transient markers of the second track. There is quite a noticeable difference, and this would allow us to play the two tracks together without them sounding "messy."

There is an argument that doing this is actually taking some of the "life" out of a song, and I would be inclined to agree, but there is one situation where this would be extremely beneficial and one group of people who, as a result, have become very fond of beat-mapping.

DJs, at least in the case of DJs who mix records in to one another to achieve a constant flow rather than just having a very short cross-fade between one track and the next irrespective of tempo, have for a very long time been able to deal with mixing tracks together that are different in tempo. By adjusting the speed of the record (or CD), they can make the tempo match, but for them to be able to mix the records properly, the tempo throughout the song needs to be consistent. It is one thing to adjust the pitch/tempo of one record to allow it to be mixed in to another, but if there were constant tempo variations throughout the song, then, no matter how adept DJs were, they simply wouldn't be able to keep up with the changes, and the resulting mix between the two would be constantly drifting in and out of sync. By being able to remap the timing of a song so that it is at a consistent tempo, even if that is at the expense of a little bit of the feel and life of the track, they now have the ability to use that song in a predictable and reliable way in their mix sets.

QUALITY LIMITATIONS OF BEAT-MAPPING

Before we move on to look at some creative ways in which we can use beat-mapping, it is probably best to consider the quality issues of what we are trying to do. The biggest problem we face with this beat-mapping technique is that no time-stretching system is totally without artifacts. Depending on the time-stretching algorithm, the actual content of the file being stretched and the amount of stretch can range from almost unnoticeable to a very clear change. The most common artifact is a so-called smearing effect, which is perceived as a loss of definition and clarity and the tendency for the component parts to blend with one another.

There are several different methods used to achieve time-stretching (see chapter 12 for more information), and each of these methods/algorithms has strengths and weaknesses. Therefore, it is always advisable to try out any different options that you might be offered in terms of different algorithms. Sometimes these are named according to what they might be suitable for, and, given that the people who come up with these algorithms generally know what they are talking about, they are often correct. However, there might be times when a file that you have doesn't fall neatly into one of these descriptions or when you just have the time to try different things out.

If we try to look at ways to reduce the number and severity of these artifacts, then we have to consider one important fact. In general, with any complex process, there is a trade-off between speed and quality. If we want to do something in real time, then we almost always have to compromise on the quality of the result. And time-stretching is a very good example of this. Computing power has increased rapidly in recent years, and it is only because of that that we can even consider doing things like time-stretching in real time. But processing power has yet to reach a point where the best quality that we have today for the stretching algorithms can be achieved in real time. And even if we reach a point where computers are powerful enough to do that, by that point there may well be even more advanced algorithms, which need even more processing power. So it is easy to believe that real-time processing will always be a second-best option when it comes to quality.

What would be a better option is to have the real-time processing, as we do now (which is actually pretty good), and all of the inherent flexibility of it, but then in addition have an option to bounce or render the file with the best-quality algorithms, once we have decided that we don't want to change it any more. There are a few synth plug-ins that are offering this option now (more-efficient CPU use during normal playback but a "high quality" mode to be used during bouncing), so there is definitely the ability to do it within the coding of the software, and, given that quality is always a concern when dealing with processes like time-stretching, it would be a good way to reduce that trade-off at the cost of an additional step of processing.

RECYCLING

Recycling uses a very different principle and is designed for a very different purpose. It still uses transient markers as a means of identifying timing references in the source material, but instead of manipulating the length of the audio between these markers, it simply cuts the audio file at these points and changes their timing from absolute (minutes and seconds) to "song position" (bars and beats). The original Recycle software (by default, anyway) created one container

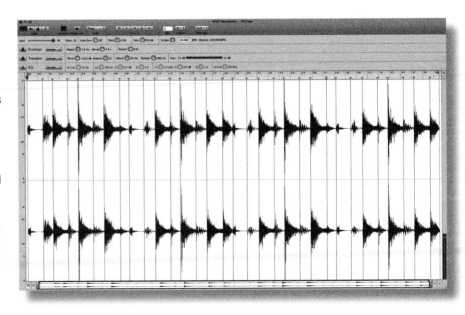

FIGURE 10.2 Propellerhead's Recycle software was the progenitor of many of the techniques we take for granted today. While it might seem quite basic compared to some of the technology that we have available today, it is still used a great deal, because it does a simple job very well, and sometimes there is no need for added complexity.

file that contained all of the audio "slices" and the timing information. There was, however, an option to export the audio information and timing information separately. The audio would be converted to a sampler patch, where each slice was mapped to a particular MIDI note, and the timing information would be created as a standard MIDI file, where each slice was triggered by the respective MIDI note at the correct time.

Given the explosion in popularity that Recycle experienced, it was inevitable that other software would come to the market that offered similar facilities. One of the limitations (if you would consider it that) of Recycle was that it was purely designed for slicing and then exporting the results for loading into another program or plug-in for playback. Some of the alternatives that followed combined the slicing features of Recycle with a playback system and wrapped the whole package up into a plug-in format. This meant that the whole process could be completed inside the plug-in, which greatly simplified the workflow. Given the importance of saving time without compromising on quality of results, clearly this all-in-one solution was a much better way to do things.

So, however we arrive at our destination, let's assume that we have our recycled file ready to use. What can we do with the file now that we have it? The main reason that we have recycled files is to allow us to change the tempo and timing of the files. In this respect it is actually very similar to beat-mapping but with one crucial difference. Because each "slice" isn't time-stretched in any way, there are no artifacts to worry about. Each slice sounds exactly the same after recycling as it does before. On the other hand, because the slices aren't stretched, there will be some potential problems if we move them or change the tempo.

When we move the slices further apart, there might be audible gaps between them. Just how much of a problem this will be depends on the source file. If we

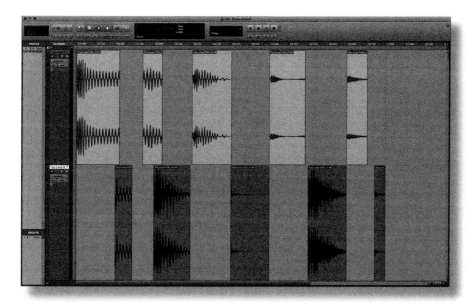

FIGURE 10.3
When an audio file is split at transient markers and then re-quantized, there can often be gaps or overlaps. The "slices" have been separated onto two tracks in the image above to make it easier to see.

have a file with legato sounds or sound with long decays, then there is a chance that the sound would naturally run into the next slice if we create too many slices. Equally, at the beginning of any given slice, there could be the end of sounds from the previous slice in addition to the sounds occurring at the start of this slice. If the sounds in the loop have a very tight sound (closed hi-hats and a short-sounding snare, for example), then we might not have any problems.

The opposite is true if we move the slices closer together. The natural decay of the sounds in each slice might be cut short by the start of the next slice. There is perhaps more scope to deal with this issue, though, because you can just play the slices back polyphonically (more than one at once). In the case of a sampler playing back a .rx2 file, you would simply have to set the sampler's polyphony to a value of higher than one to allow more than one slice to play back simultaneously. In the case of the individual slices being on an audio track, any overlapping slices can be moved to a newly created (and identical in terms of plug-ins, routing, and settings) audio track to allow the overlap to be dealt with naturally.

You could also make use of fades to deal with both these situations. In the case of gaps between slices, it might be an option to apply a fade-out to the end of each slice that was followed by a gap. This could solve the problem of the slices cutting off abruptly, but what it wouldn't deal with is the presence of the end of those sounds at the start of the next slice. If this is the case, then you always have the option of selectively time-stretching certain slices to avoid this. The problem with this approach is that, if you are time-stretching only certain slices and applying fade-outs to others, the final result can sound a little disjointed. In this situation it may be a better option to either beat-map the sample or to manually stretch all of the slices to allow consistency across the length of the file.

In the case of overlaps, the situation with fades is a little easier, because all we would be doing is returning the level of each slice to zero before the start of the next region. It probably wouldn't be necessary anyway, as we have already seen that it is very easy to play the file polyphonically, but simple fade-outs at the end of each slice might help to clean things up if for any reason you can't utilize any of the polyphonic options.

EXTENDED USES OF RECYCLED FILES

So far you might be thinking that beat-mapping sounds like a better option, and you would probably be correct if all you wanted to do was clean up the timing of a particular sample. But the possibilities of recycled files go beyond simple timing correction. One of the other key benefits of recycled files is the ability to not only change the timing of each slice to make the timing more accurate but also change the order in which the slices play back. If you are working with a MIDI region and a sampler, then it is simply a matter of moving the notes to the new positions that you want them in. The fact that a MIDI note triggers each slice means that you can actually play a new groove on a keyboard to figure out what you want to change before actually moving the notes around. You can achieve the same end result if you are just dealing with audio regions, but obviously the process is a little less straightforward.

As well as rearranging the pattern that the file plays, there is another potentially very useful thing that we can do with recycled files. If we are using recycled files within a MIDI region and sampler setup, we can use the fact that each slice is independent as a way to change the dynamics of the pattern. Most samplers will, by default, have some mapping between velocity of the MIDI note and the volume of playback of the sample. You can always change this if you don't require it, but it can be useful in our situation. By changing the velocity of the individual notes within the created MIDI region, we can change the volume of each slice as it plays back. This won't give us complete flexibility, because if there are, for example, a kick drum and a hi-hat occurring at the same time and contained within a single slice, then changing the velocity of the relevant MIDI note will cause both the kick drum *and* hi-hat to change volume. So while recycling can certainly help to either increase or decrease the dynamic range of a loop without using any kind of compression or expansion plug-ins, it doesn't give us as much flexibility as we would have if we were dealing with a multi-track drum recording.

MORE CREATIVE USES
Beat-mapping

So far we have looked at the ways in which beat-mapping and recycling were originally intended to be used. Like so many technological advancements, though, it doesn't take long for creative people to come up with new and interesting ways of using the underlying principles, and this holds true for both beat-mapping and recycling.

The main use of beat-mapping that we looked at earlier was to take an audio file and change the timing of that file to either change the tempo, make it more consistent, or radically change the groove of it. If we use it for that purpose, it can work incredibly well. But what if we really like the feel of a certain piece? Perhaps it has a certain "swing" to it that we like, and we want to take the other parts of our song (audio or MIDI) and make them fit to this feel rather than the other way round. Fortunately, we can do exactly that.

Because the beat-mapping process creates transient markers, and these transient markers exist in relation to the timing grid, we can use that information to create a quantization "groove" that, instead of simply moving everything to the nearest sixteenth note, will instead look at the timing of the audio file that we have beat-mapped and analyze the timing offsets between a theoretically perfect timing and the actual timing of the markers. Obviously these offsets would be defined in relative terms (0.03 of a beat late, for example) and could therefore be applied to any MIDI region to change the timing of that MIDI region to match the audio file.

More Creative Uses

If we want to change the timing of one audio file to match another, then we have a couple of options. If it is a relatively simple audio file (drum beat, bass line, rhythmic guitar part, etc.) then we might be able to recycle that audio file and then just apply the quantization groove we have created to the MIDI region of the recycled file. If it is something more complex, then we may need to approach it in a different way. To get the result we want here, we would need to go through a few different steps. We would beat-map the source file and create the transient markers as usual. We would then beat-map the second file to create the transient markers that we wanted to move. So far so good, but the difficulty lies in how we actually make them both line up. One way would be to cut the second audio file at the transient markers, then move each one manually to the position of the respective one in the first file (they are usually overlaid on top of the waveform display, so this isn't too hard to do), and then manually time-stretch each split region to fill the required space. Basically, we are doing what beat-mapping does (moving transient markers and time-stretching to fill the gaps) but in a manual way. This will achieve what we want to achieve, albeit in quite a time-consuming manner.

Fortunately there is an alternative, and that comes in the form of elastic audio. This is a *huge* step forward for audio editing and production and is, in fact, such a big deal that there is a whole chapter dedicated to it that will be coming up shortly. I have mentioned it here because it uses the same fundamental principles of beat-mapping as we have described here but gives us, the end-users, much more control over how the audio is processed. It is certainly a very creative use of the basic beat-mapping principles, though, and is something that many people find indispensable nowadays.

Moving on from this, we can look at another related use of the beat-mapping ideas. In addition to the subtle variations of timing within a single bar that often exist with live performances, there can also be greater and more obvious tempo changes throughout a piece. These accelerandos and ritardandos can be very effective at adding to the emotion and feeling of a song. Choruses of songs not played to a click track are quite often at a slightly faster tempo than the verses, and the final chorus may be slightly faster still. There may be a subtle drop in the tempo at the end of a verse or a chorus as well to build a slight sense of anticipation. In fact, sometimes these changes aren't subtle at all and can be deliberately exaggerated. Beat-mapping can handle all these with relative ease and allow us to get rid of them if we need to.

In certain circumstances (a club DJ needing consistency of tempo, for example) this can be very useful, but, again, what if we want to work the other way round? What if we have an audio file that has an inconsistent tempo, but, instead of getting rid of the changes, we want to make sure that the tempo of the rest of the song follows the changes in this file? Once again, beat-mapping comes to the rescue. We can use these transient markers to create a tempo map that can then be applied to the current project. Once we have done this, any MIDI regions will automatically follow the tempo changes, but we should be aware that audio files, unless prepared specifically to do so, will remain at their original tempo.

To make sure that the audio files follow the tempo changes, we can either recycle them (if they are shorter files and are not likely to suffer from audible gaps or overlaps) or beat-map them. Once this is done, then everything should follow the tempo changes that you have created from your source file.

Recycling

When it comes to being creative with recycled loops, it is fair to say that your options are probably greater in number but less drastic in effect. As with beat-mapping, you can use the location of the individual slices to create a quantization groove that can then be applied to other parts within the project in the same ways that we looked at for beat-mapping. If you have a MIDI and sampler setup for the recycled file, then it could, potentially, be even easier. You already have a MIDI region with the notes in the right place (in terms of the timing). If you look at that MIDI region, you will see that, unless you have already moved things around, each subsequent MIDI note will be one note higher up the keyboard, and each note length will fill the whole time until the next. This, as it stands, isn't very useful, but we can take those MIDI notes and, keeping their timing position the same, move them up and down the keyboard to create melodies and chords. We can change the lengths as appropriate, and we can mute any notes that we don't need. By doing this you will have a melodic MIDI part that is perfectly locked to the timing of your recycled part.

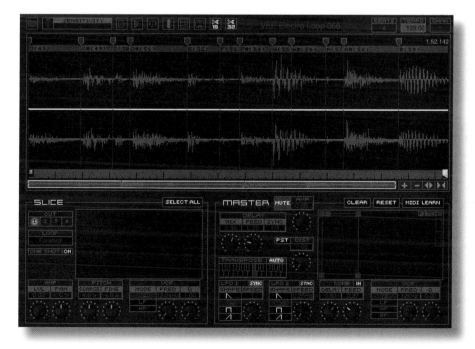

FIGURE 10.4
Building on the original Recycle concept, there are now a number of self-contained plug-ins that will not only analyze and split an audio file but also have a number of different options for playing back that file, with each "slice" having its own parameter set in addition to a number of master parameters.

Another thing you can do with recycled files is to modify the amplitude envelopes of the individual slices. This is considerably easier if you use one of the self-contained recycle-style plug-ins, as they generally offer more synth-like control of each slice (individually or globally) with amplitude envelopes (and sometimes filter envelopes too) rather than relying on fade-ins or outs. You can achieve similar results with fade-ins and -outs on audio regions, but it can take quite a bit longer, so this might be a consideration. The aim of this, whichever method you use, is to deliberately shorten the length of each slice or region with a fade-out or decay on the amplitude envelope. With drum loops, this can have the effect of making the whole pattern sound much tighter and more snappy and can be very useful to create a little more space in particularly busy arrangements without having to lower volumes or remove parts completely.

If you are using one of the Recycle-style plug-ins, then you could also have the option to have a particular loop very tight and snappy during a verse and then open it back out to its full sound in a chorus. This is simply a matter of adjusting a global amplitude envelope and setting the sustain level to zero and adjusting the decay time to give the required level of tightness. Then, when you want to open the loop back out to its full sound, you simply automate the sustain level from zero up to full. If you have set up amplitude envelopes on each individual slice, then it is also possible to open up certain elements within the loop individually. You could have one version of the loop where everything is tight, then a second version where you opened up only the snare drum hits, and then another version where the whole loop was opened back up to its full sound. Using these techniques can actually give you a lot of dynamic variation from a single source loop and could certainly be useful if you need to create interest from a limited selection of source files.

A variation on this technique is to adjust the Attack times (or fade-in times) on each "slice" or region to soften the individual hits. Again this is a way to introduce a variation to an otherwise static loop without changing its fundamental character. Simple changes like this can be very effective, as you can move a particular drum or percussion loop a little into the background by taking away the initial attack of the sound and then bring it back to the front when you need to. In many ways this technique is complimentary to the one above, and between the two you can do an awful lot to change the dominance of any given drum loop.

It's also possible to transpose either the whole loop or individual slices within the loop. Because the pitch changing in samplers or the Recycle-style plug-ins is generally based on speeding up or slowing down the sample rather than actual pitch-shifting, any change in the pitch will result in the length of each slice changing, which has all the same issues as we have already seen. But, in the event that the loop is suitable, some interesting effects can be had from playing around with individual pitches within a loop. If the loop is of a musical nature anyway, then changes to the melody of the loop might be possible in a very quick and easy way.

QUICK AND EASY MULTI-SAMPLES

While we are on the subject of melodic parts, you can also use Recycle to create quick "multi-sampled" patches for your sampler. If you have a particular combination of sounds from various synths that you like to use together, perhaps with associated compression and EQ, then you can quickly turn this into a (basic) multi-sampled sampler patch. First create a MIDI region for each of the synths that plays—in sequence and for an equal amount of time—each note, from the lowest to the highest that you want to include. The MIDI notes should have a long-enough gap between them, so that the release phase of the previous note is completely finished. Next you can bounce the combined result of all the synths playing this part. This resulting audio file is then loaded into Recycle or a Recycle-style plug-in, and, as long as the sound doesn't have a very slow attack, the software should pick up the start of each note as an individual transient. Subsequent notes will then be mapped to subsequent keys, and, perhaps with a little bit of setting up which MIDI notes the exported MIDI region should start on, you have a playable multi-sample of your combined sound. This method isn't anything approaching the best way to do this. But in instances where you need to do something quickly, it is always good to have a few little time-saving tricks up your sleeve.

RHYTHM FROM ANYTHING

The last thing we are going to look at here is a way of creating rhythmic parts out of sounds that are not percussive in nature. This isn't something that you will do all that often, and it is perhaps more limited in its usefulness compared to the other things we have looked at, but, given that it uses a similar process to the one we have just described for creating multi-samples, it seems worthwhile mentioning it here. The basics are very simple: you take an audio file (one with sounds with sharp attacks and gaps between the sounds will work best here) and then recycle it and have each separate note/word/sound mapped to a different key. It doesn't particularly matter if the key that it is mapped to is the same actual note as the pitch of the sample, as we aren't intending to use these slices to play conventional melodies. Once this is done, we now have a multi-sampled patch with different notes, words, or sounds on each key. We can now adjust the amplitude envelope to have a sharp attack, fast decay, and zero sustain, so that each sound has the kind of envelope you would expect from a drum or percussion instrument.

Then it's really up to you how to use it. You can either just play along to your rhythm track and get a feel for a groove by playing random notes, or you can take the MIDI region from another recycled file, copy it to this instrument, and then move the MIDI notes around to create a new groove that matches the other one in timing but is composed of pseudo-random snippets of vocals or other musical sounds. It may be that you want to change the pitch of some of the sounds to help them sit better melodically, or you may wish to shorten the decay time down even further, so that the pitch becomes less relevant and obvious.

As you can see, there are many ways in which you can use these tools outside of the obvious. Beat-mapping applications tend to be wider in scope and have more of an effect on the feel and groove of a song as a whole, whereas the creative uses of recycling tend to be more effective on a smaller scale. Using beat-mapping and recycling in combination, it is possible to really change the feel of a track in quite natural-sounding ways and ways that may have your clients wondering how on earth you managed to get it sounding like that.

HANDS ON

Introduction

In addition to the benefits of having transient markers in place for general editing, they also serve as a basis for a number of other more-creative editing tasks. Two of the most obvious of these are beat-mapping and recycling. In this section we will take a look at the ways in which each of our DAWs uses the transient markers to help with these increasingly useful tasks and what tools are available to help us get the most out of our audio files.

Logic

One of the most common uses of the transient markers within audio files is to enable quantization of audio files. These transient markers, in conjunction with *Flex Mode*, allow audio regions to be treated in the way that MIDI regions have been for a very long time and have a simple quantization parameter applied. Once the transient markers have been placed in the Sample Editor, you can turn Flex Mode on for the region in question, and the *Flex Markers* should appear on the waveform display of the region in the Arrange window. It might be that some of the transient markers (shown as pale vertical lines in the region in the Arrange window) haven't been converted to Flex Markers (brighter lines), and, if this is the case, you simply need to click on the transient marker to convert it to a Flex Marker. Equally, if there is a Flex Marker that you don't wish to have, you can double-click on it to remove it. Once all the Flex Markers are in place (in the sense of the markers for the transients you wish to quantize) then you can adjust the quantization simply by changing the *Quantize* parameter in the *Inspector* panel. You may need to try out different *Flex Algorithms* in order to get the best result, but the options are quite self-explanatory, and the best choice is usually obvious.

If, on the other hand, rather than quantizing a particular region to a fixed quantization feel, you wish to extract a quantization feel (or *Groove Template*, in Logic terminology) you can use the Flex Markers to do this. Select the region that you wish to use as the template, and then, in the Inspector panel, click on the Quantize drop-down list. At the bottom there will be an option called *Make Groove Template*. This extracts the timing subtleties from the currently selected region and saves this as a *Groove Template*. If you then choose a different region that has Flex mode enabled and then go to the Quantize drop-down list, you will see the name of the previous region below all of the default quantize

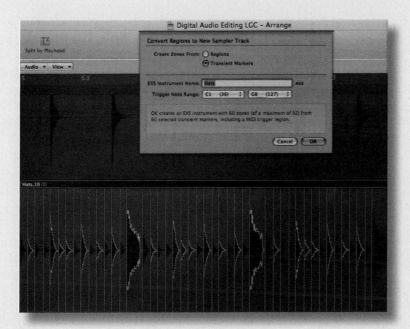

FIGURE 10.5
Logic makes it extremely easy to take an audio file and create a recycled version of it. Pretty much any audio region or file can be analyzed and split and a sampler instrument created and ready to play in just a few mouse clicks.

options. Select this option, and the current region should have the feel of the previous region applied to it. The exact quality of the results depends on the initial similarity of the two regions. If the source is a busy sixteenth-note hi-hat pattern, and the target is a more sparse and loose tambourine pattern with a triplet feel, then the results might not be 100% as hoped, but, in the majority of cases, this is a quick and easy way to match the feel and timing of two separate audio parts. It is worth noting that this Groove Template can also be applied to MIDI regions in order to really lock together the groove of, for example, a drum loop and a bass line and can be incredibly effective at tightening up the timing of a rhythm section. Finally it is also worth noting that in order for the Groove Templates to work, the regions that they were taken from have to remain in the arrangement (even if they are muted).

Another very useful feature of Logic is the ability to automatically recycle audio files based on the transient markers created either in the Sample Editor window or automatically, using Flex Mode. Once the markers are in place, you can convert any audio region into an *EXS-24* sampler instrument and companion MIDI region. Assuming you have an audio region with transient markers in place already, and you have made sure that markers exist only for the slices you wish to create, all you need to do is make sure that audio region is selected in the Arrange window and press Ctrl + E, and the *Convert Regions to New Sampler Track* dialogue box will appear. You are first asked to make a choice as to whether the *Zones* (what we have been calling slices) are created using regions or transient markers. In the context of a single audio region with transient markers, we would obviously choose *Transient Markers*, but, if you have already split a region into individual parts, you could select Regions instead. You then have the option to enter a name for the *EXS-24* sampler instrument, but the default value is the name of the audio file being converted. Finally you have the option to choose a MIDI note range that you wish to use for the resulting MIDI file. Unless you are creating a chromatic instrument, there is no real benefit to changing the default values. Once you click on OK, the audio region will be analyzed and each section between transient markers will be allocated to a single MIDI note in the *EXS-24* sampler instrument, a MIDI file created with the timing for each of the zones/slices and the original audio region muted.

Using this method you can, in just a few clicks, convert an audio region into a recycled file, and this has a huge number of uses, as we have seen in the preceding chapter. It isn't all good news, however, as there are compromises made in doing so, depending on your intended use. As we saw in the main chapter, a recycled file is great if you wish to create new variations on the original audio part by changing the order or timing of the MIDI notes, but it is only partially tempo-independent. Of course the initial timing of any MIDI notes will follow the tempo of the song but, unlike Flexed audio, the duration of each slice will stay the same, which can lead to overlaps (which are fairly easy to deal with) or gaps (which are not) depending on which way the tempo changes.

With this in mind, I would suggest that any tempo changes should be figured out before doing any recycling, but of course this isn't always practical. You also get all the other advantages of recycling (ability to change pitch and envelopes, etc.) so it is really just a question of deciding the best time to switch from Flexed audio to recycled in order to achieve what you need to. If it really becomes a problem, then you can always make the changes to the MIDI region to create the variations that you want at the original tempo and then quickly bounce the result using *Bounce In Place* (Ctrl + B) back to a new audio file, which can then be Flexed and follow any tempo changes.

Pro Tools

For Recycle-like beat slicing, Pro Tools comes equipped with the very flexible *Beat Detective*. If you wish to create a "sliced" audio file, then you start by choosing a region that you want to analyze and then open Beat Detective by pressing Cmd[Mac]/Ctrl[PC] + Num Keypad 8 or going to Event > Beat Detective. The Beat Detective tool has a number of different options, depending on what you wish to achieve, but the relevant options here are *Groove Template Extraction*, *Clip Separation*, *Clip Conform*, and *Edit Smoothing*, and we will look at each of these in order, but before we do we need to put the markers in place that Beat Detective will use for the first two options.

You first need to choose either Groove Template Extraction or Clip Separation from the *Operation* heading, and then, if you preselected a region before opening Beat Detective, the *Start Bar* and *End Bar* under *Selection* should match the region you selected. If you didn't choose a region, or if you wish to process a different

Beat-Mapping and Recycling CHAPTER 10

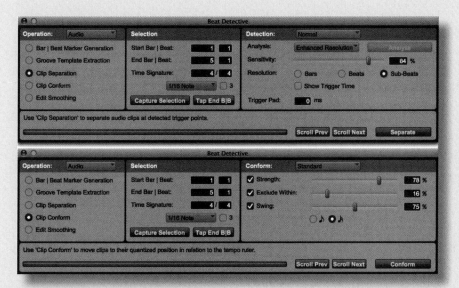

FIGURE 10.6
Beat Detective is a great option if you wish to re-quantize audio files or create Groove Templates. There are a number of different pages that are chosen on the left of the window, with each one relating to a different part of the whole process.

region, simply select the region and then click on *Capture Selection* at the bottom of the *Selection* section, and this will update the range above.

Moving on to the *Detection* section, we find three main controls. The first of these, *Analysis*, has three different options. Both *High Emphasis* and *Low Emphasis* focus on these specific frequency ranges, while *Enhanced Resolution* uses a different analysis method and in most cases is the better option. Once you have chosen an option here, click on the *Analyze* button to calculate the transient positions. *Sensitivity* is a simple slider that will increase or decrease the number of transients.

Below this are the *Resolution* controls. Choosing *Bars* will place markers on beat one of each bar and nowhere else. Equally, *Beats* will place markers at the start of every beat, but the nuances of the timing between beats won't be marked. Finally we have *Sub-Beats*, which will place markers all the way down to sixteenth notes. You can make further adjustments by manually deleting the markers by making sure the Grabber tool is selected, then moving the cursor over the marker you wish to delete, and holding down Alt while clicking. You can also reposition markers by using the Grabber tool and then clicking and dragging the markers to the new location.

Now that the analysis is done and the markers are in place, we can move on to the different processes that Beat Detective can offer. The first option, *Groove Template Extraction*, uses the position of the markers and creates a "timing map" that can then be applied to the timing of other regions to enable us to lock the timing of two separate regions together. Making sure that Groove Template Extraction is selected from the *Operation* heading, simply press the *Extract* button in the bottom right-hand corner of the window, and you will be asked to confirm the length and time signature of the extracted groove; have a text box where you can enter comments. You can then either *Save to Groove Clipboard* or *Save to Disk*. Saving to the clipboard is a temporary option where saved groove will be overwritten next time you carry out the process or lost when you close Pro Tools. Saving to disk, on the other hand, allows you to specify a file name and location to store the template for future use at any time.

The next major task that Beat Detective can perform is to separate an audio region into "slices" like Recycle does. In fact, if you have already worked on getting the markers in the right place as described above, then the only thing you need to do to carry out the separation is to choose *Clip Separation* from the *Operation* heading and then click on the *Separate* button in the bottom right-hand corner. Your one audio region will now be separated at the marker points into a number of individual regions.

If you then click on *Clip Conform*, you will see, under *Conform* on the right of the window, four controls. *Strength* allows you to vary the degree by which your regions are conformed (quantized). A value of 0% will leave them in their current position, while a value of 100% will move them to be on the quantization grid. The *Exclude Within* slider allows us to exclude certain regions from the conforming process. If a region is close to its ideal position, then it can be excluded from the process to make sure that only the most out-of-time regions are moved. The Exclude Within slider controls the amount of variation from the ideal that is allowed before confirming will take place. Finally, the *Swing* slider allows us to add a variable amount of swing to the basic conforming process, and we can select either eighth-note or sixteenth-note swing using the two buttons located below the slider. Clicking on the *Conform* button in the bottom right-hand corner will apply the process and move the regions into their new positions.

Finally we have the *Edit Smoothing* option. When you separate a region, as we have done here, and then move (conform) individual slices within that region, it is quite possible that there will be audible gaps between the end of one slice and the start of the next. Pro Tools has included the *Edit Smoothing* option (under the *Operation* heading) as a way of trying to deal with situations like this automatically. *Fill Gaps*, the first option under the *Smoothing* heading, works by moving the start point of any regions that are preceded by a gap back to the end of the preceding regions. In the event that there are still issues with the smoothness of the changeover points, then you can change to the *Fill And Crossfade* option, which will carry out the same process but additionally will apply cross-fades at the junctions of all regions. The duration of the cross-fade can be set in the *Crossfade Length* box.

Moving on from Beat Detective to *Elastic Audio*, we find that one of the most useful things it allows us to do easily is to quantize an audio file or region. First, we need to select all regions that we wish to quantize, make sure that Elastic Audio is active and that the markers are in the right places (as per the previous chapter), and once that is done, we open up the *Event Operations* dialogue by pressing Alt + 0 or Event > Event Operations > Quantize. *What to Quantize* should be set to

Elastic Audio Events, and we can leave this as it is and then move on to the actual *Quantize Grid* options. The main drop-down list gives options for all the standard note divisions as well as dotted notes and triplets, but in addition has options for *Groove Clipboard* and a list of any saved templates. If you choose one of the standard note divisions, then you also have options for *Randomize* (which will quantize notes but then apply a degree of randomization to the timing to try to maintain an element of human feel), *Swing, Include Within, Exclude Within*, and *Strength* (which are all similar to the options in Beat Detective). If, alternatively, you choose a groove template, you have options for *Pre-Quantize* (which applies a standard quantization before applying your groove template), *Randomize, Timing, Duration*, and *Velocity*. The last two of these are relevant only to MIDI regions, where they can match both note durations and note velocities as a part of the quantization process, but the *Timing* slider determines the strength of the quantization process, with higher values having a stronger effect. Any change in parameters will require you to hit the Apply button again, but it is generally quite a quick process, so this shouldn't slow you down too much.

One final way in which Elastic Audio can help us to create pseudo-Recycle files is by setting the *Track View Selector* to *Warp* and then selecting a region and using the Tab to Transient feature to move through the region one transient at a time, and then, at each new transient, use the *Separate at Selection* command (Cmd[Mac]/ Ctrl[PC] + E or Edit > Separate Clip > At Selection). This is an alternative to the Beat Detective method of splitting a region into several smaller slices.

Studio One

Once you have got transient markers in place, as we described in the last chapter, you can then start to make use of them for creative purposes. One of the most common things that you are likely to want to do with this newly flexible audio is to quantize it. There are a number of ways you can do this in Studio One, and the exact method you choose will depend on the complexity of what you are trying to achieve. We will start by looking at the ways of quantizing audio to a regular note division and then move on to look at ways in which you can match the groove of one audio part (or MIDI part, for that matter) to the groove of a different one.

The simplest way to quantize an audio part is to open the *Bend Panel* (View > Additional Views > Audio Bend or by pressing the *Audio Bend* button in the main toolbar). Over on the right of the Bend Panel you will see a heading called *Action*, where there are two controls. The first of these selects between Quantize and *Slice*. We will look at *Slice* shortly, but for now let's focus on Quantize. Below this control you will see a Strength slider and numerical readout. This determines just how much the audio (or MIDI notes) will be moved from their current position to the theoretically perfect quantized position. A value of 100% will mean that they will be perfectly aligned to the quantize grid, while a value of 0% means that they won't be moved at all from their un-quantized positions. The actual quantize grid (or feel) is chosen by changing the *Quantize Value* box, located in the main toolbar to the right of all the mouse tools. Clicking on this box will bring up a list of standard note-value quantize options along with various options that add triplets, unusual time signatures, and varying amounts of swing. Choose one of these values, and then click on the Apply button back in the *Action* section of the Bend Panel, and your audio will be quantized to the *Quantize Value* to the strength that you specified.

If you are happy with the overall quantization feel and you simply wish to quantize other parts to the same quantize grid and with the same strength, then you can simply press Q or go to Event > Quantize > Quantize, and the audio region will be analyzed, transients detected, and then the quantization applied to the same grid and at the same strength as the last time it was carried out.

If you need more control over the quantization, then you can open the *Quantize Panel* instead by going to View > Additional View > Quantize or by pressing the Quantize button located to the right of the Audio Bend button on the main toolbar. Once open, this panel gives you a number of different controls that you can use to fine-tune your quantization. To the far left are two buttons labeled *Grid* and *Groove*. Choosing *Grid* will give you an extension of the quantization controls that we saw in the Audio Bend Panel, and we will look at these first.

The first thing you will see is a series of note values ranging from whole notes (1 bar) down to sixty-fourth notes. This determines the basic timing resolution of your quantization and should reflect the minimum note division used in your project. Immediately to the right you will see choices for *Straight*, *Triplet*, *Quintole*, and *Septole*, which allow you to quantize to less commonly used time signatures and feels. Directly below these you will find a slider (and numerical readout) to continuously adjust the swing value. Whereas the Audio Bend Panel method allowed you to choose only from a number of preset swing values, this method allows you to choose an exact value (from 0% to 100%) that you want, and this is, once again, very valuable in fine-tuning the feel of quantized parts.

To the right we have a series of four controls: *Start*, *End*, *Velocity*, and *Range*. When working with audio only, the first and last apply, and Start is a control that varies the strength of the quantize by determining how much the start of the region should be moved (100% being aligned fully to the quantize grid and 0% being no movement), while *Range* is used to determine which parts are quantized. When set to 100%, all audio will be quantized, but, as the value is reduced, only transient markers progressively closer will be quantized. This can be useful in tightening up the timing of some transients while leaving others, those clearly and intentionally outside the regular quantize grid, untouched. For the most part it is advisable to start with this set to 100% and then potentially work downward if it feels over-quantized.

If we now change from Grid to Groove we will see the note division selections and the swing sections replaced with what is called the *Groove Panel*. This is a very clever feature that can analyze the rhythmic feel of any region and then store it and apply it to other regions. If you have a region that has the timing feel that

Beat-Mapping and Recycling CHAPTER 10

you want to apply to another region, then simply drag that region to the Groove Panel, where its timing will be analyzed and a timing "map" stored, which can then be applied to another region simply by selecting the new region(s) and then clicking *Apply* in the *Quantize Panel*. This method makes it almost unfeasibly easy to match the feel of an entire song to just one part of it or, perhaps, to match the feel of an entire song to the groove of another song entirely.

There is another way in which we can use the transient markers we created. In this chapter we have looked at recycling files and all the benefits associated with that idea. While Studio One doesn't offer a solution that is quite as simple as Logic, it certainly does give us the option to create recycled files if we are prepared to put in a little effort. The first step, once we have got the transient markers in place, is to make use of the Groove setting on the *Quantize Panel*. Once you have chosen the region that you want to recycle you should drag it from the main Arrange view on to this Groove Panel so that it can be analyzed and ready to be used.

The next thing to do is open the Audio Bend Panel again, and, once the transients are set at the points you wish to slice the region, you should select *Slice* from the *Actions* section on the right. There are options below for *Autofades*, *Merge*, and *Quantize*, but, for the purposes of our recycling, we certainly don't need to worry about the last two, and the first is perhaps a precautionary option you could take to avoid having little clicks and pops. When you click on the Apply button, the selected region will be split at the transient marker point to create individual regions (or slices). Once this is done, you should click and drag over all of these in the Arrange view to make sure they are all selected and then right click and choose *Audio > Send To new Sample One*. This will create a new sampler instrument and automatically map each region to a key range. This defaults to ranges starting at C3 and maps the regions only to sequential natural notes (no sharps or flats).

The next step is to create a new instrument track in the Arrange view by going to *Track > Add Instrument Track* and then opening the *Inspector Panel* (F4 or *View > Inspector*) for that instrument track and making sure that the *Output* is

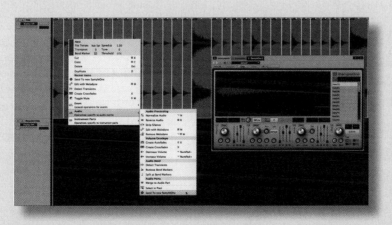

FIGURE 10.7
While perhaps not quite as intuitive as Logic in this regard, Studio One allows you to send a number of different audio files to SampleOne (its built-in sampler) that are then automatically mapped across the keyboard. This allows the creation of Recycle-type files in just a few steps.

set to the correct instrument (*Sample One*) in our case. You then open up the Quantize Panel again, click on the Groove Panel, and drag it down to the newly created instrument track. This will create a MIDI region, with the timing and velocity of the MIDI notes based on the information from the analysis of the Groove Panel. Because we used the region that we have just created the sampler instrument from, we will now have a MIDI part that will trigger the individual slices at the exact times that they originally appeared in the audio region. The only thing that we now need to do is go into the MIDI region and change the MIDI notes. Using this Groove Panel method results in a MIDI region with all the notes on the same MIDI note. We will need to go in to the MIDI region and re-map the notes, so that they start on C3 and move up one natural note (so no sharp or flat notes are included) for each new slice. This can take a bit of time if the recycled region is a few bars long or especially complex, and it certainly isn't as immediate as using some of the third-party tools available, but it does give Studio One users a way to be able to recycle files if they are prepared to go through a few stages to get there.

Cubase

Quantizing audio files is something that is, relatively speaking, quite a recent development in audio editing history, yet at the same time is something that has become pretty much indispensable. While MIDI regions can be quantized and individual notes moved around without any detrimental effects, audio regions will always have some compromise when you carry out these kinds of actions. If you are using Elastic Audio–type audio processing, then there will be time-stretching involved, and that will inevitably leave some processing artifacts. If, on the other hand, you slice any region into smaller parts, then you stand a very good chance of having gaps and overlaps that will need to be dealt with. On the whole, though, these are both manageable situations, and the choice of which approach to take is based on personal choice, a good assessment of the audio you are working with, and weighing the pros and cons of each method. Most people would agree, however, that audio quantization using *Elastic Audio* is the simpler method, because it doesn't require a great deal of user intervention after the fact, so we will look at that method first and then explore some of the slicing/recycling capabilities that Cubase has.

The first step in any audio quantization is to detect the transients or Hitpoints. We saw how to do this in the last chapter, so we will assume that this has already been done before moving forward. Once the Hitpoints are correctly in place (including any minor adjustments that you might have had to make manually to get things nice and tight), we can treat the audio largely the same as a MIDI region for quantization purposes. The Quantize Panel, which Cubases uses for audio and MIDI quantization alike, is accessed by going to Edit > Quantize Panel. Once this is open, you will be presented with a number of different options. At the very top of the panel, you will see a drop-down list of commonly used preset quantization settings. These cover all the basic note divisions along

Beat-Mapping and Recycling — CHAPTER 10

with triplet and dotted variations. Selecting any of these presets will show the quantization target points on the grid below. The grid represents a one-bar period with divisions for each beat and subdivisions for each sixteenth note. The actual position of the quantization targets is shown by the green lines positioned on the grid. To the right of the drop-down list you will also see two buttons that allow you to save your own presets and also to delete presets.

Directly below this preset list, you will find four controls that allow you to set your own quantization parameters. The first, *Grid*, determines the basic note division that you will use, and this list is the same as the presets above. Next to this you will find a *Swing* parameter that is variable from 0% to 100%. Clicking and dragging on this will change the amount of swing added to your basic note division. Any change here is reflected in the positions of the green lines in the grid below. The higher the swing value, the farther away from the standard positions some of the green lines will move. The combination of these two controls determines the basic feel of your quantizations.

The next two parameters are designed to allow for the quantization to have a less rigid and robotic feel. The first parameter, *Catch Range*, is used to determine a range either side of the quantization targets within which notes or events will be quantized. Changing this value creates a wider green bar at the bottom of each green line to show the range. Any notes or events that fall within this area will have quantization applied, and any that don't will be left as they are. This value is set as a percentage, and the greater the percentage, the wider the catchment area. However, it should be noted that a value of 0% effectively corresponds to a bypass of this parameter, and all notes or events will be quantized. The final parameter, *Non-Q*, is in many ways the direct opposite of this and allows you to specify a range either side of the quantization targets, within which notes will not be quantized. This range is shown by a red bar on either side of the green lines. The purpose of this is to be able to specify that any notes that are quite close to the ideal position (within the range of the red bar) will be left as they are, and only notes that are further from the ideal will be quantized. This will allow for a degree of natural variation while still making sure that any notes that are too far out of time are pulled back into line. You will also see a *Randomize* control that, once the main quantization has taken place, will then randomize the position of each of the quantized notes within the range specified to allow for a slightly more-natural and less-rigid feel.

Finally you will find controls for *iQ Mode* and *Audio Warp*. IQ Mode is equivalent to a strength control, in that it allows for quantization to take place to a percentage level. Clicking on the iQ Mode button will activate this mode and also display a percentage parameter that allows you to set the strength of the effect. Audio Warp, on the other hand, allows you to choose whether to quantize by moving regions/slices (which is appropriate if you have already separated your region into separate regions) or by using the Audio Warp feature with the *Warp Markers* to simply move the audio around and time-stretch/compress as necessary.

FIGURE 10.8
Groove Agent One, included with Cubase, not only allows for the very easy creation of recycled files but also packages this ability into a very easy-to-use instrument that has strong resemblances to the classic Akai MPC series.

Clicking on the Quantize button will apply the quantization based on the settings you have chosen.

If you have a particularly nice feel to an audio file or region, then you might want to take that feel and apply it to other audio (or MIDI) parts in your project. You can do this by using groove quantization. You start by analyzing the region you wish to use as a template and detecting the Hitpoints as usual. Then, while still in the Sample Editor window, click on the *Create Groove* button under the Hitpoints heading. This will copy the position of the Hitpoints onto a temporary clipboard. If you now open the Quantize Panel, you will see, at the bottom of the preset list, an additional option that has the name of the region you just used to create the groove. You will see the position of all the markers presented as the usual green lines on the grid. You can now use this to quantize other audio regions. At this point it is probably wise to save this as a new preset for recall later, and you can do this simply by clicking on the *Save Preset* button to the right of the drop-down list. If you wish you can choose *Rename Preset* at the bottom of the preset list to use a more memorable name, especially as these presets will be available in other projects, so something that refers to the name of the project as well as the region that it came from might be useful.

When it comes to recycling audio regions, Cubase has a very simple method that can carry out the process for you in just a few steps. The first thing to remember, before you get started, is that this method (like the Logic method) can accommodate up to 128 separate slices because of the limitations of MIDI note numbers. As a result, if the region you wish to recycle will have more than 128 slices, you will need to separate it and carry out the process twice. With that in mind, double-click

Beat-Mapping and Recycling CHAPTER 10

on the region to open it up in the Sample Editor window, and then click make sure all the Hitpoints are in the correct place. Once this is done, click on *Create Slices*, and this will create an audio part in the main Arrange window. You should then go to Project > Add Track > Instrument and choose *Drum—Groove Agent ONE* from the drop-down list of instruments. Make sure that the *Groups* button at the top of the plug-in window is set to "1," as this will give you the most available slices to work with. With the plug-in window still open, you should then go back to the Arrange window, select your newly created audio part, and then drag it into the plug-in window and onto the lower left of the sixteen pads. Cubase will then take each slice and assign it to the pads sequentially.

Once this is done, you need to create the MIDI region that will trigger the slices in the right order. This is simply a case of clicking on the double arrow icon in the *Exchange* section and dragging it onto the MIDI track for Groove Agent ONE. This will create the MIDI part for the recycled part, and, if you play back the MIDI region now, it should sound exactly the same as the source audio region. From here you can change the playing order of the slices, or go into Groove Agent ONE and adjust parameters for each individual slice.

CHAPTER 11
Drum Replacement

USING RECYCLED LOOPS

One additional use of recycled files that we haven't looked at so far is the possibility to take the timing information that exists within a recycled file and use that to trigger other sounds to reinforce or replace individual sounds within the file. Naturally, we could do this with a normal audio region as well. We could identify the regions or places within regions where a certain sound occurred and then create a new track and copy the replacement sound into each place that it was needed. But over the course of an entire song, this could be a very laborious process. Recycle and its ability to create MIDI regions with individual notes separated and placed at the correct times go a long way toward automating this process.

If we have a drum loop that we like but that we feel has a snare drum that is a little weak, then we simply recycle that drum loop and drop the MIDI file onto a new sampler instrument track. We can then load up a snare drum sound that we want to use to layer (or replace) the weak snare drum with and assign it to a MIDI note of our choosing (let's say we choose C3). We then locate the notes in the MIDI file that correspond to the weak snare drum, move these notes to the correct note (C3 in our case) to trigger the new snare drum in the sample, and then mute all the remaining notes. We will now have our new snare drum layered with our original one.

We may still need to do a little fine-tuning in terms of the timing if we want to layer the two sounds, simply because the transient marker that we used to "slice" the original file when we created the MIDI region may not have been 100% at the start of the snare drum. Or perhaps it was, but the start point of the new sample isn't 100% at the start of the sound. In either of these situations, there may be a slight timing discrepancy between the two. If this happens the easiest solution is to apply a MIDI timing offset to the new snare part (either a positive or negative offset, depending on what is needed) to get the two sounds to work well together. You might also need to consider the phase relationship between the two sounds. Even though they won't be identical and therefore will never completely cancel each other out, there could a great deal of cancellation between the two sounds, so either inverting the phase of one or the other or very subtle shifts in position can fix that.

This system works well if you want to work with short loops and layer an additional sound or two on top, but if, on the other hand, we are dealing with a multi-track drum recording with each drum recorded individually and with variations throughout its length rather than being a constant repeating loop, then things get a little more complicated—not so much complicated in principle but more in execution. We can still use the same principle in that we can still recycle a file that is an isolated snare drum, and we can use a file that is longer than a bar or two as the source, so that isn't a problem. The real difficulties lie in getting a "clean" snare drum (in this case) track so that we get only transients at the start of each snare drum hit and also in the fact that, given that recycle creates MIDI files for the notes and the MIDI specification allows a maximum of only 128 notes, a recycle file can never contain more than 128 separate "slices" and associated notes.

The first of these, the potential problem with getting a clean track, can be dealt with, and we have already looked at ways in which we can do this, so it's quite possible that you have already been through this process, and, having done so, have come to the conclusion that the snare just isn't right. If that's the case, then half the problem is already solved. If you haven't reached that point yet, then you can refer to the earlier chapters on how to do this. Once you have the snare drum in a reasonably clean state, you also have the sensitivity control to adjust to try to make the slices occur on only the snare drum hits and not because of any residual noise or spill.

The other issue—the 128-slice limitation—is more of an inconvenience than a problem. The solution is to create multiple recycle files to cover the length of the song. If we assume a very "straight" drum pattern with two snare drum hits per bar (on beats two and four), then a single recycle file could cover a maximum of sixty-four bars, which is quite a substantial amount of time and would mean that you would most likely need only a couple of recycle files to cover the whole song. Now of course, the more snare drum hits you have per bar (with a more intricate snare-drum pattern), the smaller number of bars a single recycle file will cover. But in all but the most consistently complex of snare patterns, you shouldn't need more than a handful of separate recycle files to cover the full song.

In the event that you do need to create multiple recycle files, you would still need to create only one new sampler instrument, as the same sound would be used by all the recycle files. You would just take the MIDI region from each recycled file and move it to the same sampler instrument track. This would be true whether you needed one recycle file or twenty. The only time you would need different sampler tracks would be if you wanted to use different sounds to layer with the original sound in different sections.

Although I have used the example of a snare drum, this technique is equally applicable to all drum and percussion sounds (all sounds, in fact, but it is most commonly used with drums), but consideration will need to be given to the number of individual slices you will end up with. A snare drum (or kick drum) with a few hits per bar could mean only a handful of recycle files to cover a whole song, but a busy hi-hat pattern could result in a single recycle file covering only eight bars or so, which means a lot more recycling to cover the whole song. Even in this case, though, it still might be far quicker than manually layering sounds.

Using this method can be quite a lot quicker than manually placing new audio files to layer with existing ones, but it also gives a much more-simple way of controlling the volume of those sounds. Now, when I say "volume," I am referring to the volume dynamics over a bar (or two or ten). Being able to use MIDI note velocity to change the level of the layered sound rather than having to draw in volume automation for each individual hit is a great time-saver. And, as we are using sampler instruments, it is also possible to map MIDI controllers to adjust the volume as well, so a modulation wheel on an attached MIDI keyboard could be used as a volume control for the layered snare drum. This might end up being a more natural way of controlling the dynamics, because it could be done by "feel" rather than just drawing lines or entering numbers. And as well as this, there is still the option to control the attack or decay times of the sampled sound, as we looked at in the last chapter.

By using this technique we can (relatively) quickly layer up a drum sound and have a good amount of control over that layered drum sound. With the right samples used, it is certainly possible to create some very natural-sounding results, but it might take a little time to record MIDI controller movements to re-create the dynamics and tonal changes of the original sound. Clearly it would be useful if we could remove some of the manual work from this to speed things up.

DEDICATED DRUM REPLACEMENT TOOLS

Given that we have a means of detecting when a transient has happened, a way of converting that into a "trigger" signal of some kind, a method of measuring a peak level of a given part of an audio file, a way of loading a sample (or velocity-switched group of samples) into a sampler, and a means of using a trigger signal to sound the sample(s) we have loaded, we already have everything that we need to make the process described above much easier. We have achieved most of these steps with the method described above, but it would be nice to be able to make the process somewhat more intuitive (and quick) by combining the various stages into one single tool.

Most DAWs have transient detection built in as a part of some other process (beat-mapping, time-stretching, etc.) but not all of them use this in the same way to aid with drum replacement. So in this chapter we will look at some third-party tools that are going to be DAW-independent and would be available

Dedicated Drum Replacement Tools

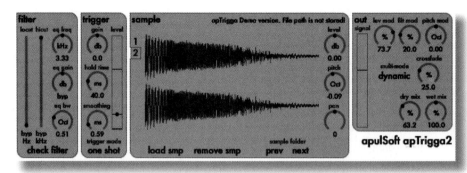

FIGURE 11.1
apulSoft's APTrigga is a simple yet effective drum-replacement plug-in that wraps up detection tools and sample playback into a single plug-in.

to everybody. All these drum replacement tools work on the same principle of analyzing an audio file, detecting when a trigger situation has happened, and using that to trigger an alternative sound within the plug-in itself. There are often options to create MIDI tracks from them as well, but, for the purposes of this discussion, I want to look at the all-in-one solution. If you were to use one of these drum-replacement plug-ins and generate a MIDI region from it, then the resulting MIDI region would be used in much the same way as the MIDI regions created from recycling, so we would have all the same options and possibilities as we have already looked at above. But let's see what these all-in-one solutions can do for us and our workflow.

The first thing we need to consider, before we get into the details of how these drum replacement plug-ins work, and what they can do for us, is the fact that, unlike the recycled file solutions that we have just looked at, these plug-ins operate in real time. They analyze the audio as it is playing, and, when they detect a transient, they trigger the replacement sound. The recycled file version analyzes the file ahead of time and creates its transient markers once the analysis has been done. Because of this, the associated slices or MIDI notes will be 100% at the same time as the transients occur. With the drum-replacement plug-ins, there will always be a latency (or small delay). It doesn't matter how fast or powerful your computer is—this delay cannot be avoided, only minimized. Naturally, the more powerful your computer is, the less this delay will be, but, if you aren't actually replacing sounds and are layering them instead, we have already seen that even the smallest delays between sounds can cause phase-cancellation issues, so this could cause more problems than just a slight "looseness" in timing.

There are two simple ways to deal with this problem, and both relate to shifting the timing of the replaced drum sound relative to the position of the original sound. The first, and the simplest, is to simply work with the slight timing lag until you are happy with the replacement sound (timing, dynamics, tone, etc.), and then bounce the replaced sound to a new audio file and nudge the timing of this bounced file a little earlier, until it is in time. This is an entirely workable solution if you are actually replacing one sound with another, because there will be no issues of phase cancellation between the two sounds, so, at worst, the timing will be a fraction off while you are still adjusting parameters prior to bouncing the replacement sound.

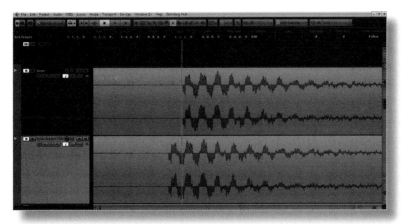

FIGURE 11.2
If the detection process means an unacceptable latency, then you can always offset the track used to trigger the detection. Here we can see the original snare drum track (at the top) and the offset copy, which is moved a few milliseconds earlier to counteract the delay in the detection process.

The second option is to actually offset the timing of the replaced sound *before* it is triggered. This might sound a little contradictory, but it is possible with a little bit of work. We have seen that there is an inherent delay in the transient detection and triggering process, so, if we are aware of this, we can get around it by creating a new audio track, copying the audio file or region that we are processing with the drum replacement plug-in to this new track, loading the drum replacement plug-in and setting everything up as we need it, and then, once we become aware of just how much the delay is, we can move the actual "source" file a little earlier to compensate for the processing delay in the plug-in. If you are doing simple drum replacement, then you should mute the original, unmoved "source" file. But if you are layering a new sound, then this method allows you to move the timing (by moving the copy of the source file used to trigger the drum replacement) of the layered sound in real time without having to bounce.

The actual amount you have to move it by will vary from plug-in to plug-in, from computer to computer, and even from one replacement sound to the next. There is no arbitrary "global" offset that you can just apply and know that it will work. It is a matter of experimentation, but, given the ease of selecting an audio file and fine-tuning its position, it isn't something that should take a huge amount of time to get right. And if, as mentioned above, you are doing full replacement of a sound rather than layering, you will need to make sure the timing is accurate only to a couple of milliseconds rather than potentially having to move the "source" file by a few samples at a time to correct a phase-cancellation issue. Now that we have that disclaimer out of the way, it's time to look at how these plug-ins work and how we can use them to our advantage.

DRUM REPLACEMENT TOOLS IN USE

There are two main parts to plug-ins of this type: the detection and the replacement. The specific features available are slightly different on each plug-in, but they share at least a few common characteristics. The main controls on the detection part of the plug-in are an input gain control and some kind of sensitivity control. The input control is used to make sure that we can maximize the dynamic range of the input sound as much as possible. Most often this gain control will be used to increase the level of the input, but, on occasion, it might be necessary to reduce the gain of an input if, in addition to the underlying audio file (which can never go above 0dB), there has been additional processing that has actually pushed the input to this section above 0 dB. In this case, having the level set so that it is clipping would actually move into the area of actually reducing the amount of available dynamic range. Once the gain has been set correctly, we can move on to the sensitivity control

In order to set the sensitivity accurately, it always helps to have a fairly "clean" signal, for the same reasons that it helps in recycling files, and to that end some drum replacement plug-ins offer additional controls to help achieve this. These could take the form of simple high- and low-pass filters to tune the frequency used to detect the transients, or could be more advanced, such as a transient designer that can help to emphasize the attack and suppress any background noise or spill or just make the overall sound more snappy and tight. Of course, even if the drum replacement plug-in doesn't actually have any features to clean up the input signal, or if the features present still aren't tightening it up sufficiently, you always have the option of additional editing or processing prior to the input of the plug-in.

Adjusting the sensitivity can be quite deceptive. After all, with very minimal controls, how hard can it be? The answer to that question is that it isn't *hard*, but it can be time-consuming. It can take a lot of adjusting back and forth between those relatively few parameters to get something that captures the subtlety of the drum performances but doesn't trigger notes in error from background spill. To help with this, there is sometimes an additional "re-trigger" control. This is very simply an amount of time that must pass once a trigger has happened before the next will be allowed. If you set it to forty milliseconds then, no matter what happens to the input signal, there will not be a trigger

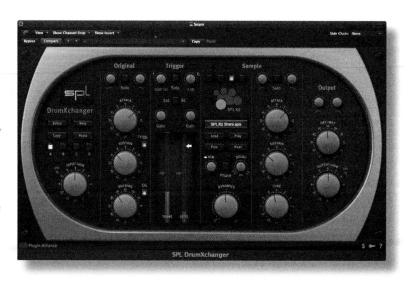

FIGURE 11.3
The SPL DrumXchanger is a very well-specified drum replacement plug-in that includes not only some very advanced detection circuits but also SPL's Transient Designer technology on both the original and replaced parts of the sound and a library of replacement sounds to get you started.

signal created until at least forty milliseconds have passed. If you know you have a pretty simple pattern with no intricate "ghost" notes, "flams,", "drag rolls," or super-fast playing in general, then you can probably set this much higher to decrease the chances of false triggers, even in a recording with quite a lot of mess and spill.

Moving on to the actual replacement part, we can see that things, however they are laid out or implemented, are basically quite self-explanatory. Fundamental here is the ability to load in samples to be triggered, and all the plug-ins allow you to do this. Many will come with their own library of sounds covering all of the common drum kit pieces, and many of these are multi-sampled, so that the sampled sound will not just simply change volume but will have *some* of the tonal change that can happen with changes of loudness in a real drum. It would be unrealistic to expect a static collection of samples to fully reproduce the tonal and dynamic range of a real drum, especially a snare drum, but as technology is advancing, larger and larger sample collections are becoming the norm, and the results are getting harder and harder to distinguish, especially if the original performance doesn't utilize a huge amount of the tonal and dynamic spectrum of the drum.

The specifics of how the samples are loaded, whether they are multi-sampled, whether you can load in more than one layer of sounds, whether there is any volume envelope editing, and whether there are then further processing effects within the plug-in will vary from product to product, and, if a particular feature is important to you, then that could be a determining factor in your choice as to which to use. Most of them do, however, allow you to use your own samples (or those from another sample library or source), but, as I have already mentioned, don't expect to get results that sound as good if you load in a single snare drum sample and expect it to represent the snare drum accurately across all volume levels.

One very clever feature is the use of "round robin" sample playback. This entails the use of multiple samples for each velocity level, meaning that, in addition to the sound of the sample changing—not only in volume but also in tone as the volume/velocity changes—subsequent triggers, even if at exactly the same level, will cycle through these alternate samples, so that no two hits in succession are ever exactly the same. Although the differences between these round-robin samples can be very subtle, the overall effect, over the length of a song, is an indefinable sense of added realism, and, best of all, all of the hard work is done for you in the creation of these sounds, so you can just set up, sit back, and enjoy! Of course things are never *quite* that simple, but we are getting closer every day.

Once you have chosen the sound you want to use, you may need to revise your settings for sensitivity and other parameters, but one thing that you will almost certainly want to do is adjust the dynamics of the new sound. The dynamics, in this context, are the rate at which the volume of the new sound increases in proportion to changes in the volume of the source sound. By reducing the dynamic correlation, you can set things up so that, even if your original source sound has a lot of variation in volume, this new sound always hits pretty hard. Alternatively, you may wish to take an original source sound that didn't have a great deal of dynamic variation and try to actually increase that. You might want to simply try to keep a linear relationship between the two, or it might be anywhere along this spectrum. The ability to actually change the dynamic correlation between the source sounds and the new sound means that, if you are mixing the new sound with the source sound, you can set things up so that the new sound really becomes apparent only on harder hits but is still there on the quieter hits; it is just that the new sound is *more* "quiet" than the source.

This brings us on to the final selection of parameters: those concerning the overall balance between the original source sound and the new sound, and the overall output level of the plug-in. Almost all the plug-ins offer a balance control that is continuously variable from only the original source sound through to only the new sound. If you have created a copy of the source sound and applied the plug-in to that track to give you the ability to nudge the timing backward or forward (as discussed above), then you can set the balance within the plug-in to be only the new sound, then adjust the balance between the source and replacement sounds in your DAW and apply separate processing or effects to each sound (something you can't do if you mix the two sounds within the plug-in itself).

That pretty much covers how drum replacement tools work from a functional point of view. Whether you would want to use these tools is really down to how you like to work and just what scope your editing job covers. There are some pretty big names from the music world endorsing some of these plug-ins, and they can certainly be a great problem-solver if you have some truly shocking recordings to work with. Or perhaps the recordings you have really aren't all that bad at all, but either the drum kit itself or the recording equipment use has just resulted in a drum track that, though perfectly competently performed, just lacks any real depth and punch. In cases like this, it helps to have a working knowledge (at least) of production techniques, as it will give you a better idea of what might be fixable by working with the original recordings and what just needs a little help and support tonally.

HANDS ON

Introduction

In the main chapter, we have looked at the ways in which recycled files and drum replacement plug-ins can be used to either reinforce or completely replace a drum sound that, for whatever reason, isn't right. Many of the drum replacement plug-ins have a number of advanced features, and many also come with custom sample sets that can hold a lot of appeal. If you don't have to do drum replacement that often, you might not feel justified in investing in these plug-ins, so we will take a look at the options available in each of our DAWs and compare them to using a third-party drum replacement plug-in.

Drum Replacement

Logic

Logic's built-in drum replacement tool, while fairly basic, can often be sufficient for simple drum replacement duties. Like all drum replacement tools, it will work much better when there is a relatively clean recording (not too much of the sound of other drums in the mic), but, because the detection options aren't as complex as some third-party tools, it is perhaps a little less forgiving. Nonetheless, it is a great place to start and may well be all you need if drum replacement is something that you do only occasionally.

The first thing to do is solo the track you are going to be working on and then go to Track > Drum Replacement/Doubling. Alternatively, you can use the key command Ctrl + D to open the dialogue box. As soon as you select this, you will notice that an Instrument track is created below your audio track and an *EXS-24* sampler loaded, ready to serve as the source for the replacement/doubling sounds. The first option that you are presented with is a choice of *Instrument*. The main effect this has is to change the default MIDI note that is used for the MIDI region. If you choose Kick, the note will default to C1, Snare will default to D1, Tom will default to A1, and Other will default to C3. Therefore, if you choose the wrong option, or you are replacing a different drum not listed, it isn't a huge deal, as you can change the MIDI note either later in the process or even after the MIDI region is created. It should also be noted, however, that it does also have an effect when you are using the *Prelisten* option (more on this below).

Next up is the *Mode* option, which can be set to either Replacement or Doubling. The only difference here is how Logic deals with your source audio track once the process is complete. If you choose Replacement, then the source audio track/regions will be muted, while if you choose Doubling, they will, obviously, remain active. But once again, whichever option you choose can be amended very easily after the process is completed by muting/un-muting the source audio track as required.

The next option, *Relative Threshold*, is the most important one, as this is what Logic uses to determine what is an actual hit that needs replacing/doubling and what is unwanted noise/spill. You can make adjustments either by using the slider, the up and down arrows on either side of the threshold numerical display, or by doubling-clicking on the threshold number and typing a value. It is useful to know that the up and down arrows can work with different increments. By default they will increase or decrease the threshold by 1 dB at a time, but if you click on a single digit of the threshold reading, it will become highlighted, and a small arrow will appear underneath it. Once this small arrow is visible, the increment of the up and down buttons will change, and only this digit will increase or decrease by a single number each time. For example, if there was a current threshold of −13.2 dB, and you clicked on the "2," then the up and down arrows would increase or decrease the threshold by 0.1 dB at a time; if you clicked on the "3," then it would be 1 dB at a time; and if you clicked on the "1," it would be 10 dB at a time.

Hands On

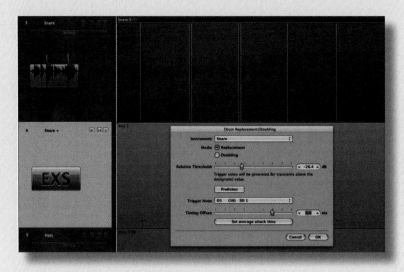

FIGURE 11.4
The built-in drum replacement tool in Logic can be quite effective, but care must be taken with "double triggers," as shown in the example above. Adjustment of the threshold level may not be able to remove these, in which case you can go into the resulting MIDI region and delete them manually.

As you adjust this threshold, whichever method you choose, you will start to see vertical yellow lines overlaid on the waveform display in the Arrange window. These lines represent the transient markers that the process will use. Lowering the threshold will create more markers, and raising it will create fewer. Many times a careful adjustment of this threshold level will create markers in all the places that you need them, but it isn't perfect. Sometimes markers won't be created where you expect them, and adjusting the threshold won't have the desired effect. In many ways it is a shame that you can't use the transient markers that you may have already placed by hand, because these would be in the right places. Instead you have to rely on the threshold method and hope that it works accurately enough.

In the event that no amount of tinkering with the threshold gives you everything that you need, don't worry, because all is not lost. You can always continue with the process and manually place additional notes in the MIDI region to cover any missed notes or delete any unnecessary notes. Of course you will have to estimate the velocity value, and having to do this manually defeats the purpose of a drum replacement tool, but at least it is a work-around if you are not inclined to purchase a more advanced third-party tool that has more options and may be able to do a better job.

Once you think that you have got the threshold level set correctly (or at least as good as it can be), you can click on the *Prelisten* button, which will allow you to audition the settings that you have chosen and the effect they will have. Once clicked, both the source audio and the replaced/doubled MIDI tracks will play, allowing you to hear if the settings you have chosen are appropriate. And here the *Instrument* setting that we mentioned above comes into play. If you chose Kick, then a kick sound will be loaded by default (and likewise with Snare and Tom), but in any case you can navigate through the default Logic library or, indeed, load any *EXS-24* settings of your own choice if you have your own sounds you would like to use. If you wish to change the MIDI note that will be used for the MIDI region, you can do so using the *Trigger Note* drop-down box, and this may prove useful if you are loading your own *EXS-24* instrument that doesn't conform to the default note choices mentioned earlier.

Finally we have the *Timing Offset* adjustment. This is included to allow minor changes to the timing of the MIDI notes. Sometimes a sample may not be perfectly truncated, and the actual sounds may start a few milliseconds after the MIDI note event triggers it. This control allows for any fine-tuning to be made (in 0.1 ms increments) prior to the creation of the MIDI file. Alternatively, you can use the *Delay* parameter in the *Inspector* panel, once the MIDI region has been created. You may wish or need to do this if you decide to change the sample(s) used after the initial MIDI region creation. While this *Delay* parameter doesn't offer adjustment in milliseconds, it does offer a resolution of 960 ppqn, which means 960 pulses per quarter note. In real terms this means that however long the duration of one beat is (60/tempo), the minimum increment for changes using this method is 1/960th of a beat.

Once you click on OK, the dialogue box will be closed and the MIDI track created using the selected MIDI note for the triggers and creating note velocity values based on the actual peak level of the individual hits in the audio track. The audio track will also be muted if you chose to replace the sound and the MIDI part ready for further editing if necessary.

Pro Tools

Unlike some other DAWs, Pro Tools doesn't feature any built-in drum replacement tools so, if you don't have access to any of the third-part drum replacement tools, you have to do things the manual way. The essence of this technique is using the *Tab to Transient* features of Pro Tools and then pasting either an audio region (sample) or a MIDI note at each transient. The key to doing this quickly and efficiently is to have the drum you are trying to replace recorded (and a fairly clean, minimal-spill recording at that) on its own track. It doesn't mean that it isn't possible to do it on a more complex track, but you certainly need to be a lot more selective with what you do, and the reasons for this will become apparent shortly.

The first step is to create a new, empty track (we will start by looking at the process for using an audio track and then discuss any differences and additional steps for a MIDI track afterward), which we will use for our replacement drum sound. For this process it makes things a lot easier if this new track is located directly below your source track for ease of navigation. Once this is done, you need to make sure all the transient markers are in the right place for the source track if you haven't done so already (see end of chapter 9 for details), because we will be relying on the accuracy of these transient markers to achieve what we want to. And finally you need to decide on the sound you wish to use to replace the drum. You should import it into the project, place a single copy on the destination track (at the very beginning is helpful), copy it, and then mute it.

Hands On

You need to activate *Tab to Transient* mode (Cmd[Mac]/Ctrl[PC] + Alt + Tab or the *Tab to Transient* button, located below the edit tool buttons at the top of the Edit window), so that you can use this feature, and you need to select a region in the source track that you want to carry out the drum replacement on.

The final thing you need to do in preparation is turn on *Keyboard Commands Focus Mode*. This is a special mode in which many of the normal modifier keys (such as Cmd[Mac]/Ctrl[PC], Alt, Ctrl[Mac]/Start[PC], etc.) are not required, and the main QWERTY keys become a very useful group of "one touch" commands. When you have to do very repetitive tasks such as drum replacement, the existence of single-key commands that achieve what we want them to do, without having to use modifier keys, right-clicks, or menus, makes life so much easier, and the benefit to you simply cannot be overstated. There are a few different places in which we turn on Keyboard Commands Focus Mode, and one of those, the one we are interested in, is for the main edit window/arrangement. To activate this mode, you need to look for a small square icon with an "a" in one corner and a "z" in the other, which is located at the top of the vertical scroll bar on the right of the window. When this mode is activated, the icon will have a yellow/orange highlight to it, and then you are ready to go.

The four keys that you are going to need for this process are **Tab** to move forward through the transients (and **Alt + Tab** if you need to go to the previous transient if you missed one), **V** to paste the copied replacement sound on the new track, **;** (semicolon) to move down from the source track to the replacement track, and **P** to move back up to the source track once you have pasted the replacement sound. From here the process is extremely simple and simply involves a very particular key sequence that you will master very easily, and you will probably be very surprised at how quickly you can progress through a track. That sequence is **Tab > ; > V > P**, which translates, in real terms, to

Move to next transient > Move down from source track to destination track > Paste previously copied replacement sound > Move back up to source track.

FIGURE 11.5
There is no dedicated tool for drum replacement in Pro Tools; nonetheless, it can be achieved pretty easily using the Tab to Transient features and a bit of copying and pasting. Unfortunately this method doesn't follow the dynamics of the source audio, so it might be necessary to use the Clip Gain control on individual drum hits to replicate the dynamics of the source, as seen in the example above.

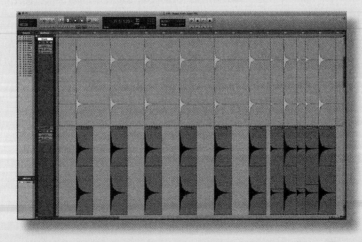

When you reach the end of a region, you may find that two presses of Tab are required to get to the next transient, as a region start is also considered a "transient" when using Tab to Transient, so you can't fully go into automatic mode yourself and will need to pay a little attention, but on the whole it is a very easy process.

As stated at the beginning of this chapter, having a recording with lots of spill or, indeed, a fully mixed stereo bounce of a drum take doesn't mean that the process is impossible, but it means more complications. As we have seen, the transient markers are the key, so, if you have a more difficult recording to work with, it will simply require you to do more work at the transient detection stage and try to make the transient markers only occur (by sensitivity adjustment or by manually placing or removing markers) when the drum you wish to replace is sounding. If you had a very complex recording, there are things you can do to help. You could, for example, make a copy of the recording and then use some plug-in processing on it (EQ, compression, expansion, and/or gating) to try to focus in on the particular sound you want to replace. You could then consolidate these regions and use transient detection on this new, bounced file and perhaps have a better chance of being able to get the transient markers in place with a simple sensitivity adjustment. There are usually ways to achieve what you want to, but some methods and situations are easier than others, so you might just need a little patience.

The biggest problem with the method described above is that each drum hit will now be *exactly* the same: the same sound, the same tone, the same level—everything the same. Unfortunately this is inherent in the process of repeatedly pasting the same file. If you want a little more flexibility, then creating an empty MIDI track (or Instrument track, if you plan on using a plug-in sampler to generate the replacement sound) might be the solution. Once the track is created, use the *Pencil* tool to create a note on the new track. The actual note (C, D, E, etc.) isn't important, because we can move this later, and in all honesty the note length isn't especially important either, but it makes sense to keep the note length fairly short, so that you can see at a glance each individual note rather than having longer, overlapping notes that are harder to make out. Once that note is created, simply cut it, and then repeat the process described above for samples. Instead of moving to the next transient and pasting an audio region (sample), you are simply pasting a MIDI note that will be used to trigger a sample when you are done.

There are two advantages to using the MIDI method. The first is that, because you are pasting a MIDI note, you can, if you wish, make adjustments to the note velocity on a note-by-note basis once the first stage of replacement is done. This would mean that, assuming you had a snare drum sample with multiple layers and velocity sensitivity, you could program variations in the velocity to try to mirror the dynamics (and often tonal change) of the source track. This is much easier to do if you have a fairly clean track to work from, because you will be able to visually identify louder and quieter hits and adjust note velocities to match. The other advantage is that it allows you to very quickly switch to a different replacement sound simply by loading a new sample, so you don't really have to worry about which sound you want to use before you start.

Studio One

While it doesn't have a dedicated drum replacement plug-in or function as such, Studio One makes the process of drum replacement very easy, owing to the *Groove* panel that we looked at the end of the last chapter. We can adapt the method that we used for recycling and instead use it as a very simple way of duplicating (or replacing) drum parts with sampled sounds.

We start the process by opening the *Quantize* panel and making sure that we have *Groove* mode selected. Then we need to select the region that we wish to carry out the drum replacement on and drag it onto the Groove panel. At this point you should note that, while the Groove panel will display the correct length of the region (however long it is), there is a limit to the number of transients that it will detect (and therefore create MIDI notes for in the next step) in any one go. As a result, if you have a very simple part, such as a snare drum on beats two and four, it might well be possible to carry out the drum replacement process on one hundred bars or more at a time, but if you have something busier like a sixteenth note hi-hat pattern, the maximum number of bars may be as little as thirty or perhaps even fewer. If you find, once you have carried out the process, that only a part of your regions was carried over into the MIDI region, then you should split your source track/region into smaller parts (based on the position of the last MIDI note in the created MIDI region) and carry out the process on only one at a time. With that in mind, once you have dragged an appropriately sized region on to the Groove panel, you will then need to create a new instrument track for the sampler that you will load the replacement sounds into. You can do this by going to Track > Add Instrument Track. Then you should click on the Groove panel and drag down to the new instrument track to create a MIDI region from the Groove panel data.

The next step is to open the *Console* panel by pressing F3 or going to View > Console. At the very left of this view, you will see a series of buttons for *Inputs*, *Outputs*, *Trash*, *External*, *Instr.* (Instruments), and *Banks*. Make sure that *Instr.* is illuminated, and then, directly to the left of all the channel faders, you will see a column headed *Instruments*. Click on the "+" symbol to the right of this, and you will be presented with a list of all of the plug-in instruments currently installed on your system. For the purposes of this chapter, we are going to use the included *Sample One*, but you could, of course, load up any instrument of your choice. Clicking on Sample One will create a new instrument, place it in the column below the Instruments heading, and also open the plug-in window. To make life easier, you can right-click on the instrument name, choose *Rename…*, and give it an appropriate name ("Kick Replacement," for example), as this will make the next step much easier.

Leaving the plug-in window open, you should now open the *Inspector* panel (F4 or View > Inspector) for the instrument track, and then, next to *Out* (default

Drum Replacement CHAPTER 11

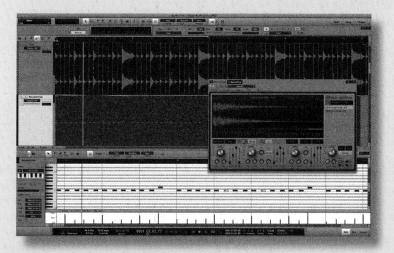

FIGURE 11.6
The combination of the Groove panel and Sample One makes drum replacement both quick and intuitive in Studio One. It doesn't have all the immediacy or flexibility of a third-party drum replacement plug-in, but if your needs are more simple, then it is more than up to the job.

value of "None"), you should click, and a list of all currently loaded instruments will appear. Choose the Sample One we just created ("Kick Replacement," in our example) and then open the *Browser* panel (F5 or View > Browser). This will open the Browser panel to the right of the Arrange view. This allows us to search for items to use in Studio One. This can include instruments, effects, and sounds, but, for what we are working toward here, we should choose *Files*. This will give us a standard file browser, which we can use to navigate to any samples we have stored on our machine. Once you have located a folder of sounds (kick drums, for our example) you can simply click on an individual sample and preview it using the Play button at the bottom of the Browser panel. By default this is set to loop play, so it will keep looping around the sample until you hit the Stop button. Usefully, though, if you click on a different sample, it will switch playback to the new sample, so it allows you to hit Play once and then switch from sample to sample without needing to hit Play each time. Once you have found a sample that you like, simply drag and drop it onto the Sample One instrument, and it will automatically be loaded and mapped to a root key of C3.

Conveniently, when we drag a Groove panel part on to the Arrange view, the MIDI region that is created has all the notes default to C3, so if we now play back the song, we should immediately hear the replacement drum part. We still need to do a little work to get things working completely smoothly, though. By default, when a MIDI region is created using the Groove panel method, each of the notes created is only about a thirty-second note in length. What this will mean is that, even though the notes and sounds are in place for Sample One, it is likely to sound very "choppy" and not natural at all, as the sample will play back only while the MIDI note is held. There are two ways we can deal with this. The first is to select all the notes in the MIDI region and make them longer, but an even easier way is to go into Sample One, go to the *Amp* section, and push the *R* (release) slider up to its maximum. What this will do is mean that the sample will play to its end each time it is triggered, irrespective of the length of the note that triggers it.

That should take care of the choppiness (and remember that you can always adjust the amp envelope settings later if you want to tighten up the sound if the release is a little too long) but we can still do a little more to improve things. Staying in the Amp section, you will notice a *Velo* knob in the bottom left-hand corner of this section. This is to control the amount of variation in volume in response to MIDI velocities. In order to try to re-create the dynamic variations in the source region as best as we can, it is normally best to turn this fully clockwise, so that there is the maximum correlation between MIDI velocity and sample playback volume.

With both of these changes made, you should have something that quite closely mirrors the timing and dynamics of the source region. You can now experiment with loading in different samples to Sample One by dragging and dropping, as each time you do this, the new sample will simply replace the old one. However, each time you load a new sample, you will need to make the changes to the envelope release and velocity controls once again. With that in mind, it might well be easiest to audition the sounds in the browser and load in only the ones (one at a time) that you think might work. You can, of course, load in multiple samples to be mapped to different keys, and there are instructions on how to do this on the website.

Cubase

Drum replacement in Cubase is made very easy in spite of the fact that it doesn't have a dedicated drum replacement function. Once again we find ourselves using *Hitpoints*, only this time we will make use of one of the other Hitpoint functions in the Sample Editor window. But before we get to that part, we first need to create a sound source to use for our replacement drum sound. To do this you should create a new Instrument track (go to Project > Add Track > Instrument) and then choose *Groove Agent ONE* from the drop-down list. We have already used this instrument to create a recycled part, and what we are doing here is broadly similar, but instead of mapping a number of individual slices onto different pads/MIDI notes, what we are aiming to do here is create a series of MIDI notes on the same note and use those to trigger a single pad/sound in Groove Agent ONE. In order to do that, we need to approach things slightly differently.

Making sure that the Groove Agent ONE MIDI track that you have just created is still the selected track, you should select the region(s) that you wish to carry out the drum replacement on and then open the Sample Editor window and start working on detecting the Hitpoints. If you have a very clean recording of the sound you wish to work on, then this will be made much easier, as the detection process will probably be largely automatic. If, on the other hand, you are working with something like an overhead mic, then you may well need to do a little bit of work in deleting unwanted Hitpoints and potentially even slightly moving some of the automatically detected ones. Once you are happy

Drum Replacement CHAPTER 11

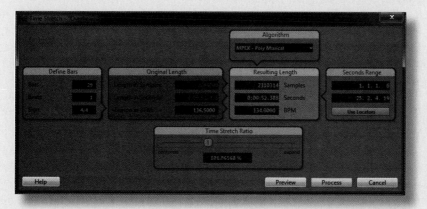

FIGURE 11.7
Groove Agent is once again used for drum replacement duties. The layout of format of the instrument makes is very useful for loading and auditioning a number of different alternatives for your drum replacement track.

that you have all of the Hitpoints that you need and that they are in the correct places, you can move on to the next step.

If you click on the *Create MIDI Notes* button, you will be presented with a number of options that determine the specifics of the MIDI region created. The first is *Velocity Mode*, which you can set to either *Dynamic Velocity* or *Fixed Velocity*. For drum replacement tasks where we would aim to replicate the dynamics as closely as possible, you should choose Dynamic Velocity, and this will scale the MIDI note velocity according to the signal levels at each Hitpoint. The next option is *Pitch*, which defaults to *C1*. The actual choice of note here isn't especially important, but you should make a note of the value here, as you will need this in the next step. You can then choose a *Note Length*, which defaults to an eighth note. Once again this isn't crucially important, as you can always change it later if need be, and in all honesty it is likely that the note length won't matter when it comes to choosing the replacement sound. The final option here is for *Destination*, which allows you to choose where the MIDI region will be created. The options here are *First Selected Track*, *New MIDI Track*, and *Project Clipboard*. We will choose First Selected Track, because this will create the new MIDI region on the first track that is selected in the Arrange window. This is the reason why we had to make sure that the new Groove Agent ONE track was the selected track before we opened the Sample Editor window. Clicking on OK will create the MIDI track, and we can then close the Sample Editor window.

If we now go back to our Groove Agent ONE track, we will see that a MIDI region has been created with the name of the audio region followed by "_midi," and it will be located in the same place as the source audio region. Double-clicking on this MIDI region will open up the piano roll editor, and you will be able to see the created notes and their velocities, which should follow the dynamics of the original audio region. You can now close the piano roll editor, and we can move on to setting up the replacement sound source.

At this point it makes sense to set up a playback loop area for the new MIDI region, so you can audition sounds in context. Obviously if you are replacing the original sound, then you should mute the audio track, but if you are wanting to double up the sound to add weight or punch, then the audio track can be left playing. Once playback has started, open up Groove Agent ONE, and we can start looking at sounds. If you left the *Pitch* parameter as *C1* in the MIDI region creation process, then you should go to Group 3, and you will see that the bottom left pad will be flashing in time with your extracted MIDI notes. If you changed the MIDI note to another value, then you can select the appropriate group (the MIDI notes for each pad are displayed on the pad) and then you should see one of the pads flashing. This is the pad that we will be assigning the sample to.

You should open one of the Cubase browsers—often the *Mini Browser* is perfectly sufficient, and this can be opened by going to **Media > Mini Browser** or by pressing **F7**. You can then use this browser to navigate to the sample that you wish to use. If you have a folder of samples, they will be listed, and you can preview each sample by clicking on it. Once you have a sample that you think might work, you just drag it from the browser and drop it on the pad that corresponds to the MIDI note you are using. Given that playback is already happening, you should immediately hear the sample in the context of the track. You can adjust a few parameters within Groove Agent ONE to fine-tune the sound, or, if it doesn't sound right at all, you can go back to your browser and drag and drop another sample onto the pad. If you find a sound that you are happy with, then this is the end of the process.

However, we mentioned in the main chapter that using a sample for drum replacement isn't always ideal if you want quite a natural result, because a single static sample will never be able to replicate the subtleties of a drum performance. Groove Agent ONE can help us in this regard by allowing us to drag multiple samples to the same pad. If you have one sample assigned to the pad and then you drag and drop another sample on to the same pad, it will be added to that pad. When multiple samples are assigned to the same pad, they are split by velocity range, with the most recently loaded sample assigned to the top velocity range. Clicking on the *Voice* button under the *Pad Edit* section will change the main display to show the different layers assigned to that pad. Clicking on *Layer 1*, *Layer 2*, etc. will switch the main display to show the parameters and waveform overview of the currently selected layer. If the layers are in the wrong order, then you can drag and drop the layer buttons to reorder the layers, and then you can adjust the specific velocity ranges for each layer. If you choose your samples well, this ability to quickly add multiple layers can add an extra dimension to the drum replacement process.

CHAPTER 12
Time-Stretching

INTRODUCTION

Time-stretching is the general name given to both stretching (lengthening) and compressing (shortening) audio files. The idea is simple enough but the execution of it is much more difficult. In the days of tape recording (and both vinyl and tape playback systems), you could lengthen or shorten a recording by slowing down or speeding up the playback mechanism. This would not only change the length of the recording but also would change the pitch: slowing down the playback mechanism not only lengthened the recording but also dropped the pitch and vice versa. If the aim was to change the length by only a small amount—perhaps to make sure a certain length to synchronize with some visuals—then the resulting change of pitch may have been acceptable. But anything more than a small change would have probably meant the results sounded too unnatural to use. To give you an idea of what kind of changes to pitch there would be with different changes in tempo/time, the table below shows some playback speed changes and the resulting changes in pitch.

As you can see, the margins for keeping any pitch change to less than one semitone are less than a 10% change in tape speed. In fact, a 5.946% increase in tape speed would result in a one-semitone increase in pitch, whereas a 5.613% decrease in tape speed would result in a one-semitone decrease. So if you needed to keep things fairly close in pitch (and tone), a 5% increase or decrease would probably not be too noticeable, a 10% increase of decrease might be pushing it, and any more would almost certainly result in the instruments or voices sounding unnatural.

The one true benefit of this way of doing things is that the waveform (and therefore the sound) is changed in no other way than simply being sped up or slowed down. The integrity of it remains completely untouched. There may be changes to the way we perceive the sound, though. We have already stated that transients and the attack of sounds can be crucial to the way in which we identify sounds, so if those transients are sped up or slowed down (even though the tonal and harmonic balance hasn't changed), it may have an effect on what

Introduction

we perceive the sound to be. At low amounts of change, this is unlikely to be a problem, but, as the amount of change increases, so do the chances of our misidentifying a sound.

With the advent of digital recording, all manner of new and previously impossible things started to become options, and the ability to separate tempo (or length) and pitch when you wanted to make changes to either was a real breakthrough. The quality of the earliest attempts at time-stretching, while revolutionary, weren't anywhere near as good as the options we have now. Advances in research and psychoacoustics have greatly improved our understanding of the way we perceive sound, and in doing so have improved the quality and versatility of the time-stretching systems (called algorithms). In addition, massive increases in computing power have meant that these techniques first became much more usable and quick in dedicated offline processing (where you have to wait for the result to be calculated), and more recently have become a reality in real-time processing. The best algorithms that we currently have still don't have a real-time implementation because of the amount of calculations required, but it is only a matter of time before processing power catches up and makes even these algorithms operable in real time.

It seems, on the surface, such an easy thing to do. After all, you are just making a sound longer, so how hard can it be? Well, in the most simple cases, it should actually be very easy. If you take a simple sine wave of any given frequency (let's say, 100 Hz) and you have a recording of that that lasts one second, then you will have an easily calculable number of complete waveform cycles. In the case of our 100 Hz sine wave, this would mean exactly 100 complete cycles. If we want to time-stretch that to, for example, 1.8 seconds, then we can calculate that we should have exactly 180 complete waveform cycles. So we duplicate a single cycle, replicate that eighty times to create the additional eighty cycles that we need, add it to the end of the previous file, and we are done.

We can, in fact, do the same with any constant waveform. If we take a recording of a sawtooth wave of 256 Hz that lasts 3.5 seconds, then we know we will have 896 full cycles. If we want to time-stretch this to 10.5 seconds in length, then

FIGURE 12.1
In theory, we could time-stretch audio by repeating single cycles, but, in anything other than a simple waveform, this would be very audible. However, developing this idea into repeating not a single cycle hundreds of times but rather hundreds of individual cycles one (or more) time, each is the basis of some time-stretching techniques.

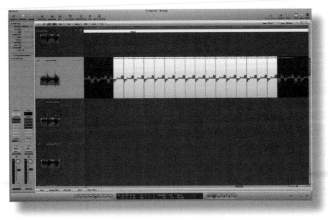

we use the same process but make sure that we end up with 2,688 full cycles. The principle works as long as there are absolutely no frequency, amplitude, or tonal variations during the whole length of the source recording. If we can satisfy all those criteria (as we have demonstrated we can with steady state "pure" waveforms), then time-stretching is a very easy process. But no real-world sounds and none but the most incredibly basic of synthetic sounds conform to that mathematical perfection. They will, over the course of a single note, demonstrate variations in frequency, amplitude, and tone, and as a result, our simplistic time-stretching method falls apart rapidly. We all know it can be done, though, so how *is* it done, and what different methods can we use?

WHY IS IT SO DIFFICULT?

In order to better understand how modern time-stretching works, we should first take look at what happens to the levels and tonal balance in a more real-world example and try to understand why there are problems and how the technology tries to solve them. To start with, and to keep things relatively simple, we can take things a step further than our previous example and start with a sawtooth wave that, over the course of a few seconds, gets gradually filtered down (via a low-pass filter) to a sine wave. The diagram below will give us an idea of what is going on.

In the top row, you can see a series of single cycles of the waveform shape from the unfiltered start to the filtered end. Below each of these is a spectrum analysis that shows the harmonic content at each interval. While our previous sine wave and sawtooth wave examples were static in tonal (harmonic) content, this one has a changing content over time. The lowest frequency (the fundamental) actually does stay constant over the whole duration, but the other harmonics reduce over time. You can also see that the higher frequencies reduce much more quickly than the low ones do. Because of this we can't simply repeat a section of this waveform and paste it on the end to change the duration. We need to look at this in a different way.

Every sound—acoustic or electronic, plucked, hit, scraped, blown, sampled, synthesized, still-in-your-head-waiting-to-come-out—no matter what the source,

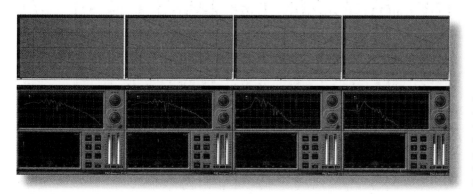

FIGURE 12.2
Even quite a basic filtered synth sound will change over time, as shown in the waveform displays and spectral analysis of the sound over its duration.

it will be composed of a number of different frequencies playing back at once at varying levels. This harmonic signature is what makes each sound unique. In theory, every single sound can be broken down into a finite number of these harmonics (plus a noise component in some cases), and the specific frequencies and levels can be analyzed and tracked over time. If we analyzed the sound in great enough detail, and we had a sufficient number of sine wave oscillators that we could tune and control the level of independently over time, we could, in theory at least, re-create any sound that exists. This serves as the basis not only for some aspects of time-stretching but also for a form of synthesis called Additive Synthesis. For those of you who are interested in a little more explanation of the processes, there is a section on the website called "Additive Synthesis Fundamentals," which will give you an overview of the process.

THE ADDITIVE APPROACH

Analyzing something as simple as a slowly filtering sawtooth wave would not present too many challenges, as each of the harmonics stays at the same frequency and simply decays in volume at a given rate. Also, in this particular case, the frequencies of each of the harmonics are relatively easy to pick out. The lowest frequency of any given sound is called the fundamental frequency, and it is this frequency that we perceive as the pitch of the note. A fundamental frequency of 110 Hz would mean that we hear the note as an A, while a fundamental frequency of 195.998 Hz would be heard as a G. Whether there is just one additional frequency layered over each of these fundamentals or whether there are hundreds, in both cases, we would hear the notes as being of the same pitch.

Additionally, the frequencies that make up a sawtooth wave are governed by a mathematical relationship and are therefore quite easy to extract. The fundamental frequency is sometimes called the first harmonic, and there are a series of numerically related frequencies that can be derived from this. They are sequentially numbered, and each number represents a multiple of the fundamental frequency. So if we have a fundamental frequency (first harmonic) of 100 Hz, then the second harmonic will be 200 Hz, the third harmonic will be 300 Hz, the fourth 400 Hz, and so on. A sawtooth wave is made of up a combination of all these harmonics up to the limit of our hearing. The actual levels of each of these harmonics usually diminish slowly as the harmonic number increases. But in any case, each of the harmonics is present at some level or other. Incidentally, just as a point of comparison, a square wave is made up of the fundamental, and *only* the odd-numbered harmonics (third, fifth, seventh, etc.).

Based on this, it would be relatively easy to determine the fundamental frequency by analyzing the audio file, and then, once that frequency had been established, the frequencies of each harmonic in the series could be calculated and a search for those frequencies carried out. This gives a good starting point, but of course any changes to the fundamental frequency (whether a subtle drift or a noticeable change) would mean that the frequencies of all the harmonics would need to be recalculated, and this would need to be assessed on a sample-by-sample basis.

The next fact that we need to consider, and which we have already spoken about, is that the transients at the start of sounds are crucial to defining the sonic identity of those sounds. We also mentioned that if we stretch a sound, including its transient, then that can cause changes to the character of the sound. Therefore, whatever method we use to time-stretch, it would be very useful if the analysis of the audio file could identify transients and separate those out, so that they aren't included in the stretching process. Naturally this complicates things a little. If our analysis of the sound determined that the first 0.2 seconds of the sound was the transient portion, then we would need to take the first 0.2 seconds of breakpoints and leave those unchanged and then change all of the remaining ones. This will change the multiplication factor for the times of the breakpoints following the transient.

To show what I mean, let's look at an example. If we have a two-second long sound of which the first 0.2 seconds was the transient region, and we wanted to make the sound four seconds long, and we weren't considering the transients separately, we would simply multiply the time of each breakpoint by a factor of two—nice and simple. However, if we are considering the transients, then we have to separate those out, which will leave us with a transient portion of 0.2 seconds and a remainder of 1.8 seconds. To make the whole sound four seconds long, we need to keep the transient portion at 0.2 seconds and make the remainder 3.8 seconds long. So instead of simply multiplying everything by a factor of two, we need to leave the transient portion untouched and multiply all the times for the remainder of the sound by a factor of 2.11 (which will make the 1.8 second portion have a length of 3.8 seconds). It's not a difficult calculation to make, but it is another step in the process.

So all in all, assuming we could get an accurate initial analysis, and subject to a number of very complex mathematical calculations that are beyond the scope of what I am writing here, we can, *in theory*, re-create a sound from a number of sine waves that have very accurate pitch and level control (along with some inharmonic or noise elements), and, by changing the speed at which we move through the different points of these envelopes, we can change the playback speed. So that should give us a very good, albeit calculation- and system-resource-intensive, method of time-stretching. But given that this is so complex, are there any other ways in which to approach the problem from a different angle?

THE GRANULAR APPROACH

One alternate technique, based on granular synthesis, is based on the much simpler idea that any changes that happen to sounds will happen over a period of time, and, if we can divide the sound up into small enough "slices" (the grains referred to in the name), then each one should be constant—or at least as close to constant as makes no difference—in its harmonic content for the duration. If we can reach this point—a series of static waveforms—then each waveform can be treated as we did with our very simplistic sine wave example at the beginning of this chapter. Each individual grain can be looped for as long as necessary, and then playback can progress to the next grain. In the case of a simple doubling of length, then we can see that each grain would be played twice before moving on to the next grain. If, as would be the case a majority of the time, the stretch doesn't involve a simple integer multiplication of length, then some form of cross-fading between adjacent grains would be necessary to smooth things out.

If we consider actually shortening a file or sound using this method, then we have additional things to consider. If we wanted to make it 50% of its length, then we would have a couple of choices: We could play back the grains in the original order but only play each one for half of its length before cross-fading into the next one. Or, alternatively, we could simply choose to play back only every other grain to give us the required length. Each of these has its problems: cross-fading will result in changes to the sound of every grain during the cross-fading period, so no single grain will play back completely throughout, while missing out alternate grains could result in a bigger tonal change between non-adjacent grains than there would have been with adjacent ones. Obviously neither of these is ideal, but, given that we are talking about *very* short periods of time for each grain, the idea is that the fact that there are so many of these every second means that the ear won't hear the repetition or truncation of the individual grains unless the stretching factor is a particularly large one.

We also need to take into account transients again, because these can be quite hard to deal with for granular time-stretching systems. The fundamental principle of granular synthesis and time-stretching is that, if we can make the grains short enough, we will be dealing with essentially static waveforms for each grain. This is quite easy to achieve with sounds that have already settled into their "steady state," but, as we have already discussed, transients consist of a lot of sound energy in a very short period of time and usually distributed over a wide frequency range. As a result, it is much harder for us to find the tonally stable grains that we need. In order to work with transients effectively with this granular technique, we need to treat them differently.

The simplest way to do this is, like before, to have a system for detecting them and to then exclude them from the stretching process. The second technique is

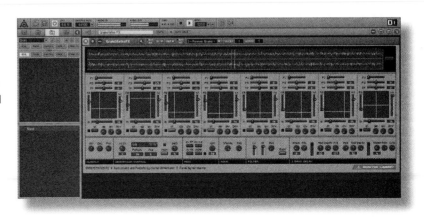

FIGURE 12.3 Granular techniques can not only offer time-stretching capabilities but also allow you to "freeze" time (or move around randomly) in your audio. GrainStatesFX is a Reaktor Ensemble that uses granular techniques for the creation of some truly unique effects.

a little different in that the transients are still a part of the overall process, but we can in effect separate them out by changing the size of the grains for a short period. If our grains are normally just a few milliseconds in length, then it is unlikely that a transient would be captured entirely within a single grain. But, by increasing the length of the grains to, say, 100 milliseconds, it may well be possible to include the whole transient in a single grain and thereby eliminate the problems with the transient crossing between grains. If we also mark this grain as one that should not be repeated even if the stretching would ordinarily require it to do so, then we can bypass the problem. This variable grain length is very useful to process drums with their obvious sharp transients but can also prove very effective on certain vocal sounds, such as words beginning with hard-sounding consonants, such as "t," "k," and "d."

One of the biggest advantages that this granular approach has is that, because it doesn't require analysis of the complete file in advance in order to figure out what is happening over time, it is much more suited to real-time applications. The downside is that, dependent on the material, the results are possibly not as smooth and accurate as the other methods we have looked at. But it is certainly an advantage to have many different methods and technologies at our disposal when we are trying to do something like time-stretching. Although it would be perfect to just have a single algorithm that covered every eventuality with equal quality (from a time and work-flow point of view, at least), having different options that we can try, even if it takes a little longer, is better than having just the one option that doesn't do everything to the same quality.

This is why there are often many different algorithms available for time-stretching. Very rarely (even in the manuals) are deeply complex technical details given of the method used, but, to make them easy to understand and use, they are often given names that refer to the material that they are best suited to: monophonic, polyphonic, percussive, simple, complex, and so on. So, leaving aside the technical details of how it is being achieved, and leaving aside the specifics of which algorithm might be better suited to which task (as the names will probably help with that), let's have a look at ways in which we can use time-stretching in our editing tasks.

HOW WE USE IT AND WHEN WE USE IT

The first thing you need to consider with time-stretching, whichever system you use and whatever you are using it for, is how you are going to define just how much you want to stretch or compress the audio file in question. And how exactly you define that will depend largely on the circumstances. If you are trying to make a mixed piece of music fit perfectly to a thirty-second commercial slot, then you will have a current running time and a desired running time (thirty seconds). In this scenario it makes sense for you to set your stretch amount in terms of absolute time. If, on the other hand, you were working on producing a song, and you had a percussion loop that you wanted to use that was currently at 120 bpm, but your project was at 110 bpm, then you would want to set your stretch amount in terms of tempo. You might have an audio file that is currently

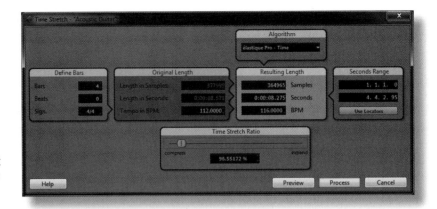

FIGURE 12.4
There are always a number of different ways in which you could define how much you want to stretch or compress our audio by. Most DAWs and plug-ins offer a number of different choices but not all offer quite as many as shown above.

nine bars and one beat long that you want to make eight bars long, and in this case, you would want to set your stretch amount in terms of length in bars and beats. You can also usually use a straightforward percentage amount and a length in samples as ways of determining the stretch amount. So with that out of the way, let's think about the ways in which time-stretching can help us.

TIME-BASED STRETCHES

One of the most basic uses of time-stretching, and one that we have already alluded to, is the process of taking an audio file and making it fit to a required length of time. The example we gave was taking a piece of music that was slightly over or under thirty seconds and stretching it so that it would fit to a typical thirty-second advertising slot. While this is a very common thing to have to do, and, while time-stretching could certainly achieve that, we should really consider if time-stretching is really necessary. If the piece is only a second or two longer or shorter, then it may be better to simply speedup or slow down the piece by changing the playback rate. This will change the pitch as well, but, if we were talking about a small amount, then it would probably be better to do it this way.

Most time-stretching software is designed to separate the time and pitch aspects from each other and to allow us to adjust one without changing the other, but, for situations like these, perhaps, they usually give us the option to link the two again and change the length by adjusting the playback rate like you would on a tape or record deck. So if the change is a small one, and *if the absolute pitch isn't important*, then this could be the best-quality result. I mentioned the absolute pitch simply because, in a situation like an advert spot where the advert will play and then stop and then another one will come on, there is no need for the pitch of the piece we are playing in the advert to match up to anything either side of it. If, on the other hand, we are trying to match up audio to a defined length in something such as a film score where there could well be music either side and any small changes in pitch will be noticeable, then this might not be the better choice.

Staying with film scores for a moment, you may have a particular piece that was scored and recorded to match a scene in a film, and that scene (and therefore the piece) runs to fifty-eight seconds long. However, owing to a last-minute change, perhaps a reedit to meet a ratings requirement, a few of the shots within the scene have had to be cut. There is another musical cue in the preceding scene and one immediately following this one. The scene has now been reduced to just fifty-three seconds, and you have to fit the file to the new length. It may be that, during the whole of the fifty-eight seconds, there are certain notes or possible certain sound effects in the musical piece that match up with on-screen action. Given that some of the shots have been edited out, it will mean that, relative to the total running time of the whole piece, those synchronized sound effects will now be in the wrong place. In this kind of situation, it makes sense to split the piece into a number of different regions, perhaps cutting the piece into regions with these sound effects as the split points. Each of those regions that start with an effect could now be lined up with the visuals, so that they were in the right place and then each of the regions stretched to the right length to make the piece flow properly.

TEMPO-BASED STRETCHES

While many of the situations where you might want to use a time-based stretch are post-production or music-to-video related, tempo-based stretches would probably be more applicable while in the process of recording or producing/mixing a song. There may be a drum or percussion loop from a sample library that you want to use in the project you are working on but the tempo is wrong. If, for whatever reason, you didn't want to recycle it, then you could simply time-stretch it. If you knew what the original tempo was, then you simply enter that and the tempo that you want it at, and the stretch should bring it into line.

Another good use for tempo-based stretches is to create half-tempo variations of audio files or, if you are slightly more adventurous, it can be very interesting to increase the tempo by 50% (for example, from 80 to 120 bpm), as this can take quite a straightforward rhythmical phrase and create a file that, when laid over the top of your other rhythm parts, creates a polyrhythm that can add a lot of interest. It is usually better to try this with things such as hi-hats, tambourines, shakers, and other similar percussion sounds. This is simply because, if you chose something like a snare drum or a kick drum, the polyrhythm created might end up a little too distracting, but with things such as hi-hats, the result can be a lot more subtle.

Creating alternate rhythms is something that can easily be done with Recycle and other software like that, but sometimes stretching audio files like we are describing here can give different results, as not only is the actual rhythm changing but, in the case of quite substantial changes such as a 50% increase or decrease, can actually lead to changes to the sound. The artifacts that we have said are an inevitable part of all time-stretching can prove to be useful sometimes, and choosing the "wrong" algorithm can also have its place, as the resulting tonal change and effect can (sometimes) be interesting and add another tonal texture, especially if used subtly.

Fundamental, though, to be able to create these tempo-based changes is to actually know what the tempo of the source file actually is. It might sound like a really stupid thing to say, because if you are recording the audio yourself, to a click track, in a project that you have set up, then of course you will know the tempo, but there are a lot of times when files might be sent to you without any indication of tempo in any written form and sometimes without a huge amount to go on to determine the tempo from the files you have been sent. If you have received any kind of drum track, then the tempo is usually relatively simple to work out. If, on the other hand, you have just been sent a vocal, then it can be more time-consuming.

LENGTH-BASED STRETCHES

If you have recordings that you know are of a certain length (in bars) but don't know the tempo, then there is another way in which you can set up the time-stretching factor. If, for example, you have an audio file that you know is sixteen bars long, and yet, when it is imported into your project, it has a different length—for example, fourteen bars and one beat long—in your Arrange window, then it is most likely because it was recorded at a different tempo than the one you are currently using. If you already know the source tempo, then you could apply a simple tempo-based stretch to make it fit. If you don't know the original tempo, the simplest method is to actually just specify the length (in bars) that you want it to be in the time-stretch parameters. You don't need to know the length that it currently is, because that will be worked out for you.

Another way this can be used is to take nonmusical/rhythmic audio clips, perhaps ambient sound recordings, and make them conform to a certain length in bars. This kind of thing is often done to match ambient sounds to a certain length in seconds for audio-visual purposes, but there might be occasions when you want an ambience to last a certain number of bars instead, and being able to stretch to a length in bars without having to work out exactly how long 16 bars is at 104 bpm is a definite plus.

These uses are all very practical and will be used by most of us at some point or another, but, in the world of everyday audio editing, it is far more likely that we will need to time-stretch on a *micro* rather than *macro* level, and at that point, all these different ways in which we can numerically quantify the stretch amount become a little irrelevant.

FILLING IN THE GAPS

Most of the uses for time-stretching that we have mentioned so far have been based on the idea of taking a recording and making it fit to a certain length for the purposes of synchronizing files either to video or to a project tempo that you are working on. These tend to be stretches to a whole file that is already pretty much how we want it except for its length or tempo. But there are many occasions, such as when we are comping vocals and making small changes to the timing of individual parts within an audio file, when we might want to stretch just a part of a file. The process is, of course, exactly the same, but the way in which we approach it is slightly different.

One of the biggest difficulties that we will have in doing this is to know exactly how much we have to stretch or compress a region by in order to fill in any gaps or remove any overlaps. If we are stretching a file to change its tempo, or if we are stretching it to make it fit a certain length (either in time or in bars), then the stretch amount is relatively easy to figure out. But if we are looking to just stretch a file or region by just a small amount in order to fill in a small gap, then the actual figures aren't quite so easy to ascertain. If we cut a file into two separate regions and then move one by a small amount (say, 1/40th of a beat), then we have to adjust the length by a corresponding amount. This is easy in principle, but it relies on us to keep a note of exactly how much we have moved it by. Given that we may move a file or region by tiny amounts more than once in order to establish a final position, keeping track of the exact amount that we have moved it by can be hard work. Naturally if you have performed a number of these changes in position, there would be a huge amount of work in calculating the exact length that each region now needs to be to avoid any gaps or overlaps.

Fortunately, almost all DAWs now give you an easier way to achieve this. Instead of having to manually enter numerical parameters for the stretch amount, you can simply click on the end of the region you want to stretch, hold down a modifier key (this varies by DAW), and then drag the mouse and the region will change length on the screen. Once the end of the region is in the position that it should be in order to avoid gaps or overlaps, you release the mouse, and the time-stretch amount will be calculated. In practice this means that you can cut an audio file into regions and move them into the positions that you want, and then, by clicking and dragging the end of each region (with the modifier key), you can stretch each region to the required length without ever leaving the Arrange

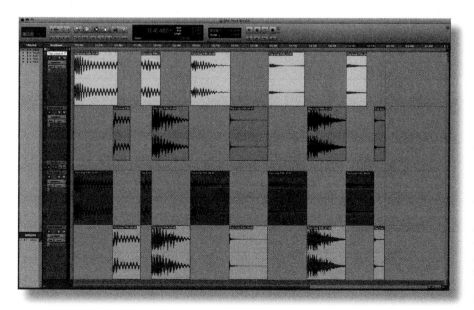

FIGURE 12.5
If we have a number of regions that start in the correct place but whose lengths don't match nicely to the gaps that we need them to fill, being able to click and drag to stretch them to a particular length can remove the need for a lot of figuring out of current and new durations for each region. Just drag and go!

window. Anybody who has ever cut up and moved audio regions in this way will appreciate the huge amount of time that can be saved by doing things this way.

"MANUAL" TIME-STRETCHING

In cases where you might need to stretch a sound by only a very small amount, or in cases where the time-stretching artifacts are unacceptable, there is another option, but it needs to come with a warning that it is time-consuming and doesn't always work as hoped. Nonetheless, it is worthwhile including. What I am in fact referring to takes us back to the beginning of the chapter, when we spoke about repeating a single cycle of a waveform. This idea is, in essence, granular time-stretching but done manually. By identifying a regular pattern in the waveform, cutting out a section of this regular pattern, duplicating it, moving the rest of the region to the right, and inserting the copied section, we can in theory carry out time-stretching. If the edits are done at zero-crossings, then it might be possible to do it without cross-fades, but they might well be needed.

It is very hard to tell just from looking at the waveform what length of region may be needed. If you zoom in enough you might see a cycle that will be related to the fundamental frequency, but a single cycle edit like this is rarely enough. So you will probably want to cut out a section of at least a few of these single cycles and possibly even more than that. The greater the length of the region that you choose to use, the more you will need to take into account any tonal or level changes. So the best bet is to get a rough idea of how much you need to insert, and then zoom out, so you get a good overview of the whole section. From there you should be able to see any cyclical variation due to amplitude changes in the waveform. Try to cut out a section that goes from "peak to peak" of one of the amplitude cycles. At that point you can zoom in and make sure that the edit points are zero-crossing edits at the beginning of a frequency cycle as well. By doing this, you stand the best chance of getting a good transition on the repeats of the section you are using.

There are a number of benefits to doing things this way, not the least of which is the fact that you can be very selective about which part or parts of the file you choose to stretch. In addition, there is a possibility that you could extend a held vocal note with vibrato without actually changing the speed of the vibrato. The overriding benefit, though, is the fact that you aren't actually stretching anything at all, so there are no artifacts involved. Naturally this comes at a price. The fact that we are repeating sections of the audio file means that there is a chance that the file will now have a slightly unnatural feel to it. This could range from an overly repetitive tonality in a vibrato to a more "metallic" overtone if the region cut is very short.

It may take a little work to get a good result with this technique, and it may not always give a perfect result, but it is certainly capable of sounding very smooth in the right situation, with time taken in choosing the right section to loop and with consideration given to cross-fading between each section. It isn't as immediate as using the time-stretching features built in to your DAW, but it does allow for indefinite stretching of regions (subject to the caveats already mentioned). For this reason, it is a technique that I would recommend you try for yourself and get used to, because you never know when it might come in handy.

HANDS ON
Introduction
Time-stretching is one of the more useful abilities of a modern DAW, simply because it allows us an extreme amount of flexibility in manipulating audio recordings to fit the often-changing demands placed upon us as editors. We may have a project that is for synchronization to a film clip. The original cut of the clip may have been a particular length, but then, owing to a create change or even a cut owing to censorship, there may be a need to change the length of that piece. Time-stretching makes this a very simple process, and the algorithms in use today are capable of amazing results. In the following sections, we will take a look at the traditional "static" (insomuch as a fixed-length change that won't automatically follow tempo changes) time-stretching tools in our four DAWs and look at some of the shortcuts to implementing them.

Logic
Logic, like most DAWs, has a number of different ways in which you can specify the amount by which a region should be stretched, but, broadly speaking, these can be broken down into the numerical and the visual. The numerical methods offer a number of project-independent ways to control things, and arguably offer more control in more situations. That's not to say that the visual methods aren't valuable, though. In certain situations, they can be just as effective and perhaps even more so, and we will look at them shortly. First, let's take a look at Logic's central time-stretching "brain" and destination for numerical stretching: the *Time and Pitch Machine*.

With an audio region selected and the Sample Editor window open, you can access the Time and Pitch Machine either by going to the Sample Editor window and choosing the Factory > Time and Pitch Machine or, if you prefer the keyboard shortcuts, Ctrl + P. Either of these methods will open the Time and Pitch Machine dialogue box, where you will be presented with a number of different options that we will look at in sequence.

The first of these, *Mode*, is used to choose whether you wish to time-stretch audio in the modern or traditional senses. What I mean by this is that the first option, *Free*, gives you control over the length of the audio that is completely independent of the pitch—in other words, modern time-stretching. *Classic*, on the other hand, replicates the effect of speeding up or slowing down a tape player, and any change in length with be accompanied by a corresponding change in pitch.

FIGURE 12.6
Time and Pitch Machine in Logic offers a very simple but well-specified way of parametrically stretching audio. On the right of the image above is the list of different algorithms available (including some third-party ones available that integrate into Logic itself).

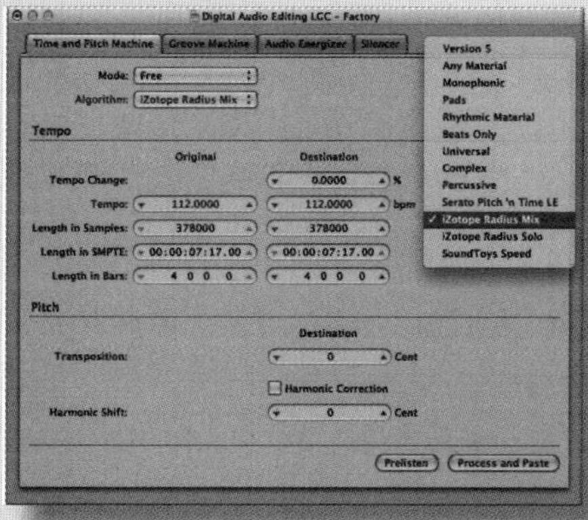

The next option for us is the choice of *Algorithm*. In the main chapter above, we discussed the reasons why different algorithms exist, so we don't need to go over that again. All we need to know is that this is the place where we get to choose the most appropriate algorithm for the material that we are stretching. By default Logic offers a choice of nine different algorithms, each tailored to a specific purpose. A short description of the uses of each of these algorithms is provided on pages 577–578 of the *Logic Pro User Manual* if you would like to know more. In addition, Logic supports third-party tools that, if purchased, will enable Logic to use additional algorithms inside of the Time and Pitch Machine without having to load up additional software.

Once you have chosen the best algorithm, you then need to define the amount or stretching or compression that is required, and this is where you have a number of choices. I say that, but the truth is you have *one* choice to make: how much to change the region. The real point is that you have several ways of defining your answer, depending on how you define the question.

If you know you need to increase or decrease the length by a percentage, then you can do this either by entering a numerical value by double-clicking in the *Destination* box to the right of *Tempo Change* and typing in the value that you want (positive numbers for stretching and negative numbers for compression) or by using the up and down arrows. Like most other Logic dialogue boxes, these work by adding or subtracting a set amount. By default this will be 1% at a time, but if you click and highlight any single digit and then use the arrows, then only that digit will increase or decrease by 1 each time.

Alternatively if you know the source tempo and the tempo you need the region to be at you should use the Tempo Change option. You can set the *Original* and *Destination* values either by double-clicking or by using the arrows. This method is ideal if you want to repurpose existing material (either your own or a sample library) to fit in a new project.

The final three options all relate to absolute lengths: length in samples, length in SMPTE (minutes and seconds), and length in bars. I have personally never used the *Length in Samples* option, simply because "samples" are not a unit I ever work in, and I have never known anybody to work in terms of "samples," but if there is ever a need for you to have an audio file or region that is a certain number of samples long—you can do it here. Exponentially more useful are the final two options. Which of these is easiest depends a lot on the context you are working in. If you are trying to synchronize audio to picture, then an SMPTE basis would be more useful, as you specify the length that you need in the units that you are working in. Equally, if you are working on a musical project, it is more likely that you will have your frame of reference as bars and beats, so this option would be better.

Whichever method you choose, the *Source* will automatically default to the current length or project tempo, so in most situations, you will need to enter only the Destination value that you need. You then simply hit *Process and Paste*, and the newly stretched audio region will take the place of the original one in the Arrange window.

Moving on now to the visual method that I mentioned earlier, what we really have is just another way of defining the amount of stretching that is required. To do this you simply move the cursor to the bottom right corner of the region you wish to stretch, and you will see the cursor change to what looks like a right-handed square bracket with arrows on either side of it. If you click and hold down the mouse button, you will see a pop-up that says "Length Change." This is the tool that you would ordinarily use to truncate or extend an audio or MIDI region. However, if you hold down the Alt key, you will see the cursor change and the left and right arrows swap sides. If you now click and hold down the mouse button, you will see that, instead of "Length Change," it now says "Stretch." What you can do now is simply drag the end of the audio file to the new position where you would like it to end, and it will automatically carry out the stretch for you (using the most recently selected algorithm) without your needing to specify a desired tempo or length.

This method makes it incredibly easy to stretch a number of files of different tempi to the same length quickly, but it does come with a couple of caveats. The first is that, in the case of time-specific files (ones with a defined bar and beat length), it helps to make sure that the source file is trimmed to a set number of bars and beats, with no silence at the end. This is simply so that it makes it easy to know how far to drag the file. If you know that the source is four beats and two bars long, but you don't know the tempo, you just drag the end of the file until it is four beats and two bars long in your Arrange window. The second, perhaps more problematic, thing to consider is that, unless you are stretching to an easily defined length (whole beats or bars), it can be difficult to get the exact required length right using this method. But in any case, in certain situations (such as stretching a number of nicely trimmed loops of different tempi to fit the project tempo, using the same algorithm), this method can be the quickest and most efficient way to do it.

Pro Tools

There are a number of ways to carry out static (as in not following the tempo or being freely moveable) time-stretching operations, but the most useful to us will be the simple numerical percentage stretch factor and the more visual drag-to-stretch approach. Let's start by looking at the numerical approach before moving on to the visual one.

To be able to work with the numerical method, you will need to be working on an *Elastic Audio*–enabled track, and the choice of algorithm plays a very significant role in the final quality of results. Pro Tools has five different Elastic Audio algorithms (*Elastic Audio Plug-Ins* in Pro Tools terminology), each of which is tailored to a particular kind of audio, and four of which operate in real time while the fifth is for rendered processing only. There is a more detailed overview of the different algorithms in the Pro Tools manual (pages 868–870), but in essence, they should be used in the following ways:

- *Polyphonic*—This is very much a general-purpose algorithm that is suitable for anything from drum loops to full mixes. It may not be the best algorithm for more specific purposes, but it is a good algorithm for general instrumental material.
- *Rhythmic*—This algorithm is designed to work particularly well on percussive material or anything with large numbers of clearly defined transients and aims to keep the tightness of timing and the tone of the sounds intact.
- *Monophonic*—Solo instruments (including voice) are the focus of this algorithm, as it contains formant-correction processing to maintain the tonal integrity and character of the processed sounds.
- *Varispeed*—This algorithm is more akin to the time-stretching/pitch-shifting techniques of speeding up a tape where pitch and tempo (and therefore time) are linked. Any time-stretching carried out with this algorithm will result in an accompanying pitch change.
- *X-Form* (rendered only)—This is the highest-quality algorithm available natively in Pro Tools and is actually based on the algorithms used in iZotope Radius. It is a great all-rounder, but, because of the complexity of the algorithm, it can be used only with rendered processing.

Once you have chosen the most appropriate algorithm, you can then enter the percentage stretch/change that you want to apply by opening the *Elastic Properties* dialogue (Alt + Num Keypad 5, Clip > Elastic Properties, or right-clicking on a region and choosing *Elastic Properties…*) and either clicking and dragging up or down on the *TCE Factor* box or by clicking and typing the value in (correct to two decimal places). This is a very useful and quick way of effecting a time-stretch if you happen to know the percentage change that is required, but that isn't always an immediate or convenient number, so we have an alternative way to do things as well.

Building on this idea, and allowing us more flexibility, is the offline (rendered) *AudioSuite TimeShift* plug-in. This can be found by going to AudioSuite > Pitch Shift > TimeShift. When the plug-in window opens, you will see four separate areas: *Audio*, *Time*, *Transient/Formant* (depending on which algorithm is chosen), and *Pitch*. Looking at the *Audio* section first, we can see a *Mode* selection that, when clicked, corresponds to the four real-time algorithms that are used by Elastic Audio. Choosing either *Monophonic* or *Polyphonic* here will also give us access to the *Range* control. This can be set to *Low*, *Mid*, *High*, or *Wide*, and this can be used to focus the algorithm on certain frequency ranges depending on the audio material being processed. There is also a *Gain* control with *Level* and *Clip* indicators to make sure that your audio isn't clipping going in to the process.

The next section, *Time*, is the most important to us here, because it allows us to set a stretch factor based on a number of different criteria. Unlike the fixed percentage/ratio that we have just looked at, this section allows us to calculate our stretch amount by entering a desired length under *Processed Length* in either bars and beats or minutes and seconds (along with time code, feet and frames, and samples—all chosen from the *Units* box on the right of this section) rather than having to make the calculations ourselves. Entering a value for the desired length will automatically update both the *Processed Tempo* and *Speed* displays to reflect the change. Equally, though, a desired tempo can be entered or a simple ratio as we looked at above. This one plug-in allows us to change the length of a region based on four different units of measurement and is, as such, very powerful.

When *Polyphonic* or *Rhythmic* modes are chosen, there will be a third *Transient* section visible, while choosing *Monophonic* will change this to a *Formants* section. The controls in these sections are quite simple but can be used to fine-tune the response of the algorithms, and more details are provided in the manual. The final *Pitch* section, while very useful, is a matter for another chapter, so we can skip over that for now.

FIGURE 12.7
The Time Shift AudioSuite plug-in is the place to go in Pro Tools for offline (rendered) time-stretching. Pitch-shifting abilities are also integrated into the plug-in, but both the pitch and time aspects can be treated separately.

Once you have chosen the settings you need, you can click on the *Preview* button at the bottom left of the plug-in window, which will give you a real-time preview of the results. If you change a value while the preview is playing, you should hear the change immediately. Once you are finally happy with the settings, you just need to click *Render*, and the results will be rendered to a new audio file.

The other alternative is to employ a more visual approach to time-stretching. This works best with regions that you know are an exact length (most usefully in full bars or, at the very last, full beats), which you can then drag to the equivalent length in your project. To do this, you need to use the *Trimmer* tool in *TCE* mode to drag the end of the file to the required length. To select the *TCE* tool, you can click on the *Trimmer* tool, and a pop-up menu of three different options will appear. The one we are interested in is the middle one: *TCE*. Alternatively you can right-click anywhere in the main project window and choose Tools > Trim Tools > TCE, and the same selection will be made. If you are working on a track that is already Elastic Audio–enabled, then the TCE tool will use the Elastic Audio algorithm already specified for the track.

You don't, however, have to have Elastic Audio–enabled on a track in order to use the *TCE* tool. As with Elastic Audio–based stretches, there are a number of different algorithms, but these are tucked away in the *Pro Tools Preferences* dialogue. Open the dialogue by going to Setup > Preferences and clicking on the *Processing* tab. At the top right, you will see a section called *TC/E*, and this is where you can make changes to the *TCE* algorithm used. If you choose the *TimeShift* option, then you are using the same algorithms as the process we described above, and so it will yield similar sonic results. The disadvantage here is that, when using the *TCE* tool, you can fine-tune the parameters in the same way as you can in the *AudioSuite* plug-in. Underneath the *TC/E Plugin* box, you will also see a *Default Settings* box, which, with TimeShift selected, gives you a number of different options for the type of material that you are processing, which, when selected, will make minor adjustments to the settings in order to optimize the results based on the type of material.

Once you have set this up, you are ready to start stretching. As I alluded to earlier, it is easiest to do this on regions that you are know are a whole number of bars or beats long, because you can use *Grid* mode and then set the resolution to either bars or beats, and, assuming that your region starts accurately on a bar or beat, the movement of the TCE tool will be limited to whole bars or beats (as chosen by you); this will help you to set the exact length that you require.

Another very useful thing to be able to do is to fill in gaps between regions (perhaps as a result of using *Beat Detective* or a similar process) by stretching regions to fit. You can use the *TCE* tool to do this, and it is very quick and easy, but it can be difficult getting the length set exactly right, as it is unlikely that all of the regions will line up perfectly on bar or beat markers. Fortunately there is a way around this. If you hold down Ctrl[Mac]/Start[PC] and then click on the end of a region using the *TCE* tool, you will be able to extend it all the way to the start of the next region but not any further. This can make it a lot easier to use this process to fill in the gaps, because you can just click and drag regions and not have to worry about the exact position that you are dragging to.

Studio One

On the surface, Studio One doesn't look to have as many time-stretching options as some other DAWs, but, owing to the ways in which it deals with audio, it really doesn't need as many. Most DAWs provide options to deal with time-stretching by inputting a source tempo and a destination tempo. Studio One circumvents this somewhat by allowing you to enter an original tempo for any audio file, at which point it will automatically be time-stretched to match the tempo of the current song or project. To do this, you need to open the *Inspector* panel by pressing F4 or going to View > Inspector and then entering the correct tempo in the *File Tempo* box toward the bottom of the Inspector panel. Once you have done this, you also need to make sure that the *Tempo* parameter (at the very top of the Inspector panel) is set to "Timestretch," and then the imported region should be in sync with your project. As well as the obvious benefit of being a hugely simple way of stretching file to a given tempo, this also has the very significant advantage of meaning that all of your audio files will follow any changes in tempo without any further intervention from you.

Of course this does rely on you actually knowing the tempo of the files that you are importing, which won't always be the case. In the event that you don't know the tempo of the file but do know its length (in bars and beats), then you can import a region and then manually time-stretch it to match a given length in the Arrange view. If, for example, you imported a file that you knew was exactly two bars long, but, because of the tempo difference, it looked to be two bars and three beats long in your project, you can stretch it by making sure that the *Arrow* tool is selected (press 1 or the Arrow tool button in the main toolbar) and then moving the cursor over the right hand region boundary. The cursor (that we use to adjust the length of a region) should change to a vertical bar with left and right arrows on either side, but if you hold down Alt while this cursor is showing, the left arrow will change to a small clock, and you now have the ability to drag the end of the region and time-stretch it. In our example, you would drag the end of the region from being two bars and three beats long to exactly two bars long and, irrespective of what the original tempo was, it will now be at the correct two-bar length. Once you have done this, you can go to the Inspector panel and enter the current project tempo in the File Tempo box, and, even though that wasn't the original tempo of the region, now that it has been stretched to its correct length, you can define current tempo as being the original tempo, and the region can be made to follow the tempo changes like any other region does.

FIGURE 12.8
The Inspector Panel in Studio One has a number of different options for time-stretching algorithms as well as a place to define the original tempo of the material (if known) and a "Speedup" factor that can stretch or compress the audio by a fixed percentage in addition to any change that may have been affected automatically by following the tempo of the song.

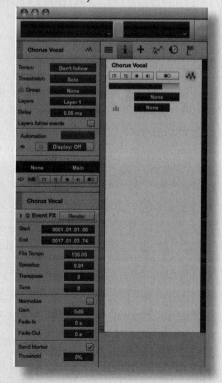

If you use this manual stretching method and then look in the Inspector panel, you will see a parameter, under File Tempo, called *Speedup*. This is essentially a simple playback speed ratio where a value of 1 represents 100% playback rate (130 bpm for example), a value of 1.5 would represent 150% playback speed (195 bpm), and so on. If you have used Alt and clicked and dragged the end of a file to stretch it, then there will be a value other than 1 (the value for an unstretched file) in this box. You can, if you wish, enter a simple playback speed percentage (to two decimal places) in the Speedup box, but, unless you have a very convenient ratio, it is often easier to use the visual method. However, if you have a whole collection of regions that you know are at the same source tempo and you work out the Speedup value for one, then you always have the option of typing in the value for the others rather than using the drag method, if you prefer to work that way.

One other advantage of using the manual drag method or the Speedup method is that it can apply a stretch to an individual region even if that region is just one instance of a particular audio file. If, for example, you import a drum loop from a sample library that is at, say, 110 bpm into a project at 100 bpm, set the File Tempo to be 100bpm, and then make multiple copies of that region throughout your song; each copy will be tied to the File Tempo that you specify, and changing the File Tempo for any one region will change the tempo for all of them, as they all have the same source file. So if you wanted to slow one copy down for some reason, then you could bounce just that one region to a new file (Cmd[Mac]/Ctrl[PC] + B or Event > Bounce Selection) and then stretch that one region separately; alternatively, you could use either the drag method or the Speedup parameter to create a different playback speed/tempo for just that one region, even though the source file is the same as for the other copies of that loop.

If you do bounce a new region, then the source tempo of the new file will be the tempo that it was bounced at. There is no real advantage in doing this for the project you are working on (other than for a particular purpose such as the one we just described), but it might be useful if you wish to create a new version of a particular audio region for purposes outside of the current project (or for use in another DAW).

No discussion about time-stretching would be complete without mentioning the fact that Studio One, like most DAWs, has a number of different algorithms available for time-stretching and Elastic Audio purposes. And like most DAWs, these algorithms are named in line with the kind of material they are best suited to. Here are the four different options that Studio One has for time-stretch algorithms and the kinds of material they are most suited to:

- *Drums*—This algorithm is optimized for drums and percussion or, in fact, any material that is transient heavy. By focusing on transient preservation, this algorithm will give the best representation of these kinds of sounds and will be most likely to preserve the timing of the three main time-stretch algorithms.
- *Sound*—This is very much the general-purpose algorithm that Studio One offers. It gives a good balance between *Drums* and *Solo*, and it the most suited for complex polyphonic material and mix stems or complete mixes. It doesn't have the advance transient awareness of Drums or the formant awareness of Solo, but it a very good algorithm for most material.
- *Solo*—Designed to be used on clearly recorded monophonic instruments or voice, this algorithm is very much format-aware and will apply automatic format correction when used for pitch change/transposition (more on this in Chapter 14). Like all three of the main time-stretching algorithms, you can use this algorithm to process any audio, but the results are optimized toward the monophonic.
- *Audio Bend*—This final algorithm is focused specifically on the Audio Bend processing that studio one offers. It actually separates the transients out from the remainder of the signal, so that, when moving the Bend Markers around, the duration of the transient portion remains untouched, and only the post-transient portions of the sounds are stretched or compressed to accommodate the length change as a result of the Audio Bend.

To change which of these algorithms is used on any given track, you need to open the *Audio Bend* panel (View > Additional Views > Audio Bend or press the *Audio Bend* button on the main toolbar) and select the desired algorithm from the *Timestretch* control under the *Track* heading. This algorithm can be changed at any point, and, depending on the material, the differences can often be quite noticeable. While the algorithm names are a good guide, and while in many cases, the material you are stretching will lend itself to one of the three main algorithms, there is no harm in trying a different algorithm, as it is all nondestructive, and it could yield a slightly better (or even pleasantly unexpected) result.

Cubase

Cubase has quite a few different options when it comes to time-stretching. Some are destructive and parametric in nature in that you have to enter in a numerical value for the stretch amount, while others are nondestructive and are based on visually selecting the stretch amount. Both have their place, and it would be impossible to say that one method is inherently better than the other, but there will be times when one is more applicable than the other. We will start by having a look at the destructive, numerical methods and then move on to look at the nondestructive methods. But before we do, that we will have a quick look at the different choices of algorithms that Cubase offers, as these algorithms are shared between both methods of time-stretching, and there is a degree of overlap between the two.

There are three main algorithms available for time-stretching, and each of these has variations that are focused on processing certain kinds of sounds. The first main algorithm is *Élastique*, and this has three modes—*Élastique Pro*, *Élastique Pro Formant*, and *Élastique Efficient*—and each of these is then further subdivided into *Time*, *Pitch*, and *Tape* variants. Both of the Pro modes provide the best quality, while Efficient is set up to provide a slightly lower CPU load (important in real-time processing) at the cost of a slightly lower quality. Of the two Pro modes, the Pro Formant mode is the same as the Pro mode, only with built-in formant preservation (more relevant when pitch-shifting, which we will look at in a later chapter). The Time variant places more emphasis on maintaining rhythmic accuracy, while the Pitch variant, naturally, places more emphasis on maintaining the correct pitch. The Tape variant works by linking tempo and pitch, so any increase in one will create a corresponding increase in the other and vice versa. The choice of which of these algorithms to use is often quite a simple one. Efficient, even though it is still a very good algorithm, should be avoided, unless CPU load is really an issue for you, and the choice between Pro and Pro Formant really depends on the choice of material. If you are processing acoustic sounds, then Pro Formant is probably a better choice, as it will deal with the natural resonances in acoustic instruments, but for electronic sounds (which don't have any natural resonances), you could probably stick to Pro. The choice of variant is pretty simple as well. If you are stretching rhythmically complex material, or any material with very clear and sharp transients, then the Time variant would be the obvious choice. For solo instruments such as voice, then Pitch would be the better choice. Tape is something that can be useful, but it takes away the naturalness of the result, and would, as a result, most commonly be used as a special effect.

The second main algorithm, *MPEX*, also has a number of different variations designed for different purposes. Of these, *Preview Quality* is, as the name implies, meant solely for previewing purposes and will not give the best results. *Mix Fast* is also meant as a preview algorithm but optimized for complex material and whole mixes. Both *Solo Fast* and *Solo Musical* are algorithms meant for solo instruments, with the Musical version giving a higher quality at the expense of a higher CPU load. Equally, *Poly Fast* and *Poly Musical* are optimized for more-complex musical material, with Musical again being the better-quality option. Finally, we have Poly Complex, which is the best quality version of the *MPEX* algorithm but which also comes with quite a heavy CPU load. This is the best setting to use for complex or important material, and is perhaps best applied in an offline, destructive way to avoid loading your CPU too heavily with a real-time stretch. Additionally, when we come to look at pitch shifting, you will see that there is also a Formant version of each of these variations that will preserve formants during the pitch-shifting process.

The final algorithm, *Standard*, is the most CPU efficient of the algorithms, as it is designed for real-time processing. There are seven variations here, which are named according to their intended uses. *Drums*, *Plucked*, *Pads*, *Vocals*, *Mix*, and *Solo* are all optimized for particular instrument types, while the final option, *Custom*, is the only one of the algorithms that allows you to change the parameters. You can adjust parameters for *Grain Size*, *Overlap*, and *Variance*. This type of granular time-stretching is capable of both very natural-sounding results and some stunning and otherworldly results, and, by adjusting these three parameters, you will be able to get quite a lot of variation in the final results. If you have the time to experiment, then it is definitely worth giving this mode a try.

So, moving on to the actual time-stretching process itself, the destructive stretching process is started by going to Audio > Process > Time Stretch. The window that opens up includes a lot of different options and a lot of ways of defining the actual stretch amount that you require. If you know the duration (in bars and beats) and the time signature of the file you are stretching, then you can enter that on the left-hand side under *Define Bars*. Doing so will then adjust the *Tempo in BPM* value under the *Original Length* heading. Alternatively, if you know the original tempo, you can just enter the tempo here. You can then choose the algorithm that you wish to use from the full list of algorithms above.

Then we come to the *Resulting Length* and *Time Stretch Ratio* sections. These are the important ones that allow you to define how much you wish to stretch the file by. The length in samples and in seconds of the source file will already be known by the software, so you can define the new length as a length in samples or as a length in seconds if you desire by clicking in either the *Samples* or *Seconds* boxes under the Resulting Length heading and typing the new value that you require. If you are working to a particular tempo, then it might be easier and

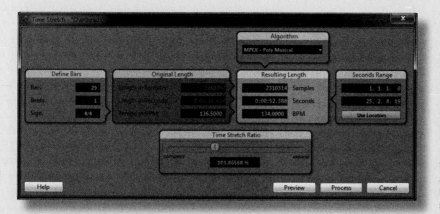

FIGURE 12.9
The Time Stretch window in Cubase gives you a large number of different ways in which you can define the amount you wish to stretch by, and all the different methods are linked. If you change the length in seconds, then the BPM, length in sample, length in bars, and beats and ratio will all update as well.

more useful to simply enter the resulting tempo that you wish to have by double-clicking on the *BPM* box in this section instead. Or, if you need to have a specific ratio of stretch, you can use the Time Stretch Ratio box at the bottom of the window to enter a stretch percentage between 50% and 200% of the original. Changing any one of these options for the resulting length will automatically update the others.

When you have entered the values that you wish to use, you can click on the *Preview* button to get a sample of what the result will be, and you can then click on the *Process* button to finish things up. You could receive a warning stating that the project is using the same audio material and giving you the option to apply the processing only to the selection by creating a new version. Clicking on *New Version* effectively renders the stretch result as a new file, while just clicking on *Continue* in this warning box will write the changes to the existing file, which will affect not only other regions in this project but also, potentially, other projects. If in doubt, you are probably safest to render the results to a new file each time.

The major advantage of this destructive method is that it places no load on your CPU at playback time, as there is no real time-stretching happening. The biggest downside, of course, is that the audio is then fixed at that tempo, and if you wanted to try a different algorithm, you would have to *Undo* the process and carry it out again with a different algorithm, which doesn't lend itself particularly well to experimentation with either song tempo or algorithm choice. If you apply the real-time methods of time-stretching, then you don't have either of these problems, as not only are the methods used more flexible, but, more importantly, any changes that you make to settings or algorithms will always be applied to the original source file without the need to undo or go back through other processes.

We have already stated that the process is a very visual one, but it is also an incredibly simple one. To time stretch an audio file using this method, all you need to do is select the *Sizing Applies Time Stretch* version of the *Object Selection* tool (either by clicking on the down arrow below the main Object Selection tool and choosing a different version or by pressing 1 repeatedly to cycle through the different Object Selection tool versions) and then move the mouse to either of the bottom corners of the region that you want to stretch. When you do sot the cursor will change to left and right arrows with a clock face below. This indicates that you are ready to resize the region by time-stretching it. Just click and drag the end of the region to the position that you want it to be in, and the stretch will be applied. If you wish to change the algorithm that is being used, you just double-click on the stretched region to open the Sample Editor window and then, at the top of the window near the right-hand side, you will see an option to change algorithms. You can change the algorithm in real time while the file is playing, and this enables you to perhaps change the algorithm in order to get a better result.

CHAPTER 13
Elastic Audio (Time)

DEFINITION OF ELASTIC AUDIO

The whole concept of elastic audio is a relatively recent idea, which has only been made possible (or practical at the very least) by recent huge advances in computer processing power. The ideas behind elastic audio are very basic and are really just an extension of the ideas that we have looked at in the last few chapters. In essence, elastic audio, in one sense at least, is another way of carrying out time-stretching. The revolutionary aspect of it is that it can do all of this in real time and in a nondestructive way, which allows us to try out new ideas quickly and easily and, most importantly, safely. The time-stretching that we have been discussing relies on us committing to a certain amount (or ratio) of stretch and then processing the file to create that stretch. If we then decide that we need to change things, we will need to either locate the original untouched file and apply the new stretch, or, worse, apply the new stretch to the already stretched file, which results in a further loss of quality. This is neither a very practical nor elegant way of working.

By being able to time-stretch audio without committing to any particular stretch amount and without having to worry about having to undo edits and then do further off-line processing, we have a lot more freedom to try out ideas and be generally more adventurous. Just to reiterate, though, elastic audio isn't so much a new editing possibility as a much more intuitive way of doing something that we are already used to doing. This in itself is a great thing, but there is a little more to it than that.

Audio that is truly elastic will have two aspects of flexibility: time and pitch. Some DAWs and software have functionality in one of these aspects, some in the other, and some on both; for the sake of clarity, it is best to separate them, as they do require a slightly different mind-set to get the best out of and, in some cases, different software as well. In this chapter, we will look at the time-related aspects of elastic audio in greater detail, and in the following chapters, we will cover the pitch-related aspects.

DIFFERENT ALGORITHMS

As with traditional time-stretching, most implementations of elastic audio have multiple algorithms that are optimized for different types of audio. The principles behind the different algorithms are exactly the same as for normal time-stretching. Some will be optimized for more percussive material and will be more transient-aware, some are better suited to sustaining material, others are better suited to monophonic or "solo" instruments, but the actual algorithms used may not be the same ones that are used for the normal time-stretching.

What all the elastic algorithms have in common is a need for the file to be pre-analyzed. Whenever the elastic audio function is turned on for a particular file, it will need to be analyzed to prepare the file for manipulation. Once this analysis has been completed, it will be used by whichever algorithm is chosen, and the file won't need to be re-analzed if the algorithm is subsequently changed. When it comes to actually using these different algorithms, there is one major thing to consider that differentiates them from the algorithms used in the normal time-stretching. Normal time-stretching will stretch the whole file and, with the exception of the transient detection and subsequent exclusion, will treat every part of the file the same. All these elastic audio algorithms work in the same basic way, but they have the additional factor of the markers used to identify important points in the sound.

These markers are the key to the flexibility of the system but can also prove to be a cause of problems. If these markers aren't put in the right places—or, more importantly, are put in places where they don't need to be—then the algorithms can misidentify transients, and, because those transients are treated differently, this can result in the stretched audio, sounding "jumpy" and uneven. Choosing the correct algorithm will always help, and adjusting any parameters that are available for the algorithm can further improve things, but the most important factor is the markers themselves. The actual way in which you can control the amount and placement of these varies from DAW to DAW, but they do all offer some degree of control; you might need to deviate a little from the "default" settings in order to get the best results and it is always advisable to try to get the markers set correctly *before* you actually try to move anything.

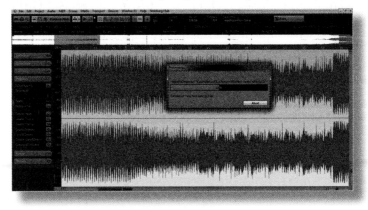

FIGURE 13.1
In order for the Elastic Audio functionality to work, audio will need to be analyzed initially. It may be that this is done behind the scenes, or it may, as is shown above, be something that you have to choose to enable, at which point the analysis will take place.

ADVANTAGES AND DISADVANTAGES COMPARED TO REGULAR TIME-STRETCHING

Fundamental to the idea of elastic audio is the inherent ease of use. You don't need to worry about the actual tempo or the actual length, and you don't need to worry about calculating ratios or percentages. It is very much a "drag and drop" approach to time-stretching. It hugely simplifies the process of a number of very common editing tasks, but it can also have a negative side to it, because there are some compromises with the idea in general and some with the quality aspects.

When looking at the actual work-flow, it is clear that the fact that the regions are forced to be continuous means that there will be no audible gaps, but, at least without intervention from the user, this can mean that regions may be stretched or compressed unnecessarily. Let's consider a vocal with quite a legato style. When the words are flowing into one another, this way of working with elastic audio is a great time-saver. We move the start of a region (word), and the region immediately before it is compressed, and the region we are moving is stretched to compensate for the movement. This is probably what we would do if we were doing the stretching manually. But what happens if we are at the end of a line of vocal, and there is a reasonable gap before the next word?

With this automatic stretching and compression, any movement applied to this last word would mean that the region would be stretched or compressed when it probably wouldn't need to be. If we moved the region slightly later, there is a good chance that there would still be enough of a pause after the word for us to not have a problem. Equally, if we moved the region slightly earlier, then we wouldn't necessarily have to stretch it. The stretch itself might be only very minimal, but, given that no stretching algorithm is totally free of artifacts, and especially given that the algorithms used by elastic audio are generally focused on being able to operate in real time and therefore aren't necessarily the absolute best algorithms available, it seems wasteful to stretch a region that doesn't need to be stretched.

Another possibility is that you might have a single, held vocal note that has a pitch bend at the start. Because the note is continuous, the analysis will probably pick this out as one region, and therefore, any stretch would be applied to the whole note. If you lengthen or shorten the region, then you will be lengthening or shortening the duration of that pitch bend as well. This might be fine, but it would be better if you separated the start of the note (with the pitch bend) from the more constant pitch of the second part of the note. You can do this by manually adding a marker at a point after the pitch bend, but, even then, you are still required to do a little more work. If you have a marker at the start of the note and one at the start of the constant pitch (after the bend), and you then move the start position of the note as a whole, all that will happen is that the section between the very start of the note and the constant pitch part of the note will be stretched or compressed. This is, in effect, the exact *opposite* of what you want to do! It can be remedied by then moving the second marker to be the same relative distance from the moved first marker, as it was before we started, but this isn't really an intuitive way of working.

The other potential problem with this method of time-stretching is that, dependent on the specifics of the algorithms used in both the traditional time-stretching and the elastic audio parts of your DAW, there is a good chance that the quality of the time-stretching, when done the elastic way, might not be quite as good as when done the traditional way. It may be that at some point in the future, even the most elaborate and advanced time-stretching algorithms that we have today can be adjusted and process in real time in the way that would be needed for them to work with elastic audio. But, if that happens, it may be that new algorithms have been developed that are more advanced, so we could find ourselves in the same situation that we do today, albeit with better quality all round.

QUANTIZATION OF AUDIO FILES

Aside from the obvious use of being able to move individual sections of audio files around to adjust the timing, elastic audio offers us a number of possibilities for adjusting things for us. One of the more obvious ones is that it gives us the ability to quantize audio files in the same way that we would quantize MIDI regions. In a way this is very similar to the functionality that recycling an audio file would give, but, as we saw in the chapter on recycling, one of the things that can be a problem with recycled files is the presence of gaps at the end of each of the "slices" if we move them from their original positions. We have just been looking at the ability for elastic audio to eliminate these gaps when we are manually moving sections of audio around, so it makes sense that this can be applied to sections that are automatically moved when we quantize the audio. In effect, what we are doing is very closely related to the beat-mapping that we looked at in chapter 10.

But it is worth considering that the transient markers that the DAW will insert won't always be exactly where you want them and that not everything will actually benefit from strict quantization. Sometimes it would be better to have things sitting just a little off the beat to take into account slightly slower lead-ins to particular sounds. Therefore, to get the very best out of this process, it would be ideal if we had the option to select parts of the file and mark them as nonstretching/quantizing parts (as we talked about earlier in this chapter), as this would give us the option, quickly and easily, to be selective about what we quantize and not risk completely ruining the feel of a part.

FIGURE 13.2
By using Elastic Audio, it is possible to automatically re-quantize audio files without worrying about having gaps or overlaps, as each section of audio between the markers is automatically stretched or compressed to reflect any change in position.

AUTO-CONFORM AND FOLLOW TEMPO

Another process that some DAWs offer, which is linked to the technology behind elastic audio, is the option to have any files that are imported into the project to be automatically analyzed and mapped to the project tempo. That way, as soon as the files have been imported, they will be available to preview at the project tempo. This can be a time-saver if you want to audition a number of different files and don't want to have to stretch them when they are placed into the Arrange window. Features like this, although incredibly simple, can soon become indispensable. Whether or not you choose to eventually replace the elastic version of the file with a manually chopped and stretched one is a matter of individual choice and circumstance. Perhaps you might choose to recycle the loop rather than stretch it. In any case, this is a great feature to have available in almost every instance where you are using rhythmic audio files.

One of the great things about this automatic tempo matching is that the files treated in this way will not only match to a projects tempo when they are imported but will also adapt to any changes in tempo either as a global tempo change or as tempo change midway through a song. In chapter 10, we discussed the idea of extracting the ebb and flow of the tempo from other songs and then applying that as a tempo map in a project of our own, and, if any audio files we have will follow any tempo changes automatically, it will save us a great deal of work in manually stretching regions and files to match a changing tempo.

As well as audio that has been imported into a project, when recording audio files directly within the DAW, in many cases, as soon as the recording is done, analysis will be carried out, and the file will instantly respond to tempo changes. The usefulness of this is that, once again, we could record to a click track (if that is the preferred method for the project) and then, at a later stage, could decide to incorporate little changes of tempo just to manipulate the energy of the song a little more.

The biggest concern with these methods is to do with the artifacts associated with time-stretching. I realise that no time-stretching can (currently) happen without these little artifacts, and, especially with the very latest algorithms, these artifacts are becoming more and more subtle. However, they are still there and they can certainly be audible even within a full mix. My concern is not to do with the artifacts themselves as such but more to do with them changing during the song. Let me explain what I mean a little more clearly.

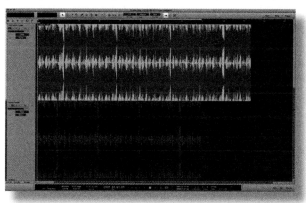

FIGURE 13.3
The image above shows duplicate audio files, with the one on the bottom track being set to follow any tempo changes and the one at the top being set to remain at its original tempo.

If we as an audio editor (or engineer or producer) are presented with a perfectly recorded set of tracks to work with, it would seem a shame to subject them to any processing that would leave behind these little tell-tale audio signatures as a result of time-stretching. But we are in a unique position in that we will hear them before stretching, while the consumer will hear the end product only in the stretched state. Unless the artifacts are really bad—and if they are, then we shouldn't let it through our own quality control like that—then the listener will just perceive that as being "how it sounds." And rightly so, because that *is* how it sounds. However, if the stretching amount or ratio is changing over the duration of the song, then the artifacts present at each point will change as the tempo changes. And it is this specific point that could present us with a problem.

It doesn't take long for our auditory system to get used to patterns. And the repetition that is present throughout most modern music (to a greater or lesser degree, depending on the style) means that we will very quickly recognize those patterns. Therefore, any changes to those patterns will be quickly identified. And it doesn't need to be major changes, either. Small changes in the tone—perhaps a slightly "ringy" or "metallic" overtone creeping in here and there, or perhaps a softening of a transient or a slight "phasey" quality—all these things can be enough to register as a change. And when you consider the fact that all the audio files that are being made to follow tempo changes will have these subtly shifting artifacts, it could very quickly become noticeable.

This aspect of the tempo-matching will improve over time as computers get (even) more powerful and the algorithms and processing techniques improve, but it's just something to be aware of now. It is very easy to think that all these different techniques for, and uses of, time-stretching mean that we can take any piece of audio and use it anywhere, anyhow. We aren't at that point just yet—but we are certainly well on the way to that.

THE BEST OF BOTH WORLDS

One way in which the current implementations could possibly be improved in future software versions is to incorporate a dual-algorithm system. The file would be analyzed and prepared as usual for elastic audio, and an algorithm chosen that was appropriate. This algorithm would process any changes made during normal playback. However, in addition to this algorithm, there would be an option of a high-quality/off-line algorithm as well. This would be one of the algorithms used for normal time-stretching that could provide a better quality, and this would be used when the file was bounced/frozen/rendered. When this option was chosen, the file would be cut, moved, and stretched exactly as we described above, but all done automatically and behind the scenes.

In order to make this system usable, there would need to be some kind of preview function, so that you could hear the result of using the off-line option and make sure that the correct algorithm had been chosen. One way of doing this might ns be to select a region (or regions) and then have a preview button that processes the region and creates a temporary file that the user then listens to, to make sure that the correct algorithm has been chosen. It would also be necessary for this preview to be played back through any plug-ins that are on the channel for that region to ensure that the comparison between the elastic and "high-quality" algorithms was an accurate and representative one. With this comparison in mind, it might be good to have the preview feature able to do quick A/B comparisons between the elastic version and the off-line version as the preview is playing back to make sure that everything is as you expect it to be.

Whether or not this way of working would suit everybody, I don't know. Equally, there is no way of knowing if DAW makers would ever incorporate a feature like this, but, given how useful I believe it would be, and given that the technology to do it is already there, I see no reason why it couldn't be incorporated at some point.

HANDS ON
Introduction

It should be clear from this chapter just how much of a big deal Elastic Audio can be and how much time it can save for some of the more detailed editing processes. For quite some time we have been able to change the length of digital audio files, but it was always done destructively, and it would have been very tedious to just try things out for the sake of trying things out. However, all the DAWs that we are featuring in this book have their own versions of this concept and, as a result, allow us the freedom to do just that (try things out) without having to spend huge amounts of time manually stretching individual parts or regions. We will take a look at how each of them work in practice and list out any important things to remember when using them.

Logic

Flex Time Editing is the terminology used in Logic for what we have been calling Elastic Audio, and it offers a large number of benefits. Not only does it allow us to make *Flexed* audio files follow tempo change automatically but it also allows us to get inside the files and change the timing of individual notes within the recording either manually or automatically (through quantization). The first thing we need to do in order to use *Flex Mode* is to enable the Flex view, and, as always, there are a number of ways of doing this. By default the Flex View button above the Arrange window should be visible, and clicking this toggles the Flex view on and off. Alternatively you can use the menus at the top of the Arrange window and choose View > Flex View and, finally, you have the option of using the Cmd + F key command.

Once Flex view has been enabled, you will see an additional drop-down list in each track header at the left of the Arrange window. By default this is set to *Off*, but clicking on this will show the list of the different Flex modes available in

Logic. The next step is to decide which of the algorithms to use. As with the time-stretching algorithms, most of them are named according to the intended source material, but if you wish to know more about each algorithm and their associated settings, you can find a full description on pages 529–531 of the Logic Pro manual. If you aren't sure, you can always try one that you feel might be best and change it later. In addition, sometimes using the "wrong" algorithm can lead to some very interesting results, so, if time is not as issue, it is always worth trying a few options rather than going for the immediately obvious one. Note that many of the Flex modes have additional parameters that can be changed (details are in the Logic Pro manual as mentioned above), so if the results are not quite as smooth as you like, it might be worth making small adjustments to these parameters (if your chosen mode has any) before trying a different mode.

Now we move on to the issue of Flex Markers. While these markers are related to transient markers, and will often be placed in the same places as transient markers, they don't necessarily have to be the same. When you initially select a Flex mode, there will be a short pause while the audio file is analyzed, and, once this is done, you will see faint vertical lines appearing in the positions where Logic has determined there to be transients. These are the markers that, in the absence of any created Flex Markers, Logic will use as a basis for the time-stretching. Once the analysis is done, if you change the tempo of the project and then hit Play, you should find that the newly Flexed audio file will automatically change to the new tempo. However—and this is a very important thing to remember—it is always best to change the project tempo to match the audio file you want to Flex *before* you carry out the analysis and then change it back to its original tempo.

If, for example, you imported an audio file that was 100 bpm into a project that was at 160 bpm, it would play back at 100 bpm. If you then carried out the *Flex Mode Analysis* by choosing one of the modes, Logic would assume that the file being Flexed was at 160 bpm. Once the analysis was complete, the file wouldn't magically change to 160 bpm but would stay at its original tempo. It *would* follow any tempo changes but in proportion only. If you changed the project tempo to 144 bpm (a 10% drop), then the Flexed file would drop to 90 bpm (a 10% drop). In order to get the tempo to follow correctly, you need to change the project tempo to match the audio file and then carry out the Flex Analysis at its original tempo. Once this has been done, you can change to any tempo, and everything should work as expected.

The other thing we need to do is set our Flex Markers. If all we want to do is make audio follow any tempo changes, then we really don't need to worry about the Flex Markers as the default anaylsis, and transient markers should do the job adequately. But if we want to manually move the timing of individual parts of the file, then we need to create Flex Markers as well. If you move your cursor over an audio region with Flex view enabled and a Flex mode chosen, you will see the normal cursor change to a vertical line instead. If you then move the cursor over one of the transient markers, you will notice a further change to a vertical line with what looks like a down arrow at the top. Both of these are tools that we use

Elastic Audio (Time) CHAPTER 13

to create Flex Markers. The fundamental difference is that the plain vertical line is used to create Flex Markers between the transient markers, while the vertical line with the arrow is used to convert transient markers into Flex Markers. If you click using either of these tools, it will create a Flex Marker, which is denoted by a brighter vertical line with an orange "anchor" at the top of the region.

If you now move the cursor over one of the Flex Markers, you will notice that the vertical bar with the arrow at the top now has two additional left and right arrows, and this indicates that we can now move this Flex Marker. As you move this marker, you will see the waveform overview to the left and the right of the marker stretching and compressing, depending on which way you move the marker. You will also note that dragging the background of the overview will change on one side to green and other the other orange. This is to indicate that the areas are being compressed (green) or stretched, so that you can see at a glance which areas have been affected. When you release the mouse button, the background goes back to the normal color, but the waveform overview itself has a colored border in line with the above.

And that is pretty much all there is to setting things up. Once this is done, all that remains is to move the Flex Markers as you see fit. There is the option to automatically quantize the markers, and this is done in the same way that you would quantize a MIDI region: by selecting a value for the *Quantize* parameter in the *Inspector* panel. You should be aware that quantizing works on all the markers created. It isn't (currently at least) possible to choose only a certain range of markers within a region and quantize those. If you wish to do such a thing, you will have to create a separate region by splitting an existing one into multiple parts. Once this is done, you can selectively apply quantization only to the markers in a particular region.

If you prefer you can manually move the markers to get things into the right position, and this is often the preferred method as it retains the most natural "feel." Naturally it is more time-consuming, but, like most things, the extra time

FIGURE 13.4
The coloring of the outline of each section of a Flexed audio track will give you a visual indication at a glance of which parts have been stretched and which compressed.

spent can lead to more rewarding and satisfying results. One good practice tip is, where you have a note you wish to move that is preceded by a period of silence, to create a Flex Marker in the silent space so that you minimize unnecessary compression of the preceding section. The creation of this kind of marker can help to maintain the integrity of the audio, as using Flex Mode does inevitably lead to some artifacts, and any compression or stretching that isn't absolutely necessary should be avoided.

Pro Tools

Because of the deep level of integration of Elastic Audio into Pro Tools, there are, aside from the usual audio-warping capabilities, some very intriguing and useful possibilities. Aside from the (now quite common) ability to have audio regions follow tempo changes, Pro Tools also has the ability to analyze an audio region and make it automatically fit to your current tempo. The exact process that you follow depends on whether it is a region that you have already imported or if you are planning on importing new regions.

If you have an audio region in your project already, then you can make it conform to tempo by enabling Elastic Audio, choosing an appropriate algorithm (see chapter 12 for details), and then going to Clip > Conform to Tempo. The region will be analyzed, and a tempo and length estimated, and then, based on that, the region will be stretched or compressed to match the current tempo. If you are looking to import new regions, then things can actually get even easier, but you need to use a slightly different method of importing regions than you might be used to.

You can import audio regions by pressing Cmd[Mac]/Ctrl[PC] + Shift + I or going to File > Import > Audio, but there is another way of importing that gives you additional options. If you press Alt + ; (semicolon) or go to Window > Workspace, you will open a Pro Tools Browser window. Much of the layout of this is similar to a typical operating system file browser, so it should be fairly familiar. However, there is an especially useful feature built into this browser, which comes in the form of the *Audio Files Conform to Session Tempo* button located at the top of the browser window next to the level meters. When pressed, this button turns green, and the audio regions in the browser window will be previewed at the current project tempo. By default you can preview a file/region by clicking on it and pressing the Space Bar once to start the preview and again to stop it. As this process uses the same algorithms as the other Elastic Audio processes, you will find a drop-down list to the side of the *Audio Files Conform to Session Tempo* button that will enable you to choose which algorithm is used for the previews.

The ability to preview the files at the project tempo is extremely useful, but, continuing on from there, if you then drag and drop a region from the browser window into your project, it will be automatically conformed to the correct tempo, so it will work in your project without you having to do anything further. However, in order for your conformed files to automatically follow tempo changes, you may have to carry out an additional step.

Elastic Audio (Time) CHAPTER 13

If your regions were already in your project and you conformed them using the first method, then all you have to do is make sure that the track is a *tick-based* track rather than a *sample-based* track. The difference simply refers to how Pro Tools itself determines the position of a region. In earlier versions, Pro Tools was always sample-based, meaning that the position of a region was given in absolute time (minutes and seconds). Changing the tempo of a project wouldn't make anything move, because a position of three minutes and twenty-two seconds is the same whether the tempo is 100 bpm or 150 bpm. Tick-based tracks, on the other hand, determine the position based on bars and beats. A position of Bar 3, Beat 1 might be ten seconds into the song, but if you change the tempo that could change to eight seconds or twelve seconds in. As such, *tick-based* tracks are required in order to allow regions to follow any tempo changes.

Fortunately it is very easy to change a track from one type to the other. If you set your track height to *Medium* or larger, you will see, to the left of the *Elastic Audio Plug-In Selector*, the *Timebase Selector*. This allows you to easily change between the two options by clicking and then, from the pop-up menu that appears, choosing either *Samples* (which has an icon which looks like a blue clock face) or *Ticks* (which has a green metronome as an icon). Changing between the two has no immediate effect on any playlists or edits that you may have made and simply changes the reference scale that Pro Tools uses. Changing a track with Elastic Audio enabled to a tick-based track (if it wasn't already) will mean that any tempo changes—either a constant change for the whole project or tempo changes throughout the song—will be followed by any regions on that track.

If, however, you have dragged and dropped regions from the browser when they might be conformed to the current project tempo, they won't necessarily be set up to follow tempo changes. And it all depends on exactly where you dragged them to. If you dragged them onto an empty track, then, by default, they will just adopt the settings for the track (Elastic Audio–enabled or not; sample-based or tick-based), in which case you will need to make sure that the track is set up as needed, as detailed above. If, however, you drag and drop the region onto an empty part of the project below any existing tracks, then a new track will be created for the region, and it will already be Elastic Audio–enabled and tick-based. You might still need to change the actual algorithm for the best results, but, if the track is created as part of the drag-and-drop import process, then the region should not only be at the right tempo to start with but also follow any tempo changes that occur later.

The above techniques undoubtedly give massive flexibility when it comes to using audio regions. In many ways, especially if you consider that you can quantize them as easily as if they were MIDI regions (see chapter 10), audio has now become very malleable, and we have the ability to automatically correct (in the loosest sense of the word) audio regions. But sometimes we might not want to apply a blanket quantization to an audio region. Sometimes we might just want to nudge a particular part or two of it that could have been tighter timing. And that is where Elastic Audio can help us out once again.

We have already seen that transient markers are created when we enable Elastic Audio for a track, and we have also seen how to add, delete, and move these markers. But in order to be able to have full audio-warping capabilities, we need to look at Warp Markers. In essence, these are very similar to transient markers, and, for the most part, they will be in the same locations as the transient markers, but we still need to actually create them before we can freely move audio around within a region. If you set the *Track View Selector* to *Analysis*, you will see the transient markers, and if you then set it to *Warp*, those transient markers will still be visible. In order to convert a transient marker to a Warp Marker, you need to do one of two things: With the Grabber tool selected, you need to position the cursor over an existing transient marker (vertical line overlaid on top of the waveform overview) and then double-click or press **Ctrl[Mac]/Start[PC]** and click. When you do so, the gray vertical line will change to a block vertical line with a small triangle at the bottom. This is now a Warp Marker.

As you move the cursor over a Warp Marker, it will change from the standard Grabber cursor to a horizontal line with arrows at each end. Using this cursor, you can now move this Warp Marker earlier or later and move the audio as you do it. Moving a Warp Marker will stretch or compress the audio on either side in proportion to the amount that you move the marker. However, it should be noted that it will move all the audio between the Warp Marker that you are moving and either the next Warp Marker or the end of the region. Therefore, if you have created only one Warp Marker in the middle of a region, perhaps to adjust the timing of a single note, moving this Warp Marker will adjust the position of *all* the audio from the start of the region to the Warp Marker and from the Warp Marker to the end of the region. If all you want to do is move a single note, then you have to create three Warp Markers: one at the position of the note you wish to move, one at the start of the preceding note, and one at the start of the following note. Doing this will make sure that only the note you wish to move is actually moved, and all the rest of the audio (barring the preceding note, which will be stretched or compressed to compensate for the movement of the note you are working on) will remain untouched. You can, of course, create

FIGURE 13.5 Clicking and dragging on the Warp Markers allows you to change the position of individual sections of the audio, whether that involves a subtle shift or a complete change of groove and feel.

Warp Markers that aren't tied to existing transient markers. To do this you simply hold down Ctrl[Mac]/Start[PC] and click at the position you wish to create the new Warp Marker. Equally, you can delete existing Warp Markers by moving the cursor over the Warp Marker and then holding down Alt while clicking.

Studio One

We have seen, in previous chapters, that Studio One is one of the more intuitive DAWs when it comes to audio manipulation, and it builds Elastic Audio–type functionality into the very core of its audio file handling process. You can set the preferences to automatically stretch new audio files to the song tempo by opening up the Studio One *Options Panel* (Studio One > Preferences or Cmd[Mac]/Ctrl[PC] + ,) and then clicking on the *Song Setup* button at the bottom of the panel. From there click on *General* at the top of the panel, and make sure that the *Stretch audio files to Song tempo* check-box is ticked. This sets the default behaviour for all newly created songs in Studio One, and for many people, this will be a very good option to have selected in order to speed up the whole work-flow.

The actual Elastic Audio process revolves around the use of Bend Markers. We have seen in previous chapters how we can go about detecting transients and then adding, deleting, and moving individual Bend Markers, so there is no need to revisit those techniques here. If you need a reminder, then look back to the Studio One section at the end of chapter 9. We will move forward on the assumption that the Bend Markers have been added and checked to be in the right place, and that the *Audio Bend Panel* is open. Finally, you need to make sure that the *Timestretch* mode (located in the *Track* section of the Audio Bend Panel) is set to *Audio Bend*.

In addition to straightforward quantization of audio regions, which we have already looked at, the Audio Bend functions of Studio One allow you to manually reposition individual notes or sounds within an audio region by clicking and dragging on the relevant Bend Markers. You will need to have the Bend tool (press 7 or the Bend tool) button on the main toolbar in order to be able to move the Bend Markers around, as using the Arrow tool and trying to click and drag on a marker will result in the entire region being moved.

As you click and drag on a Bend Marker, you will see the audio either side of the marker (between this selected marker and the two markers either side) being stretched or compressed, depending on which way you move the marker. To help give you a quick visual reference as to what you have changed, any parts of the file that have been shortened (compressed) will be colored green, and any that have been lengthened (stretched) will be colored red. This helps you very quickly identify what kinds of changes you have made.

If you wish to automatically constrain the new positions of the markers to a quantization grid, then you should make sure that *Snap* is selected by pressing the *Snap* button beneath the *Snap Timebase* control in the main toolbar or by

FIGURE 13.6
When the Bend Markers are moved in Studio One, the waveform overview changes color to indicate whether each particular section has been stretched or compressed. This is very useful in determining which parts of a file have been adjusted if something doesn't feel quite right with the new timing, and you want to adjust back toward the original position.

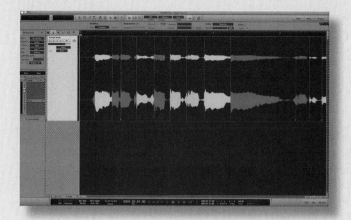

pressing N. Doing this will mean that markers can be dragged only to positions that the current *Quantize Value* defines. Using this method is quite a simple way to quantize only certain points within a region. You could, for example, turn *Snap* on and then drag only the *Bend Markers* that represent notes or sounds that should be on the beat (whole notes). This will mean that the loop has very tight timing when it comes to the solid and definable beat of the song but allows for much more freedom of groove on the crucial eighth or sixteenth notes that fall between the beats.

One *very* important thing to remember when using Audio Bend in Studio One is that moving a Bend Marker will stretch or compress the audio either side of the selected Bend Marker between the marker itself and either the adjacent marker or the end of the file. Notice that it is the end of the file and not the region. If you have imported a sixteen-bar file, and you cut this into smaller sections—perhaps a pair of eight-bar sections representing a verse and a chorus—then if there is no Bend Marker at the end of the first (verse) section and you drag a Bend Marker in this section, not only will the audio between the Bend Marker and the end of the first section be stretched but also all the audio throughout the second section as well. This can be slightly confusing, as it is quite different from the behaviour in some other DAWs, but it is easy to work around as long as you are aware of it happening. You can either place a Bend Marker right at the end of the first section, or you can bounce the region (section of the file) to a new file by pressing **Cmd[Mac]/Ctrl[PC] + B** or going to **Event > Bounce Selection**. This will create a new file that is only the length of the section you bounced, and, therefore, any manipulation of the Bend Markers in this new file cannot stretch beyond where you might expect.

If you have multiple tracks which have been grouped together then the Bend Markers on each of these tracks will be linked however there is a slight variation on how they function on grouped tracks. Ordinarily, if *Snap* mode was active and you moved a Bend Marker, the marker would move to the nearer point on the quantization grid. This is still true to an extent if tracks are grouped and there are Bend Markers very close to each other on different tracks. In this situation, the timing relationship between Bend Markers on the different tracks will remain static. What this means is that if you had, for example, four different tracks, and the positioning of the Bend Markers was slightly different for each of the tracks (as would be natural with any live performance), then moving a Bend Marker on one of these tracks—let's say the first track—would mean that that particular Bend Marker snapped to the quantization grid, while the markers on the other three tracks would maintain their positions relative to the first one. If you then moved the Bend Marker on the second track so that it snapped to the quantization grid, then the relative positions would still remain the same, and the marker on the first track would now move away from the quantization grid.

While this might seem like a flaw in the system, it is actually far from that. It is, in fact, essential if the phase relationship between the different tracks is to be maintained. If the Bend Markers were not locked together in this way on grouped tracks, not only might you have issues with timing because of "ghost notes" from the spill into other mics, but, perhaps more problematic, the direct source (the snare drum mic, for example) might be stretched or compressed in comparison to the snare spill into the overheads, which, under certain circumstances, might cause phase-cancellation effects or even "flanging" effects. So by maintaining the timing offsets between the Bend Markers on different tracks, Studio One avoids these situations. Of course you can, if you wish, ungroup the tracks and move the Bend Markers individually, in which case any of the problems mentioned might occur, but at least the software does what it can to avoid these things happening by mistake.

One definite downside of using Bend Markers on grouped tracks is the fact that if you move a Bend Marker on a single track within the group and there are no markers on the other tracks that are in close proximity, then Studio One will create Bend Markers on those tracks that don't have them and move them accordingly. If you are working on phase-locked drums, then, again, this makes sense, but if you have tracks grouped that weren't multi-mic recordings of the same instrument—perhaps groups of vocal harmonies—then you might not want to stretch individual notes unless absolutely necessary. A sustaining note on one track would not need a Bend Marker created midway through it, just because one of the other harmony lines had a change in pitch and therefore a detectable (and movable) transition from one note to the next.

None of these things really cause problems—at least not problems that can't be overcome—and the reasoning behind it makes sense in most situations, but it is worth being aware of some of the limitations as well as the numerous advantages of any Elastic Audio–type system. In the cases we have been discussing here, it isn't so much a limitation of the underlying technology as a limitation of the way in which it is applied (grouped tracks).

Cubase

In this section, we will look at two distinct aspects of the Elastic Audio features in Cubase. The first will be the ability to automatically keep audio files in sync with any tempo changes, and the second will look at selectively moving certain parts of audio either to subtly change the feel of a rhythmic part or to correct minor timing errors in a performance.

If you have imported audio files or regions into your project, and you want them to adjust to a change in tempo, then you will need to open the *Audio Pool* by going to Project > Pool or by pressing Cmd[Mac]/Ctrl[PC] + P. In the Audio Pool, you will find a list of all audio files and regions that are in the current project. Across the top of the Audio Pool window, you will find a number of column headings. The ones we need to make use of here are *Tempo* and *Musical Mode*. The Tempo parameter should be set to the original tempo of the file. If the original tempo is not stored in the file metadata, then Cubase will make a guess at what the tempo should be. In the event that the tempo is wrong, you can double-click on the value and type in the correct tempo—assuming, of course, that you know it. If you then make sure that the Musical Mode check-box is ticked, you will have done all you need to to not only make sure that the audio files you have imported are automatically matched to the current project tempo, but also that they will follow any future changes in tempo, whether that is a change in the fixed tempo of the song or changes due to a tempo track.

If you don't know the tempo, then the file can still be made to follow tempo changes by checking the Musical Mode check-box, but the absolute tempo may still not be correct. For example, if you have a project at 80 bpm and an audio file of unknown tempo, then increasing the tempo to 100 bpm will result in a 25% increase in tempo for the audio file. This will not give you an absolute tempo for the file now, but it will, at least, adjust the tempo of the file in line with other changes in the tempo of the song. This can be useful if you have some kind of free-form sound (perhaps a sound effect or a spoken-word recording with no clear tempo) that is designed to last a certain number of bars. Because it doesn't have an absolute tempo defined by a rhythmical pattern, you won't need to specify a tempo. But if you change the tempo of the project, then you will still want it to last the same number of bars, so enabling Musical Mode will allow you to do this even if there is no clear tempo or you just don't know what the tempo is.

Now that we have got all our audio files locked to our project tempo, we can look at ways in which we can go in and manipulate the timing of individual notes or sounds within the file. This is what most people think of when they hear the term "Elastic Audio." The audio quantization that we looked at the end of chapter 10 is an automated form of doing this, but there will be times when we don't want to simply apply a quantization groove on top of a region or when we simply want to correct the timing of a particular part of an otherwise great performance. In situations like these, you want to be able to get in there and move things around manually. And in order to do that, we can once again go back to using Hitpoints. If you are in the Sample Editor window and you have already detected and potentially edited your Hitpoints until you are happy that they accurately represent transients or other important musical events, then you can click on *Create Warp Markers* underneath the Hitpoints heading. This will create one Warp Marker for each Hitpoint, which you can then use in the *AudioWarp* section of the Sample Editor window.

Clicking on AudioWarp reveals a number of different controls, but, more importantly, shows the Warp Markers overlaid on the waveform overview. You will notice that the Warp Markers, while in the same position as the Hitpoints, are represented by vertical yellow lines instead of gray ones, but they actually work in a very similar way. You can manually delete them and move them, and, unlike Hitpoints, you can manually add them. But before we get to that, we need to make sure that Musical Mode is selected under the AudioWarp heading. When the button is illuminated, the mode is active. You may find that making the mode active removes all of the Warp Markers, in which case, with Musical Mode still active, you will need to return to the Hitpoints section and click on *Create Warp Markers* once again before returning to AudioWarp.

You can apply audio quantization here by setting *Resolution* to an appropriate value for the material in question and then adjusting the *Swing* slider. As you do this, you will see the Warp Markers moving to reflect the changes in the quantization values. The effect here is exactly the same as if you had applied the quantization using the method described in chapter 10, but here you have the advantage of being able to further adjust the position of individual Warp Markers post-quantization. In order to do this, or to move the audio around without applying a fixed quantization first, you will need to click on the *Free Warp* button.

When you do this you will see, if you are zoomed in to a high-enough level, that there is a number just to the left of each Warp Marker. This number represents a stretch ratio for the audio preceding the current marker. A value of 0.75 (for example) would mean that the audio between the previous Warp Marker and the current one had been compressed to 75% of its original length, while a value of 1.34 would mean that it had been stretched to 134% of its original length. This is similar in principle to the red and green coloring applied by Logic and Studio One but, while more accurate in telling you the amount of stretch that had been applied, is perhaps less intuitive, as it doesn't show at a glance what has been changed. Nonetheless, it is there, and it is a very useful feature to have.

Now that we are in Free Warp mode, you will notice that the cursor has changed. If the cursor is positioned between Warp Markers, it will look like left and right arrows with a clock below and a vertical line between the arrows. Clicking with this tool will create a Warp Marker at the current position. This is very helpful if the automatic detection process has missed a Hitpoint or Warp Marker and you need to add one, but it is also useful in create Warp Markers to serve as anchors. Moving the Warp Markers will move all the audio between the current Warp Marker and the preceding and following ones. If there is quite a bit of space between the current note or sound and the previous one, then moving the current one to the left would mean that all the audio directly to the left would be compressed. As there would be sufficient space (in this situation), then it wouldn't actually be necessary to compress the preceding audio. In this

FIGURE 13.7
Warp Markers can be moved and created independently of any transient markers or Hitpoints, and this gives a lot of flexibility if you wish to change the timing in the middle of a legato phrase that doesn't have any distinct transients, or if you want adjust the tonal evolution of a longer sound.

situation, we can create a Warp Marker in the silence after the end of the previous note, so that, should we move the following note, only the space between the note and the marker created in the silence would be compressed, and the preceding note would be left unstretched.

Moving the cursor directly over a Warp Marker sees the cursor change to the left and right arrows with the clock but without the vertical bar. This is the tool we use to actually change the position and timing of the audio. As we move the Warp Marker with this tool, you will see the audio to either side of the marker being stretched or compressed, and the waveform overview updating in real time. You will also see the stretch ratio number at the top of the Warp Marker changing. You will see light gray vertical lines in the background of the waveform overview, and these represent the position of the standard timing grid of sixteenth notes. As you drag a Warp Marker close to one of these gray lines, it will automatically snap to its position, but you can easily drag the marker away from that snapped position if you want to.

If you now move the cursor to the very top of a Warp Marker, you will see the cursor change once again to simple left and right arrows. When this cursor is active, it allows you to reposition the Warp Marker itself without actually moving the underlying audio. If you zoom in and find one of the markers is placed slightly ahead of where it should be, then you can reposition the marker in this way by dragging it to the correct position. As you do this, the area between the current position and the new one will turn orange, and, upon release of the mouse button, the audio will slide over, so that the marker stays in the same position relative to the song position, but the audio has realigned.

Finally, if you wish to delete a Warp Marker for any reason, all you need to do is to position the cursor over an existing Warp Marker, then hold down Alt (the cursor will change to an eraser), and then click on the marker. Unlike with Hitpoints, deleting a Warp Marker in this way doesn't leave any indicator as to its position prior to being deleted. You can, of course, add another marker back at that position, but you will have to place the marker by eye rather than being able to simply click on a triangle to reestablish the marker, as you could with Hitpoints.

As with any Elastic Audio processing, you will get the best results when the changes are quite small. Any significant repositioning of notes or sounds should be carried out another way. If at all possible, it is often best to split the region and move any badly out-of-time parts manually and then make good any gaps or overlaps using other techniques, but for simple, subtle timing correction, Elastic Audio is a revelation.

CHAPTER 14
Pitch-Shifting

INTRODUCTION

Pitch-shifting is, in many ways, a natural counterpart to time-stretching. The easiest way to change pitch on a tape recorder is to change the speed of playback, as we already know. However, in the context of an ordinary tape recorder, this would also change the tempo/duration of the piece being played back, so it wasn't pitch-shifting in the sense that we know it today, where a change in pitch has no effect on duration. That would require something a little more complex than a straightforward tape speed adjustment. However, it didn't need, as many people thought it would, a digital revolution to make it possible.

A BRIEF HISTORY OF PITCH-SHIFTING
Analogue

Analogue circuitry was used in electronic keyboards in the 1960s to create a "divide down" technique that allowed signals an octave (or two octaves or more) down to be created from any input signal. This technology would later evolve into stand-alone effects, some of which were adapted to allow creating a signal an octave up as well. Although primarily designed as guitar effects, there was no reason why other signals couldn't be run through them, although the results would hardly be considered to be accurate. In any case, the creation of an octave up or down is a little limiting in terms of pitch-shifting, so there was a need for something more adjustable. And this was where digital technology came in? Actually, no. There was a fully analogue, tape-based device that not only allowed for fully adjustable pitch changing over a near half-octave range in either direction but also had a time-stretching feature that would change the speed *without* changing the pitch. This marvelous piece of studio equipment was called the Eltro Information Rate Changer, and it was used on a number of records and was even used on the voice of HAL in the film *2001: A Space Odyssey*. Perhaps most surprising, though, is that the basic principles used by the Eltro can actually be traced back to patents as early as the 1920s, so this pitch-shifting had been a long time coming. If you are interested in finding out more about how the Eltro unit worked, there is a brief description of the technology and principles at work on the accompanying website.

Digital

Nonetheless, even though the technology was available to carry out pitch-shifting in the analogue domain, the whole concept of controllable pitch-shifting was kick-started in 1975, when Eventide released the H910 "Harmonizer." This digital unit was built on the same ideas as the Eltro and other tape-based pitch-shifters and, thanks to some high-profile users, quickly established digital pitch-shifting as a real possibility. This was followed up in 1977 by the H949, an improved version of the H910 that claimed to have fewer audible artifacts. The fact was that these early digital machines from Eventide gave pitch-shifting something that wasn't available in the analogue versions: precise and repeatable control.

The actual shifting amount was presented as a simple pitch ratio, which meant that it wasn't overly intuitive and musical in its operation. If you wanted to shift a sound upward by one semitone, you would, in theory, need to apply a ratio of 1.059463:1 to the input signal. The controls on these early Eventides were adjustable to two decimal places, so the closest you would have been able to get for a one-semitone shift would have been 1.06:1. Similarly, for a two-semitone shift, the theoretically correct ratio is 1.122462:1, which would have equaled 1.12:1 on the Eventides. These figures would have been plenty close enough though as any deviation from the theoretical ideal ratio would be very hard to discern and would lie within the window of what would normally be considered to be "in tune" anyway.

While these Eventide units would pitch-shift any incoming audio signal, they would certainly get their best results with monophonic, clean, "solo" sounds. This isn't really surprising, because the sound is much less complex, and therefore a better and more consistent result will be achieved. This fact pretty much set the standard for digital pitch-shifting, and good-quality pitch-shifting would remain monophonic for quite some time.

Other manufacturers rapidly caught on to the popularity of these early devices and introduced alternatives capable of digital pitch-shifting, but in 1986, Eventide changed the game again with the introduction of the H3000, which was the first pitch-shifter to work "intelligently" and have the pitch changing and harmonizing defined in terms of actual musical scales rather than simple pitch ratios. Instead of choosing a static pitch shift of, for example, +4 semitones (a major third interval), you would specify the key that the song was in, then specify a shift up of a musical interval of a third, and then the H3000 would know whether that should be a major third (+4 semitones) or a minor third (+3 semitones) based on the musical scale and the actual note you were shifting. It also had the more-traditional option to set the pitch shift in semitones and cents. There was, though, another development that increased the creative potential for the H3000. The relatively recent introduction of MIDI (in 1983) meant that the parameters of the H3000 could be controlled over MIDI and, therefore, recorded into sequencers. It was now possible to not only change the shift amount in real time (which it always had been since the H910) but also actually record those changes, so that they played back in sync with the rest of the sequenced parts.

Computer-Based

With these new possibilities, studio owners and producers all around the world began to use pitch-shifting effects more and more. But one less commonly referenced aspect of the H3000 would form the basis of a plug-in that would become a worldwide phenomenon when it was released in 1997. The H3000 had a "quantize" feature, which automatically tuned the output of the processor to the nearest correct note. Interestingly, the H3000 user manual states that this "quantize" feature was normally off and that it had to be specifically turned on if wanted. This simple feature formed the basis of Antares Audio Technologies' now (in)famous Auto-Tune, which aimed to provide automatic correction of pitch to either chromatic or diatonic scales of monophonic signals. While a monophonic source may have been recommended for the best results with the Eventides and similar hardware units, with Auto-Tune it was actually specified that it worked correctly only on monophonic inputs. So this ability, in itself, was nothing new, but its use on the 1998 song "Believe" by Cher, with its heavily corrected vocals, led to a worldwide awareness of the effect. This so-called Cher Effect has been used (and some would say abused) on a huge number of records ever since that day.

Auto-Tune does actually possess a graphical mode that allows for much more-subtle retuning of notes to make them closer to what they should be without necessarily being 100% accurate in tuning. It also allows for gentle pitch bends into notes without assuming that these are "wrong" and automatically correcting them. One other factor that has to be taken into consideration when looking at the almost-ubiquitous nature of Auto-Tune (and similar effects) is the fact that, prior to the release of Auto-Tune, any pitch-shifting would have been done

FIGURE 14.1
Antares' AutoTune allows both automatic and manual pitch correction, but, even though it is actually capable of quite subtle and natural results, it is most known for the creation of so-called hard-tuning effects used as a part of the signature sound of some artists.

on very expensive hardware that would be able to operate on only a single track at a time. The plug-in nature of Auto-Tune meant that multiple instances could be used simultaneously, and the (much) lower price meant that it was inherently more affordable even for smaller studios. Both of these probably had a lot to do with the rapid spread and use of Auto-Tune.

Things took a step forward in terms of technology and simultaneously backward, away from mathematical perfection, with the 2001 introduction of Celemony's Melodyne. In its initial incarnation, Melodyne was a stand-alone product. Audio files would have to be imported into Melodyne, manipulated, and then exported to a new file before bringing them into your DAW. The numerous options that Melodyne had for correcting or completely changing pitch, preserving or changing formants (more on this shortly), and time-stretching and quantizing were a revelation at the time, and Melodyne seemed far more adept at preserving the subtleties of a performance than Auto-Tune. The work-flow could prove problematic, though. Many changes like this are best done in context while listening to the rest of the track. Because Melodyne was a stand-alone application, it made this difficult. Not only would you potentially have to export any comped tracks to single, continuous audio files, but you would also have to export a "rough" mix of the rest of the track to work against. This stand-alone version of Melodyne did support multiple tracks, so it was certainly possible to have your "rough" mix playing in the background while you made changes to your vocal/melody sound, but I don't think anybody would have said it came close to Auto-Tune's immediacy and ease of use.

As well as actually tuning a note, Melodyne also allowed users to modify the amount of pitch modulation. This meant that overly strong vibrato could be tamed or subtle vibrato could be emphasized. The actual pitch glide between notes could be changed too, to allow a quicker or slower transition. The tonality and character of the audio could be changed by moving the formants, and the volume of individual notes within a part could be changed. And, of course, the position of a note could be changed with the length of the note itself and those on either side of it being adapted accordingly. Many of these features were very compelling, but Melodyne perhaps didn't get as much attention and use as it deserved, owing to the relatively complexity of using it in practice.

At the end of 2006, Celemony released a true plug-in version of Melodyne, which meant that there was no need for a complicated work-flow any more. Audio was still recorded as needed and then analyzed, but this all took place within the plug-in itself. All of the familiar Melodyne tools and options were there as expected, but there were some additional benefits other than just a simplified work-flow. Time-stretching had always been possible in Melodyne, but it had been a manual process to some extent. Now, because Melodyne was a tempo-aware plug-in, any audio process by Melodyne would follow any tempo changes automatically. No intervention was required. There were still tools to manually adjust the timing and position of individual notes within a passage, but the passage as a whole simply followed any tempo changes applied by the DAW. Another useful benefit was that the instances of the Melodyne plug-in were saved as part of the DAW project, so file management now became a lot easier, and there was less of a requirement to actually export any Melodyned files, because the Melodyne plug-ins would simply reopen, with all of the necessary analysis data and files, when you next open the project. It still wasn't quite perfect, though.

Then, late in 2009, Celemony returned with a product that defied all expectations and did something that many had been asking for but few, even those well acquainted with the deeper workings of pitch-shifting technology, had believed was possible: polyphonic pitch-shifting. Now it was possible, thanks to the newly developed DNA (Direct Note Access) technology, to change the pitch and timing of an individual note within a chord. Using this idea, it was possible, with a few clicks, to change an audio part from major to minor or to a number of more exotic keys. It was possible to transpose polyphonic parts, and it was possible to actually copy and paste individual notes within a polyphonic part. In fact, if you think of a MIDI region and all the major things you can do with that (moving timing and pitch, adjusting volume, etc.) then you are probably pretty close to the capabilities of this new version of Melodyne.

Even with all of this, we still aren't quite done. For quite a while Melodyne and Auto-Tune were the undisputed leaders in terms of manipulating the pitch, but in 2012 newcomers Zynaptiq released Pitch Map, which promised an equal degree of manipulation for polyphonic audio files. In one respect, Pitch Map promises a combination of both Auto-Tune and Melodyne. It offers the ability to process polyphonic audio files in the same way that Melodyne does, but it processes them in real time without the need for preanalysis in the way that Auto-Tune has. Once again the actual technology behind it is a closely guarded secret, and all Zynaptiq is saying is that it is based on perceptive modeling, which tries to analyze the audio input in the same way that we ourselves perceive sounds. It all sounds very complex, but the results do actually seem to be pretty good.

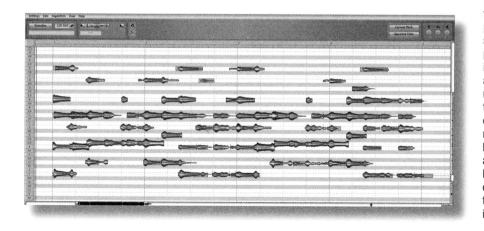

FIGURE 14.2
The latest versions of Melodyne allow for the retuning of polyphonic audio, something that was previously thought impossible. As long as the changes are relatively subtle and the audio itself not overly complex, the results can actually be very convincing, and it allows for last-minute creative changes, even after the recording session is over.

Pitch Map has an interesting work-flow that certainly helps speed things up. Instead of setting an arbitrary shift amount (as was the case with early pitch-shifting), or choosing a scale and having the input automatically "snap" to the nearest note (as was the case with Auto-Tune's automatic mode), or moving notes around manually (as with the usual Melodyne operating method), you simply route a MIDI input to the plug-in and play the note or notes that you want the audio to be shifted to and the plug-in does the rest. It really cannot be overstated how much help this can be. If you are working with a track that has chords outside of the usual major and minor chords, and you aren't especially musically knowledgeable, you don't need to spend time working out which notes you should correct a part to—you can simply use an existing MIDI part to specify the notes, or, if you are working completely with audio files, you could use a tool like Melodyne to extract MIDI note information and then use this as a MIDI input to Pitch Map in order to shift the notes of one audio region to match the notes of another audio region.

Now, it may be that your needs don't extend as far as the latest and greatest software that we have just been talking about, or it may be that your budget doesn't, but, in either case, there are many other plug-ins that deal with pitch-shifting and correction; better still, modern DAWs are fairly well-equipped straight out of the box to deal with the most common pitch-shifting tasks and some go much further than that. So let's go over some of the different ways in which pitch-shifting could be achieved, consider some of the complications that we might be faced with, and then take a look at ways we can put this technology to use.

DIFFERENT APPROACHES TO PITCH-SHIFTING

There are a number of different ways in which pitch-shifting can be done, from the fiendishly clever Eltro system to the proprietary (and top-secret) algorithms used by the new breed of pitch correction and manipulation plug-ins. At the most basic level, digital pitch-shifting can be achieved by using a combination of traditional and modern time-stretching techniques. We can use the traditional "resampling" method of time-stretching, which changes the playback speed of the sample in the same way that a tape machine or vinyl record would, linking a change in pitch with a change in tempo. We would adjust the playback rate to give us the desired pitch change, and then we could apply any one of our modern time-stretching techniques to return the file to its original length.

This method is, in fact, exactly what the Eltro systems did. The rotating tape heads meant that even though the speed of the tape running through the machine was constant, by rotating the tape heads at different speeds or in different directions, it was possible to change the *relative* tape-to-head speed. This would increase or decrease the pitch of the track and change the speed and therefore length of the recording. But the genius part was that, by having a number of reading heads rotating, it was possible to read the same part of the tape over and over. This repetition of sections of the recording is very similar to the way in which granular time-stretching works, and it would allow, theoretically, for the sound to actually be "frozen" in time. Of course, there would be audible effects with this method, but it should be pretty obvious by now that none of these methods are totally without audible effects and artifacts.

Granular techniques could also be used to achieve pitch-shifting. The pitch of a sound is determined by its fundamental frequency and that frequency is the number of waveform cycles that are completed every second. So, if we can make each of the "grains" a single cycle in length, then we can achieve pitch-shifting simply by changing the number of the grains that are played back in a single second. If we decrease the number, then the pitch will drop, and if we increase the number, the pitch will rise.

Finally, going back to our multiple sine wave resynthesis model that we spoke about in chapter 12, we saw that, if we could split a sound into a number of sine waves, each of a defined pitch and level with changes to that pitch and level tracked over time, we could time-stretch this sound by changing the rate at which the sound moved through those changes. Importantly, when it comes to pitch-shifting, the actual frequencies of these sine waves aren't as relevant as is their numerical relationship to each other. A very simple sound with a fundamental frequency of 440 Hz and harmonics at 880 Hz and 1320 Hz could be defined as having frequencies of n, $2n$, and $3n$. If we wished to pitch-shift the sound, so that the fundamental frequency (the perceived pitch) was now 260 Hz, then the ratio would stay the same, and the harmonics should be shifted to 520 Hz ($2n$) and 780 Hz ($3n$). Pitch-shifting using this method should, as with time-stretching, provide us with a near perfect result; however, the reality is that we can't realistically create an unlimited system, so there are compromises to be made in terms of the number of sine waves we could realistically use, the accuracy and extent of the envelopes, and, most important, the accuracy of the initial analysis to determine the pitches and levels and changes over time.

Then, as we have already seen, there is extensive research going on in psychoacoustics and the perception of sound, and the results of this research are starting to manifest themselves in remarkable new technologies and new products with bewildering and (perhaps deliberately) obscure acronyms for names. One thing that is abundantly clear is the fact that the people who are behind all these ground-breaking products clearly aren't out of ideas yet, so it is safe to say that things will continue to improve over the coming years.

SIZE MATTERS

There is one issue that needs to be considered whichever technology or approach is used to actually carry out the pitch-shifting. All acoustic sounds are made by the vibration of a physical object. The nature of that object varies from instrument (including voice) to instrument, but it is, in all cases, a physical process carried out by a physical system. And because the system is made up of physical objects, it will have physical properties: material, density, mass, rigidity, and so on. There are ways in which we can manipulate the system these objects operate in to create different pitches, but they are ultimately all governed by a set of rules.

The first rule that we should consider is the fact that, when it comes to acoustic sound generation, "bigger" equals "lower" and "longer." No matter what instrument we consider, if we were to increase its size, the notes we get from it would get proportionately lower in frequency and last longer in duration. In order to illustrate this, we only need to look at a grand piano. The sound-generation system for a piano starts with the hammers and strings. The hammers are the cause of the vibration in the strings, but they have nothing to do with the actual pitch of the note generated. That is determined by the length (and thickness) of the strings, and, even without knowing the physics behind it, it is easily seen that the thicker and longer strings at the left of the piano are the ones that make the lower frequency notes. If each note is played and allowed to decay naturally to silence, you will hear that there is a substantial difference in duration of notes as they rise in pitch as well. A typical grand piano might have the lower notes lasting as much as 50 seconds, while the higher notes can last as little as 0.3 seconds.

To be fair, the relationship isn't quite as simple as "bigger" means "lower," because there are other factors to take into account, such as the materials used, but our argument holds true ceteris paribus, meaning that if only the size were changed, then we would see the correlation we expect. The same is true of guitars, albeit in a slightly less immediately obvious way. There are two ways you can increase the pitch of the note you are playing on a guitar. The first is to select a higher fret, which shortens the length of the string (not the absolute length of it but rather the length of the part that is allowed to freely oscillate), and the second is to choose a thinner string. As you moved across the fret board from the thickest string to the thinnest, just playing the "open" strings (no fretting used), you will hear that the pitch rises. The strings are all (more or less depending on the exact positioning of the bridge saddles) the same length, and the tension in all the strings is reasonably consistent and certainly consistent enough to not be responsible for the large variation in pitch. So from this, we can see that shorter lengths (smaller) and thinner strings (smaller) both contribute toward a higher pitch. Actually playing a guitar will yield a similar result to a piano in terms of the decay time of the notes. The lower notes will last longer than the higher ones, although in this case the differences are nowhere near as pronounced.

If we move on to instruments that have sounds generated by wind (breath), we will see a similar relationship between the size and the pitch. In order to make the visualization easier, let's consider a pan flute. This incredibly simple instrument consists of a series of tubes, closed at one end, of increasing length (and sometimes diameter) that generate sound by air being blown across the top of the open end of the tubes. Again we see this correlation: longer tubes give a lower pitch. The relationship between size and duration is a little more complex in the case of these wind instruments, because the sound isn't created by a single event (a hammer hit on a piano or a pluck of a guitar string), but rather a continuous one. Even so, when the continuous excitation source (the air movement across the top of the tubes in this case) is ceased, there will be a natural decay back to silence, and once again this time is longer for the longer tubes than for the shorter ones. This decay to silence is also similar for bowed stringed instruments: once the bowing has ceased, there will be a longer duration of note decay with longer or thicker strings.

We have stated that we don't want the pitch-shifting process to change the length of the sound, and there are many occasions when that is true. If we are pitch-shifting a vocal phrase or perhaps a string-section phrase, then changing the overall length when pitch-shifting would mean that the melody would now be out of time. But, if we consider a piano part for a moment, although we need the overall length to stay the same for timing and rhythm reasons, not changing the length of the individual notes within that part could lead to a slightly unnatural sound. If we were to play a particular passage on a piano and record it, each note within it would last a certain duration (if we were talking about legato notes). If we pitch-shift that recorded part up an octave, then not only would the whole part be the same duration, but also the length of each note within it would be the same. If, on the other hand, we were actually to replay that piano part an octave higher, we would probably see that each of the notes was actually slightly shorter, owing to the decay characteristics of the notes varying with pitch. Perhaps we are talking about really tiny details here that wouldn't really be noticeable unless the pitch-shift amount was extreme; nonetheless, it is something to consider.

FORMANTS

Something that is also related to the physical properties of an instrument, and that can be much more audible and discernible, is the subject of formants. Formants are peaks in the spectral/frequency response of a sound that is caused by natural resonances in the physical body and mechanism used to create the sound. In the case of stringed instruments, this would be the hollow "body" of the instrument, while in wind instruments (including brass instruments), it is the "tube" through which the air flows. Drums and many percussion instruments have resonances caused by the "shells" of the instruments, and pianos have the physical body of the piano combined with the soundboard. The human voice, while capable of much more variety and expression, behaves in basically the same way as any other wind instrument and has its own set of "tube" resonances, but this is further complicated by the fact that the mouth, tongue, and nasal cavity all have a complex interaction with the basic throat "tube" and can change and move these resonances. And these resonances also follow the "bigger" is "lower" rule. The resonant frequencies of a larger sound-producing mechanism will be lower than those of a similar but smaller mechanism.

Formants

The most important thing to know about these formants and resonances is that they are not related to the pitch of the note. They are consistent with the physical structure of the instrument, and the frequency of these formants doesn't change as the pitch of the note changes. If we consider this in the context of everything we have looked at so far, we will see why this could be a problem with pitch-shifting. When we pitch-shift a sound, we are multiplying all the frequencies by a constant fact. A multiple of two would produce a shift of an octave up, for example. So while the relative harmonic balance will be consistent (assuming perfect pitch-shifting), we will have shifted the frequencies of the formant peaks as well. If the frequencies of the formants are higher, but the spectral balance of the sound is the same, then the effect will be that the sound is still recognizable as being what it is, but it will sound like it has been shrunk. Equally, a pitch shift down will move the formant frequencies lower, which will make it sound like the instrument has been made physically bigger.

This is noticeable in all acoustic sounds to a greater or lesser degree, depending on the relative strength and complexity of the formants but it is especially noticeable with the human voice. One of the reasons for this is simply that we have evolved, as a species, with our hearing systems being especially finely tuned and sensitive to the sound of other voices. This has meant that our ears are very good at picking up even very minor changes in this frequency range and tonal balance of these sounds. A greater factor, however, is the flexibility and adaptability of the human voice and the range of sounds it can produce. The formants in our voice are much more complicated than in any other instrument and are much more moveable. Each different vowel sound, for example, is formed by a different combination of positions of our jaw, tongue, and mouth. As a result of this, each vowel sound has a different resonance, as the size and shape of the overall sound-generating system have changed.

The frequencies of, and the spacing between, these formants is what give vowels their unique sound. You could, without stopping the note at all, hold a vocal note and change the sound from "aaah" to "eee" to "ohhh" to "ooo." The fundamental frequency of the note wouldn't change, and therefore the pitch would remain the same, but the formants would be changing as you changed the sound. If you were to record this and then look at the frequency distribution during each of the sounds, you would see clear peaks (other than the peak of the fundamental frequency of the note) that moved as the sound changed. If you now recorded yourself holding an "ahhh" sound but changing the pitch of the note, and then you looked at the frequency distribution, you would see that those really prominent peaks would stay in the same place, even though the other frequencies were moving as you changed note. To further complicate matters, if you were to record somebody of the opposite sex repeating your experiments with the same pitches used and then compared the frequency distribution, you would most likely see the formants in different places. And if you were to repeat this again with a child, you would see the peaks in yet another place.

These formants are, therefore, crucial not only in determining the actual character and intelligibility of the sounds being made but also can vary depending on size and gender. If a simple pitch shift is carried out on a vocal sound, the resulting shift in formants would, at the very least, change the perceived tone of the singer and could risk sounding like a gender or age change had happened along with the pitch-shifting process. There is also the possibility that certain vowel sounds could become slightly confused. Naturally, if we are only talking about fine-tuning a note that was just a little flat to get it in tune, then the amount of formant-shifting taking place would probably be minimal enough to not be an issue. But if we are considering actually shifting the pitch up or down by a few semitones or more, then we run into greater risk of unwanted tonal changes.

The good news is that these formants are dealt with very effectively by most modern pitch-shifting systems. Some give you control over them independently of pitch, while others simply automatically correct and re-map the formants to the positions they were at before the pitch-shifting occurred. In fact, formant-shifting has actually been used on its own, not in relation to any kind of pitch-shifting but as a part of a voice-modeling application. The idea that the frequency of these formants can determine perceived size and gender has led to the idea that, by manipulating the formants (among other things) of a signal without touching the pitch, you can quite substantially alter the tone of a person's voice. Antares Audio Technologies, makers of Auto-Tune, actually released a plug-in called Throat, which used "vocal tract modeling" to allow you to alter the length, width, and other characteristics of the vocal tract to offer subtle or radical changes to the vocal tone.

The fact that formants can be automatically put back in their original positions following a pitch-shifting process, or that they can be moved independently, gives us much more control over the final result. In a very simple way, this movement of formants is similar to the tonal matching we spoke about back in chapter 6, but rather than matching the exact balance of a wide range of frequencies, we are instead just looking for those few really obvious peaks in the preshifted sound and looking to shift the corresponding peaks in the post-shifted sounds to these positions. The actual ways in which this is achieved vary widely among the different methods used for achieving the pitch shift, but they all aim to achieve the same end result.

So a "simple" pitch change of a vocal track is actually a three-stage process: play back the audio at a different rate to achieve the pitch shift, apply formant correction to recover the correct tonal balance, and then apply time-stretching to correct the change in length. Given that formant correction is an integral part of most pitch-shifting systems, the formant correction will also be applied to any sound that is processed and the correction applied automatically. While the effects wouldn't be so obvious in other acoustic sounds—and not necessarily present at all in synthetic sounds owing to there being no physical mechanism used to generate the sounds—the fact that there is a system in place, checking and making sure that there aren't any horrible formant-shifting surprises at the end of the pitch-shifting process, certainly means that we can push things farther than we ever would have done before when it comes to pitch-shifting sounds.

USES OF PITCH-SHIFTING

The earlier forms of pitch-shifting were good at either chromatic correction (moving the pitch of the incoming signal to the nearest note) or the creation of simple, fixed interval harmonies, but it is really the introduction of the later technologies, and consequently being able to move notes around in the same way that we would MIDI notes, where we really started moving into the areas of elastic audio from a pitching sense. With the integration of Melodyne technology deep in the structure of PreSonus Studio One, and with many DAWs offering similar functionality as a part of their built-in feature set, it is getting to the point where the pitch of our audio is automatically elastic from a few moments after we stop recording or importing. As a result, there is a very fundamental change in the way that we think about audio.

Correcting a slightly off-key vocal is likely to be one of the main tasks that this technology will be useful for. Although the fully automatic options are highly tempting, it is almost always better to do things manually, because you can just keep a little more realism in the performance as a whole. While a fully automatic algorithm will take a note that is cents flat and tune it to the exactly correct pitch, if you choose a manual option, you can still get it close (say, five cents flat), but the minor differences from note to note will make it sound somehow more authentic and genuine rather than heavily processed and uniform. Admittedly it isn't *as* quick, but even re-tuning every single note of a vocal performance wouldn't be a mammoth task with the ease of use that the software has today. Correcting pitch is usually as simple as selecting the right tool, clicking on the note you want to re-pitch (perhaps holding down a modifier key to select a "fine" mode), and moving the mouse up or down while watching the display readout to see how far from the perfect tuning you are. You adjust, and then you move on to the next note.

It would be best to adjust the tuning by ear, of course, but if you were in a hurry, you could probably move to within five cents, perhaps even ten at the most, of the target pitch without even listening to the result and be quite safe. Even so, once you were done adjusting the pitches "blind" like this, it would be especially prudent to have a listen through and check everything, making note along the way of any words that still sounded like they needed a little more work. Melodyne actually has a "Correct Pitch" function which will automatically move notes to the nearest chromatic note (or note within a scale, if you have defined one), but this is different from the fully automatic pitch-shifting, because it has a slider that allows you to gradually change from 0% (no correction) to 100% (perfect pitch for every note), so you could easily select every note in the performance and then adjust this one parameter to something like 75%, and then 20 cents sharp would move to 5 cents sharp, 33 cents flat would move to 8 cents flat, and so on. From there you could listen through and adjust notes that needed further work individually.

The obvious next step from this idea is to create copies of tracks and to actually create vocal harmonies from a single vocal track. The notes of the lead vocal can be moved around at will to create whatever harmonies you want. Most pitch-shifting software still works only monophonically, so complex harmonies would need to be built up over a number of tracks. Each track could have the actual note changed to the new harmony note; the fine-tuning of the new note changed, so that not all the harmonies for a particular word were sharp or flat by exactly the same amount; the timing moved slightly; and even the tone changed a little (after the initial formant correction had been done). Once all these steps had been done, you could end up with quite complex harmonies from a single vocal recording. However, no matter how great the possibilities for pitch (and even tone) manipulation are, the one problem with this method of creating harmonies is that they will all have exactly the same delivery. The phrasing will be the same, the pronunciation will be exactly the same, any vibrato will be at the same rate, and even the subtleties of the accent will be the same, so harmonies created in this way will never have the richness, depth, and complexity of those recorded by different takes and preferably different singers. Used carefully, though, and perhaps as a backup to harmonies already performed by a backing vocalist, they can add a lot of interest to a well-recorded but perhaps rather limited vocal.

A variation on this, of course, is to use the ability to change the notes completely to change only certain notes to create a new melody. This could be to cater to a last-minute moment of inspiration that called for a different chord change after the vocal had been recorded, or it might be something as simple as changing one note within a phrase at the end of a final chorus of a song just to give a sense of something being a little different for this final repeat. Simple little touches like that can really add a sense of interest to a song, but, if you have been put in a position where you have only one tidied up and comped chorus that was intended to just be copied and pasted to all locations it was needed,

FIGURE 14.3
Vari Audio, as used in Cubase, makes it very easy to create harmony lines from a single vocal. In the image above the original vocal (bottom) and created harmony (top) are both shown, and you can see that while the pitch has changed, the subtleties of the micro-tuning are mirrored to some extent.

then small changes like this can go a long way toward making each chorus slightly different and individual and reduce the sense that it has just been copied and pasted out of laziness.

There are many uses for elastic pitch other than simply enhancing a recorded performance. One spectacularly useful one, especially if you have access to polyphonic pitch-shifting, is the repurposing of existing audio files. It may be that, while working on a project, you have a sudden moment of inspiration and think that an acoustic guitar arpeggiated chord, perhaps similar to the one you used on your last project, would sound perfect with this—obviously not the same, because this track is much slower and in a different key, but along those lines. But it's late, and you are working alone. You don't want to forget about the idea, though, and you have a very specific thing in mind that you might not be able to capture by simply writing it down. So you load up the acoustic guitar part in question from the previous project. When you have imported it in and tempo-matched it to the current project, you load it into your pitch-shifting software and then move the notes around to fit the chords of the project you are working on. Perhaps you mute a few of the notes and change the timing feel of the ones that are left. You listen back, and you realize that it isn't perfect, and that you couldn't use it anyway, because the band that you recorded for the last project probably wouldn't be happy with your using parts of their recording on this project. But it doesn't matter; in those few simple steps, you have created something that is much closer to the effect that you want to create than you could have easily put down in words. And next time your guitarist friend is nearby, you can get him to pop in to the studio, and you can play for him the exact riff and pattern that you have in mind, and he can work directly from that and be 90% of the way toward getting you what you need before he even plays a note.

And then, finally, you can always use a Pitch-to-MIDI feature (if one is available) to extract MIDI note and timing information from, for example, the bass guitar, and then load up a really warm and "subby" synth bass sound and just layer that underneath the bass guitar recording, so that you have a separate track that can be used to add a little "weight" if needed when it comes to the final mixdown. It might be possible to compress and EQ the bass, so that it has more depth to it, but, given that there is only a certain amount of EQ-ing that you can do on any given sound, having this synth bass "double" available is certainly worth the time and effort that it would have taken to create it.

HANDS ON

Introduction

As we have seen in this chapter, pitch-shifting has evolved from being a simple ratio-based technique, through a chromatic interval-based technique, and finally on to much more intelligent systems that give audio files much of the flexibility of MIDI regions. While many DAWs offer advanced pitch-shifting capabilities,

Pitch-Shifting **CHAPTER 14**

the specifics and implementation of the systems varies quite substantially, so in this section, we will take a look at the ways in which the DAWs themselves, without the use of any third-party software or plug-ins, can manipulate the pitch of audio recordings.

Logic

Quite surprisingly, given some of the very advanced audio editing features of Logic, its pitch-shifting capabilities are actually quite simplistic. The only real options that you have are using Logic's *Pitch Shifter II* or *Pitch Correction* plug-ins (for real-time, nondestructive changes) or using the Pitch section of the *Time and Pitch Machine* (for destructive changes). Each of these methods has its advantages and disadvantages, and it is often the case that a combination of the two will give the best results. We will take a look at the plug-in methods first and then the Time and Pitch Machine method before finally looking at a way to use the two methods together.

The choice of whether to us Pitch Shifter II or Pitch Correction will depend on the task you have to complete. If you are aiming to correct small tuning inaccuracies and just bring things into line with the key of the song, then Pitch Correction will be the right choice. It should be noted, however, that while you can play polyphonic audio in to the Pitch Correction plug-in, the process (unlike the latest versions of Melodyne) will work properly only with monophonic audio parts. The easiest way to use this plug-in, if you know the key of the song, is to choose the appropriate key/scale using the *Root* and *Scale* controls at the top of the plug-in. It comes with a good selection of common (and not-so-common) scales, including all the usual major and minor scales and a number of choices that limit the notes to only those used within the selected chord.

For example, selecting a C major scale will set the keyboard display to highlight the notes C, D, E, F, G, A, and B. That means that any incoming audio will be tuned to the nearest of those notes and will then be in key. However, choosing a C major chord from the Scale box will only highlight the notes C, E, and G. Using this setting, an incoming note of A, which would technically be in key, would be re-tuned to a G, so that it was not only in key but also within the component notes of the chord itself. Alternatively you can manually select and deselect notes in the keyboard display directly, so you can create custom scales or restrict tuning to a single note if you desire.

Once the scale has been chosen or created, the main control that you will use will be the *Response* control on the right of the plug-in window. This is variable between 0 ms and 999.99 and works by determining the amount of time that it will take for the plug-in to correct the incoming audio. For a natural-sounding result, the default value of 122 ms is a good starting point. Reducing this time, potentially all the way down to 0 ms, will lead to what is often thought of as the classic Auto-Tune effect. Equally, increasing it up toward 999.99 ms can lead to a more subtle result, but this is really advisable only if the source file were fairly close to the required pitch; otherwise, there would be a very noticeable bend in pitch.

Just to finalize matters, the *Range* control offers normal and low ranges, but, according to the Logic documentation, the low-range option should really be used only for audio that has frequencies below 100 Hz, as these low frequencies can sometimes cause pitch-detection problems. The *Detune* control, located on the right of the plug-in window, is used to apply a global tuning offset to the audio after the pitch correction has been carried out.

Moving on now to the Pitch Shifter II plug-in, we have a much simpler affair. The two tuning controls, used to define the static pitch shift, allow you to specify a *Semi Tones* shift (+/− 12 semitones) and a *Cents* shift (+/− 100 cents). Below these two controls, there are three buttons to choose between different algorithms. Of the three choices, *Drums* is optimized more to maintain the timing and groove of the source, *Vocals* is optimized more toward maintaining the intonation of the source, and *Speech* is a good balance between the two. The naming of these options isn't perhaps overly descriptive, but, once you know what they are each used for, it is good to have different algorithms to choose from. Finally, there is a *Mix* control that allows you to vary the balance between the source and processed audio.

However, this brings us to the biggest problem with this plug-in: latency. If you set up an appropriate shift and then move the *Mix* control between 0% (only source audio) and 100% (only processed audio), you will be able to hear a clear delay. The extent that this is a problem depends on the type of material. Very percussive and rhythmic material will clearly suffer more than more-legato and sustained material, but in any case, it can be very distracting. You can mitigate the problem to some extent by using the *Delay* setting in the *Inspector* panel, but it can be a lot of messing around. As a result I often use the Pitch Shifter II plug-in to figure what tuning amount is required in a real-time scenario and then move over to the *Time and Pitch Machine* to carry out the final shift using arguably better algorithms and no resulting delay.

We have already looked at *Time and Pitch Machine* at the end of chapter 12, and have discussed the functionality of the *Mode* and *Algorithm* controls, so we don't need to address those again. Equally, we don't need to look at the controls relating to the tempo/time-stretching, as those aren't relevant to what we are doing here. In fact, the only two controls that we need to focus on are the *Transposition* and *Harmonic Shift* controls and the *Harmonic Correction* button. Transposition is totally self-explanatory, and the only comment I will make here is to make sure that you remember to enter the transposition amount in cents. To shift upwards by 7 semitones you

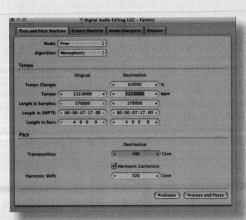

FIGURE 14.4
Time and Pitch Machine is also the place to go for off-line pitch-shifting in Logic. Note the independent settings for Transposition and Harmonic Shift. They can be tied together for the most natural-sounding results, but having the option to set different values for both allows for more fine-tuning of the result.

would need to enter 700 cents; to shift down by 4 semitones and 24 cents you would need to enter −424 cents.

The other two controls are related to dealing with formants. The *Harmonic Correction* button will engage Logic's built-in automatic formant correction, which will improve the realism of the resulting sound at the cost of an increased processing time. The Harmonic control works with the formants and allows, in conjunction with the Harmonic Correction button, to control the formants independently of the transposition. A value of 0 equates to full harmonic/formant correction, and a value equal to the transposition value equates to no harmonic/formant correction. Any value (+/− 3600 cents/36 semitones) can be applied to the transposition and harmonic controls, and they can be processed separately. It is, for example, possible to apply a transposition, and then, at a later time, apply harmonic correction if desired.

Pro Tools

ProTools, as standard, doesn't include any pitch-shifting plug-ins that enable real-time pitch-shifting on a track-by-track basis. Some might argue that this is quite an omission, but, in its defense, it actually goes one stage further and offers real-time pitch-shifting on a region-by-region basis. This ability is a part of the Elastic Audio engine that we have already looked at in depth. In order to change the pitch of a single region, you first need to make sure it is Elastic Audio–enabled and then open the Elastic Properties dialogue (right-click on the region and choose *Elastic Properties…*, Alt + Num Keypad 5 or Clip > Elastic Properties). At the bottom of this dialogue, there is a *Pitch Shift* box that allows you to enter a semitone value (+/− 24 semitones) and a cents value (+/− 99 cents). This incredibly simple method means that a pitch shift can be applied, in real time, for each individual region, which means that, depending on how much you have separated your recording into regions, each note could theoretically be pitch-shifted by a different amount.

While this method doesn't offer the visual fluency of some third-party tools such as Melodyne, it does, in a way, provide much of the functionality of Melodyne in the sense of being able to correct each individual note without being forced to use destructive editing. What it won't do is automatically correct your region to a given note or scale, and if that is what you need to do, you will need to look at a third-party application or plug-in. If, however, your needs are simpler, and you just need to carry out manual tuning of certain notes or passages, then this method can enable you to do that with minimal effort and interruption to your work-flow.

As with all real-time pitch-shifting processes, though, it is worth remembering that the algorithms used aren't necessarily going to give the best possible results, because there is always a compromise between immediacy (real time) and quality (off-line/rendered). Real-time, and in particular region-by-region, pitch-shifting is an amazing tool for you to figure out what needs to be changed and by how much, but, once you have this figured out, it could be a better option to commit those changes by rendering them using an *AudioSuite* plug-in. Not only would this give you arguably better quality, but also it would free up system resources. Any processes that are applied in real time (such as time-stretching and pitch-shifting) will place a load on your CPU. An occasional use of such things probably wouldn't make much of a difference with modern computers, but, with the ease of use and huge benefits of Elastic Audio processing, it is extremely easy to find yourself using it in more and more situations, and, as such, the CPU load can quickly mount up. Off-loading some of that by rendering changes should make your whole system more responsive, so it should definitely be considered if the changes made are ones you are fairly certain will be final.

Another quick way to carry out pitch-shifting, if you have an audio track that uses the *Polyphonic* algorithm, is to open the *Event Operations Transpose* dialogue by pressing Alt + T or by going to Event > Event Operations > Transpose… and, using the *Transpose by* sliders to set an amount you wish to pitch-shift by, then pressing *Apply*. This will apply the chosen amount of pitch shift/transposition to the region(s) that you had selected. It is still a real-time process, so it can be removed by deactivating Elastic Audio for that track, but that would, of course, also remove any other Elastic Audio editing that you may have carried out. If you aren't happy with the results then, you can either undo the process, or, if it has been a while since you applied the transposition, simply transpose again by an equal but opposite amount to get the audio back to where it was.

At the end of the chapter 12, we looked at the using the *AudioSuite TimeShift* plug-in to carry out time-stretching, and we noted that there was a section in there called *Pitch*, which wasn't relevant at the time but now most certainly is, so it's time to revisit that now. Select the region(s) that you want to pitch-shift, and then open up the plug-in by going to AudioSuite > Pitch Shift > TimeShift. Here you will see the four sections that we discussed in chapter 12, but this time, we will be using the pitch-shifting aspects of the plug-in, so the middle *Time* section isn't relevant. The first thing we need to do, as is so often the case, is select the appropriate algorithm from the *Audio* section at the top.

While it is possible that you might wish to pitch-shift drums and percussive parts, it is most likely that you use pitch-shifting more on melodic parts, so the choice will often come down to either *Monophonic* or *Polyphonic*, depending on the nature of the material. Monophonic will give arguably better results on solo instruments, because it is optimized to preserve formants and thereby retain a more natural-sounding character. However, if your material isn't actually monophonic, then such formant preservation is unlikely to work correctly, so it is better to stick with the Polyphonic algorithm.

If you use the Monophonic algorithm, then the third section will change from *Transient* to *Formant*, and this will allow you to independently adjust the formats without adjusting the pitch. The main chapter above discusses formants and why they are important, and having a separate control means that you can either adjust the formants without adjusting the pitch or, alternatively, can use the *Shift* control to fine-tune the sound of the pitch-shifted audio. The pitch-shifting process will automatically apply formant correction to try to avoid "munchkinization" effects, but often a little adjustment of the formants can help to make a sound feel like it sits a little better in the mix.

Finally moving on to the *Pitch* section, we have just one control but three different ways of adjusting it. For the most part, the *Transpose* control will be the one that you use, because it allows you to click on the readout and type a shift amount in semitones (+/− 24 semitones). If you need to include cents, then these are just the decimal part of the number. For example, 5 semitones and 58 cents would be equal to 5.58 semitones. If for any reason you would prefer to calculate the pitch shift in terms of a %age chance, you can do that by clicking on the Shift control and typing in a %age change (to two decimal places). And finally you can use the large knob to the right of the section to adjust the shift amount. If you hold down Alt while clicking on any of the controls, the shift value will reset instantly to zero.

As with the time-stretching examples, you can preview the results by clicking on the *Preview* button at the bottom left of the plug-in window, and then, once you are happy with the results, click on *Render* at the bottom right to commit the results to a new file.

FIGURE 14.5
Time Shift also allows you to create pitch-shifting effects along with (subject to the choice of algorithm) an independent formant shift control.

Studio One

Studio One is one of the best-equipped DAWs when it comes to pitch manipulation of audio files. In addition to very flexible region-by-region pitch-shifting, there is also the very tantalizing option of deeply integrated Melodyne audio processing. As we saw in the main chapter, Melodyne offers the ability to move audio around in the same way that we can MIDI notes, and this really does start to push forward the boundaries of what can be achieved with audio files. Studio One Professional includes a fully licensed copy of Melodyne Essential, while the Artist and Producer Editions include a demo version. It is worth pointing out that the version included does not include the Melodyne Editor DNA (polyphonic) option, but, considering that it is included as a part of the Studio One Professional package, this isn't really anything to complain about. If you do happen to have the latest version of Melodyne Editor, then, in addition to working as a plug-in, it also works with the Studio One ARA implementation, which we will look at shortly.

Before we get to that, we will take a look at the more traditional Elastic Audio pitch-shifting capabilities that all versions of Studio One have. Any audio file that is imported will automatically be ready for transposition. There are two ways to access the transposition parameters. The first is to open the Inspector panel by pressing F4 or by going to View > Inspector. Toward the bottom of the Inspector panel, you will see all the event parameters relating to tempo, gain, transposition, and fades. The two parameters that we are interested in here are *Transpose* and *Tune*. The *Transpose* parameter has a range of +/−24 semitones (whole semitones only), while the *Tune* parameter can be set between +/− 100 cents. The other way to access these parameters is to right-click on an audio region, and then, in the pop-up box that appears, you will see these same event parameters at the top. In either case you can either click and drag up or down on either parameter, or, if you prefer, you can double-click on either parameter and type in a value.

At the top of the Inspector panel, you will see a parameter for Timestretch, and, while this might not seem relevant, it can have an impact on the final result of any pitch-shifting that you carry out. We looked at the different algorithms at the end of chapter 12 and noted that each of them was optimized for a particular purpose. One of them, *Solo*, had built-in format correction for better results when pitch-shifting, so we should choose this one if we are looking to carry out any pitch-changing on a monophonic track. Unlike some DAWs, there is no direct control over the amount of formant correction, and you can't carry out formant shifting independently of pitch-shifting. While for most purposes this won't be too much of an omission, there could be times when you would like to shift the formants without changing the pitch, so, if this is the case, you will need to look to a third-party tool to achieve it.

One of the best things about the Studio One method is that the changes are nondestructive, so that you can quickly and easily change the values without having to worry about rendering time or without worrying that you will be transposing an already transposed region. If an off-line (destructive) method were used, then each subsequent pitch change would add artifacts to the previous change, but this method always applies any change to the original file, so a number of changes will have no more of a detrimental effect than a single change. In addition, there is the very significant benefit that this is done on a region-by-region basis. As a result you can split a region into as many parts as you want and apply different amounts of pitch-shifting to each part. You could, in theory, apply a different amount of pitch-shift to each note in a performance and create an entirely new melody.

But if this is what you want to do, then there is a much easier method that comes from the very deep integration of Celemony's Melodyne. While Melodyne has been around for quite some time, and it's widely regarded as being the pinnacle of audio pitch manipulation, the way in which it is integrated into Studio One marks a new step up in terms of functionality and ease of use. The plug-in version of Melodyne was fairly easy to use, but it always acted in a way that was layered on top of your audio. You could edit audio with it, but the edited audio existed inside the Melodyne plug-in. Now, with the development of the ARA (Audio Random Access) plug-in standard, the Melodyne editing acts much more like it is a part of the audio itself. With the plug-in version, you would have to record audio into the plug-in in a given location, say, between bars thirty and thirty-six. Once you had made the changes, they would take over from the original audio. If, however, you wanted to move or copy the audio between bars thirty and thirty-six, you would have to bounce down the Melodyne version; otherwise, simply moving the audio region in the arrangement would have no effect, and the original Melodyned audio would remain in place.

With ARA this is no longer the case, as the Melodyne changes are tied to the region itself rather than a specific location. This means that you can freely copy, move, and split regions, and the Melodyne changes will remain correct. This freedom to manipulate the edited audio in this way, combined with the fact that the editing takes place in the Editor panel, makes the whole process feel much more like a part of Studio One rather than being a plug-in or an add-on. It might seem like a small detail, but the fact that you select an audio region and press Cmd[Mac]/Ctrl[PC] + M or go to Audio > Edit with Melodyne, rather than having to load a plug-in and then record the audio into the plug-in, just makes the whole process much easier and more intuitive.

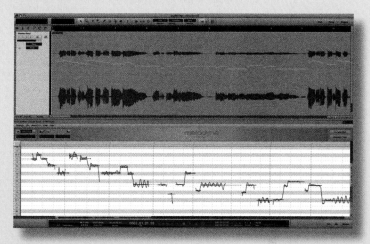

FIGURE 14.6
The integration of Melodyne into Studio One takes native pitch-shifting and pitch manipulation to a whole new level. The version of Melodyne included with some versions of Studio One allows for only monophonic processing, but it can be upgraded to allow for full polyphonic pitch manipulation to be a part of and integrated deeply into a DAW for the very first time.

It doesn't end there, though, because the Studio One/Melodyne combination has one more trick up its sleeve. While this isn't strictly pitch-shifting as such, it is related to the pitch-manipulation process, so it deserves a special mention here. When a region has been edited with Melodyne and the pitch analysis carried out, you can use this pitch-detection data to create a MIDI part that mirrors the melody and timing of the audio region. Pitch-to-MIDI isn't a new concept, but the ease of use here is simply stunning. To transcribe an audio part to MIDI, you simply need to select the audio part, press Cmd[Mac]/Ctrl[PC] + M or go to Audio > Edit with Melodyne, allow the analysis to be carried out, check that melody has been detected correctly (making any small adjustments to the detection as are necessary), go to Track > Add Instrument Track, then simply click on the Melodyned audio region (it will have "piano roll" notes over the top of the waveform overview, showing you it has been analyzed), and drag it on to the MIDI track, where the pitch data from Melodyne will be converted to MIDI note information. And that is all you have to do. This makes it also inconceivably easier for somebody to be able to quickly sing a melody in and then convert that to, for example, a bass guitar part. As a result, this is a tool that goes beyond simple correction and opens up all kinds of creative possibilities.

Cubase

When it comes to pitch-shifting, Cubase offers some very advanced features that fully embrace the Elastic Audio concept. Like Studio One with its Melodyne integration, Cubase features the ability to not only pitch-shift an entire region in real time but also go into the region and change things on a note-by-note basis. This includes not only changing the root pitch of the note but also controlling what Cubase calls the *Micro-Pitch Curve*. This is the detail of the minor pitch fluctuations within an individual note, such as subtle pitch bends, glides from note to note, and, of course, vibrato. In fact, *VariAudio*, Cubase's pitch-manipulation tool, features a great deal of the functionality of the monophonic version of *Melodyne*, so its inclusion is a significant advancement on earlier Cubase versions.

In order to work on any pitch manipulation, you will need to open the Sample Editor window and click on the VariAudio button. After an initial analysis, you will be presented with what looks like a combination of the piano roll editor and a normal waveform overview. To the left of the window, you will see a piano keyboard with associated note names, and, overlaid on top of the waveform overview, you will see horizontal bars that represent the notes and durations of the analyzed audio. The detection process is usually quite accurate, but you may need to may a few small changes before you can really get started.

If you need to adjust the length of any of the *Segments*, then you should click on the *Segments* button under the VariAudio heading. If you now move the cursor over the top half of one of the Segments, it will change to a pair of left and right arrows with a small vertical line between them. This tool is used to reposition or change the length of a Segment. If you click with this tool in the center of a Segment, then you can click and drag to move the entire segment backward or forward in time, but only up to the edge of another segment. Positioning this cursor at either end of a Segment and then clicking and dragging will allow you to lengthen or shorten a segment, but, again, only extending it up to a maximum length, where it will be adjacent to the following section.

At this point, you may be wondering what happens if you shorten segments so that there are gaps between them. The answer to this is simply that the segments represent parts of the audio that are processed by VariAudio. If there is a space between segments, but there is audio in that space (as in, not in an area of silence), then, for the duration of that gap, the original unprocessed audio will be used. Using this idea, it would be possible to carry out the VariAudio analysis and then delete all the Segments apart from one or two, and then work on only those segments. The result would be that you would hear the original audio in all areas apart from the Segments, which would play back the processed audio. If authenticity to the source is a priority, then using Segments only where you actually need to correct or change the pitch will give you the result with the fewest artifacts.

Another thing that you might have to change is the actual separation of the notes themselves. Each note is represented by a Segment, and, for the most part, each separate note will have its own segment. There could, however, be cases where Cubase has only one segment, which covers more than one note. This may be because there is a subtle glide from one note to the next or because the transition is not immediately obvious to the analysis. In a case like this, you will have to manually split the segment into the individual notes. If you now move the cursor over the bottom half of a segment, you will see it change to a "scissors" cursor. Clicking with this cursor will split the Segment at the point where you clicked, and this can allow you to separate out glides or other situations where the note separation wasn't picked up accurately.

If you are unsure exactly where you will need to split the Segment, then you can zoom in so that the area around where you think you will need to make the split is quite detailed. Once you are zoomed in, you can select the *Play* tool from the toolbar. This tool allows you to audition any part of the audio by clicking on the point at which you want playback to start. The audio will then be played from that point until the mouse button is released or the end of the region is reached. In order to help you visualize where playback is, using this tool will color the background of the waveform overview as it plays back, so you will have a clear indicator of what part of the waveform is being played back. Using this method, you should be able to quite easily pinpoint exactly where the note transition occurs, and then you just have to click on the Segments button again and use the scissors to cut the Segment at the appropriate point. Once the Segments are properly in place, you can start working on the pitch of the notes.

Clicking on *Pitch & Warp* allows us to actually start moving things around, and, in many ways, it really is a case of drag and drop. Moving the cursor over the bottom half of a Segment will change the cursor to a finger, and this can be used to move a note to another note by clicking and dragging. As you do this, the whole note, including the *Micro-Pitch Curve*, will be moved to the new note. If the original note was slightly sharp or flat, then moving it this way won't change that. If the original note was a slightly sharp C and you move it up two semitones, then you will be left with a slightly sharp D. This method doesn't apply automatic tuning of the notes. If you want to automatically tune the note as you move it, you can hold down Cmd[Mac]/Ctrl[PC] as you drag, and it will not only move the root note but also correct any slightly out-of-tune pitch as well. If, on the other hand, you wish to move a note by only a small amount and don't want to jump to the next semitone, then you can hold down Shift while you drag, and this will allow very fine tuning.

If you do decide that you want to apply an AutoTune-esque automatic pitch correction to an audio part, then you can do so using the *Quantize Pitch* control under the VariAudio heading. Adjusting this slider allows you to go from no pitch correction all the way to 100% accurate pitching. At the 100% setting, you have very much a classic AutoTune effect, where everything is completely in tune. However, as theoretically ideal as this is, a perfectly pitched performance can often sound extremely unnatural. The use of the Quantize Pitch slider allows you to get progressively closer to that and still be able to stop some way

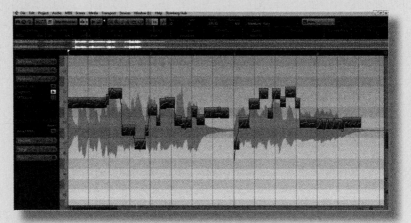

FIGURE 14.7
While it may not offer the polyphonic option that Melodyne does, the Vari Audio technology in Cubase is significantly more intuitive than almost any other software out there and, when combined with the chord track and other aspects of Cubase, offers great creative and compositional options as well that go far beyond the realms of simply pulling things into tune a little.

short of perfect pitch. It may be that you use this slider to get most of the performance sounding great, but there is still a note or two that needs additional work. Rather than pushing the slider further over, you can simply go in and fine-tune any individual note(s) in the manner we have just described.

Below the Quantize Pitch control, we have the *Straighten Pitch* control. Obviously, the center pitch of any note is the most crucial thing when it comes to determining if something is in tune or not, but it is often the small pitch fluctuations (or *Micro-Pitch Curve*, in Cubase terminology) that contains the true emotion and expression of the performance. Vibrato, for example, can be extremely powerful in expressing emotion, but the very wide vibrato that some singers have can be a little too much at times. Adjusting the Straighten Pitch slider will minimize vibrato and any other small pitch deviations. Taken to extremes it can sound very robotic and unnatural, but used carefully, it can give you a great deal of control over the subtleties of the performance.

The key here is that one word: "subtlety." Timing quantization, pitch (center) quantization, and micro-pitch flattening are things that are all too easy to just apply and "set and forget," which will almost inevitably result in a very sterile and unnatural-sounding finished product. But if we apply these modifications selectively and carefully, or even if we apply the same principles of timing and tuning correction but move everything manually, the end result is always worth the effort.

SECTION 3
Restorative Editing

CHAPTER 15
Editing in the Third Dimension

INTRODUCING THE IDEA OF SPECTRAL EDITING

So far, when it comes to actual hands-on editing and not the behind-the-scenes processing of time-stretching or pitch-shifting algorithms, we have been dealing with standard "two-dimensional" editing. Perhaps that concept isn't familiar to you, but what I mean by it is simply that we work mostly on no more than two aspects of a sound at any given time. When we copy or move a region, we are moving the whole sound—all frequencies and all amplitudes—along the time axis. So that is what I would consider to be a "one dimensional" edit. When we change the gain of a particular region as a whole, that would also be a one-dimensional edit, as we are changing the amplitude of all frequencies for the whole duration of the region. If, however, we were to use a volume automation change to create a volume increase or decrease over time, then this would be a two-dimensional edit, because we would be affecting all frequencies equally, but the gain would change differently over time. Using a conventional EQ would be a two-dimensional change, because we would be changing the gain (first dimension) of only certain frequencies (second dimension), but this would be constant over time (forgetting automation for the moment). Time-stretching would be one-dimensional, because we are changing only the time component, and frequencies and amplitudes wouldn't change (artifacts aside), and pitch-shifting would also be one-dimensional, because we would change frequencies but not amplitudes or the time dimension.

From this I hope you can see that my definition of "three dimensional" editing (where the three axes are pitch, volume/level, and time) involves making changes to a *specific* range of frequencies at a *specific* amplitude over a *specific* period of time. In order to do this, we don't necessarily need specialized tools. We have already seen that an EQ change is two-dimensional, because it changes the amplitude of a specific range of frequencies (the cut or boost amount at the specified center frequency), but this change is a constant one, because it is applied to the whole track for its duration. In order to add to this, we can automate the cut or boost amount or the frequency, and, in doing so, it becomes a three-dimensional change, because the change

Introducing the Idea of Spectral Editing

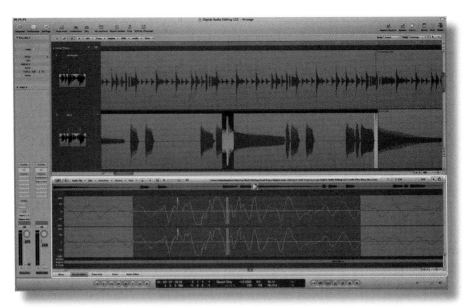

FIGURE 15.1
An unfortunate impact between a leg and a microphone caused a loud "thump" on an otherwise great take of a bass guitar recording. You can see the impact highlighted in the image above. It might be possible to choose a section from a different take, but if we have only a single take, then it would be great if we could try to fix this.

isn't constant over time. So now that we have one example of how to achieve a three-dimensional edit, what kind of uses could we put this to?

A very simple example might be in a bass guitar recording. The bass is going through an amplifier, which is then mic'ed up, and the bass player is standing nearby. During the course of the take, she turns toward her amp to check a setting, but in doing so her foot hits the microphone stand and causes a low frequency "thump" on the recording. If, for whatever reason, there isn't an alternative take to comp with this one, then we have to try to make the best of the one we have.

If we can isolate the frequency at which the majority of the weight of the "thump" happens, then there is a chance that we might be able to use an EQ to get rid of it or, at the very least, diminish it. There is a chance that, in doing so, we will pull out some important frequencies from the actual bass guitar itself, but, if we are careful enough with the EQ, we can certainly minimize the damage. The problem is, though, that if we simply apply a static EQ to correct this problem, that quite-deep EQ cut will be present throughout the whole track and will disrupt the sound of the bass in other places. The easiest way to get around this would be to set up the EQ to get rid of the thump by looping around a very short section either side of the "thump" and adjusting the EQ until it was as good as you could get it, and then putting automation in place to keep this EQ bypassed until fractionally before the thump, and then bypass it again fractionally after.

It would be possible to make this change a little more subtlety by automating the cut amounts for each of the frequency bands that you used, so that it wasn't a simple on/off change, but almost like the change faded in and faded out (albeit very quickly). This way would involve keeping the plug-in active and starting with all bands at 0 dB gain, and then, maybe over the course of

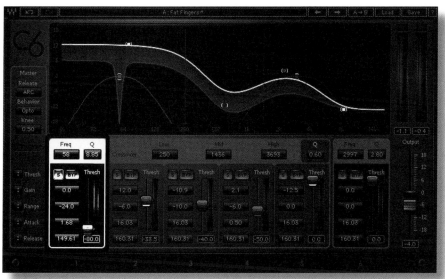

FIGURE 15.2
Using a very narrow frequency band on a multiband compressor, it might be possible to isolate the main body of the thump and then automate the Threshold control, so that, just before the thump occurs, the threshold level is pulled way down, and the body of the thump is compressed heavily before returning the threshold to 0 dB to minimize the effect it has on the overall bass sound.

50 milliseconds or before the thump occurred, you could drop the level of each used frequency band from 0 dB to whatever amount of gain reduction you had it set to. Once the thump sound had passed, you could automate the level of each band back to 0 dB over a similar time scale. You might need to adjust the exact timing and possibly even the shape of the automation in order to get the best result, but the theory is sound.

Another way in which you could approach the problem would be by using a multiband compressor. You could use a very narrow band centered on the frequency of the thump and set the Threshold to be very low, and the compression ratio very high, so that that frequency was essentially squashed, to the point of being almost inaudible. Once again, though, without automation, to make this only a temporary effect, it would be a "two-dimensional" change, which would have an adverse effect of other parts of the track. You would need to either bypass the effect completely on either side of the thump, but you would, again, run the risk that the transition from no compression to compression so abruptly would sound quite nasty; or you could use automation as described above for the EQ approach. In this case, you could either automate the Threshold from 0 dB down to whatever level you had it set at, and back up again, or perhaps you might be able to leave the Threshold control set to its very low value but start with a ratio of 1:1 (which would give no compression even if the Threshold level were crossed) and automate that up to 20:1 or whatever the highest ratio you had available was and back down again. These two options, either individually or perhaps even in combination, would smooth the transition from unchanged to changed and back again.

As I stated, this is a very simplistic and perhaps not all-that-common situation where this kind of editing might be useful, but it is far from the only situation. In a recording of a live gig, there could be a cough or shout from an audience member that interferes with the mood of a quiet piece. And in any kind of dialogue recording for film or TV, there could be any number of unwanted sounds in the background, which we might like to get rid of in order to have a clear voice recording. In cases like these, the automated EQ or compression approach might not be detailed or subtle enough. These methods are great for very short bursts of sound in a narrow frequency band. They might be usable, depending on the situation, to get rid of an audience cough or shout, but they wouldn't be especially useful for removing something like a police siren from a field recording or a news report. In cases like that, a simple EQ cut wouldn't do the job. It might, theoretically, be possible to remove a police siren from the background of a field recording by using a very narrow frequency band (or more than one) and reducing the gain substantially, and then automating the frequency of the EQ band(s) to track the pitch change of the siren, but this approach would be very time-intensive and unworkable in all but the most simple of scenarios.

TAKING THINGS FURTHER

All the situations we have looked at so far are essentially the same, in that we are looking to remove a simple sound—but what happens if the sound is more complex? Perhaps it has a broad frequency range that won't respond well to EQ cuts, or perhaps it is a harmonically complex sound that would require a very large number of EQ bands to even get close to removing it. And what happens if, instead of wanting to remove a small part and leave the majority unchanged, we want to extract only a small part and isolate it, thereby getting rid of the rest of the sounds present?

If there is a sound like a siren that we want to remove, and we can successfully track its pitch and automate our EQ bands to follow that pitch change, then we stand a good chance of being able to remove it. However, all but the very simplest of sounds (sine waves) are composed of multiple harmonics. Using our narrow band EQ, we may well be able to track and remove the lower order (fundamental, second harmonic, third harmonic, etc.) frequencies and remove them, but it will be much harder to detect and track any higher harmonics that are at a much lower level. And even if we could, we might not have enough frequency bands on our EQ to track them all. Beyond that, even if we had an EQ with unlimited bands, the task of trying to draw in automation curves to track multiple harmonics over a prolonged period of time would be, if not impossible, then certainly beyond challenging and tiresome.

Fortunately, if we can get rid of or greatly reduce the presence of not only the fundamental frequency but also a good number of the lower-order harmonics, then there is a good chance that the higher-order harmonics will simply get lost in the background of whatever other sounds are present. Technically, they will be there, but, because of the presence of the other sounds, they will be masked to some extent. If, on the other hand, we are trying to isolate a sound from within a recording, then things become far more complex. The first difference that we have to deal with is the fact that all of those high-order harmonics, the ones that we said were more difficult to track because of their lower level, are the ones that generally give a sound its unique character and also its clarity and definition. If we can't successfully extract those, then at best the sound will be dull and, in the case of speech, perhaps hard to understand; and at worst, the actual tone of an instrument could be so compromised that it is very hard to tell exactly what the instrument is. The second problem is simply that it is far easier to mask low-level harmonics in a recording as part of the general background and noise than it is to somehow re-create those harmonics in a very empty sonic space.

In order for us to really be able to stand a chance of doing more-complicated, three-dimensional edits, we need to carefully *look* at the problem from a different perspective. There are two very common ways of representing sound in a three-dimensional way: the waterfall plot and the spectrogram method. Both of them aim to present a view of frequency response that isn't instantaneous (like the spectrum analyzer common in EQ plug-ins) but rather shows the evolution of frequency response over time and does so in different ways. By showing how a sound evolves over a specific time window, they can help us to get a far better picture of how a sound evolves over a period of time, and both methods are very adept at helping us literally *see* into the sound. While both methods aim to, and do, achieve the same end result, they do it in very different ways so it is probably helpful to look at each one individually.

FFT ANALYSIS AND WATERFALL PLOTS

Almost all sound visualization techniques, including the relatively humble spectrum analyzer in EQ plug-ins, are based on the Fast Fourier Transform (FFT) technique. And that is, in turn, based on Fourier Analysis. What, you might be wondering, is Fourier Analysis? Well, you remember, back in chapter 12, when we spoke about re-creating a sound by layering a large number of sine waves, and we mentioned that we needed a way to analyze the input sound and break it down into its component harmonics? That is exactly what Fourier Analysis and the associated Fast Fourier Transform does. It breaks down a complex sound into a cluster of simple sine (technically, sine and cosine) waves that give us a new way of looking at sound. This simple idea forms the basis of the majority of spectrum analyzers, both hardware and software, which give us an instantaneous picture of the frequency balance and relative levels within a sound.

One of the well-known ways of actually presenting this data over time is in what is commonly known as a "waterfall" plot. An easy way to conceptualize it would be to imagine taking one screenshot of your spectrum analyzer every, let's say, 1/50th of a second. Imagine that each of these is printed on a piece of cardboard and the top of the cardboard cut away to leave only the area beneath the spectrum analyzer curve. Each "snapshot" would then have a decidedly mountainous

Fft Analysis and Waterfall Plots

FIGURE 15.3
Waterfall plots provide one way of looking at the evolution of a sound over time. In the examples above, the start of the sound is at the back, with the sound "flowing" toward us, and lower frequencies are on the left and higher on the right. While these plots can be very informative, we can't make changes on the actual plots themselves.

look about it. If you then took each of these and lined them up one behind the other, you would start to form a three-dimensional shape, and it is exactly this kind of pseudo-three-dimensional shape that a typical waterfall plot will produce. There are a few examples of these kinds of plots below. Just for fun, could you take a guess at what each one is?

These kinds of representations of sound in a pseudo-three-dimensional space are quite easy for us to visualize and therefore understand. The big peaks and hills will represent the louder parts of the sound, and the dips and flat areas represent quieter parts of the sound, or, in the case of the flat regions, silence (in those particular frequencies at that particular time). The time-density of these graphs can, of course, vary, and the time interval between snapshots will usually be adjustable to allow both short and long sounds to be analyzed in this way.

These three-dimensional displays can often be rotated so that time "flows" from front to back, back to front, left to right, right to left, or anywhere in between. This is because, depending on the harmonic content of the sound, from certain angles there will be the chance that important details could be hidden behind larger peaks. By rotating the view as we have just described, it should be possible to see everything that you need to from one angle or another. Most people working with these kinds of waterfall plots will have a preference as to how they like to view the sound and will more than likely always start from this position, but having the ability to "fly around" the sound in this way is very helpful.

To further complicate matters, there is the issue of static displays and dynamic displays. While they are based on exactly the same idea and analysis, they are used in different situations and for different purposes. Static displays are just that—an overview of the evolution of a particular sound over a specified period. They can be rotated to change the viewing perspective, but the data comes from an analysis of a sound (or part of a sound), and, once it is displayed, it doesn't update. This can be very useful if you are looking to identify a particular problem in a select region of a sound, but uses other than investigatory ones are pretty limited. These static displays are almost disconnected somehow from the audio files that they represent.

Much more useful in some ways are the dynamic versions. These present the same information, but they are updated as the audio file (or "live" input, if you are using them as a plug-in insert) is playing. The plot will still show a defined period of time—let's say, three seconds—but instead of you choosing a period of three seconds from within the file and then having that shown, the display simply updates to show the last three seconds. We spoke about it being possible to imagine these as a series of static snapshots of a spectrum analyzer, each printed out onto a piece of card and then arranged in order, front to back. In a static display, these snapshots are presented in order, and that's that; but with the dynamic displays they are shown in an order that much is the same, but after the passing of a certain time interval (defined by the software), a new snapshot is taken and placed into the display, and the oldest one removed. This results in a constantly updating display of the harmonic content in something that looks like you are flying over some very rugged terrain but which is, in effect, a standard spectrum analyzer with some "persistence" to the display. It never shows you what is coming up, but it does show you what has just happened, and this can be very useful in identifying where potential problems might lie.

What we haven't really dwelled upon so far is that, while these waterfall plots are actually very effective tools in helping us to figure out what is going on over a period of time inside a particular sound, they are pretty much useless in helping us to make changes to the sound itself. We would still have to rely on manipulating the audio file with EQ, dynamics, and level processing in order to isolate any particular part of the sound and either increase or decrease its prominence. They are purely analysis tools. In order to make them meaningful, active editing tools, we need to change the approach a little.

SPECTROGRAMS

These waterfall plots aren't the only way that we can represent frequency and amplitude variations over time. There is another way we can view our audio to achieve this. While perhaps not as immediately obvious in a visual use, spectrograms can also provide the crucial third axis that we need to do this. They are, in fact, very close to waterfall plots in how they achieve this. You can see from the examples of waterfall plots earlier that they represent frequency in one horizontal direction and time in the other horizontal direction, with amplitude represented in a vertical direction. We view this information from an angle that creates the sense of looking at terrain. As we have said, this is a very visually comfortable way of looking at the data. But if we stick with the terrain analogy for a moment, there is another way of presenting terrain data that doesn't rely on a 3D projection.

Topographical maps that use contours are very common. With these maps, we look directly down on to the terrain, and there are lines drawn on the maps that represent the contours of the terrain. All points that lie on that line have the same altitude, and adjacent lines represent an increase or decrease in altitude. To make these maps easier to read, they are sometimes colored so that there is a color change as the altitude increases. The highest areas on the map might be colored red, and then the colors would gradually change as the altitude dropped, from red through orange, yellow, green, blue, and down to purple. While this

272 Spectrograms

view of the terrain isn't as instantly recognizable, it is still quite an informative view, and one that is commonly used. So now let's go back to our audio realm and see how we can use this.

Instead of using different colors to represent different altitudes, let's use them to represent different *amplitudes*. Red colors would represent the highest amplitudes, orange the next highest, and so on down to purple for amplitudes approaching zero. If we use this to "map" our audio, we can have frequency and time represented along our normal x and y axes, with time running horizontally along the x axis, as this is the way we are used to perceiving it in our DAWs, and frequency along the y axis. We can then color-code the amplitude measured at different frequencies, as we have just described. Low frequencies (close to the bottom of our y axis) with high amplitudes would be shown by areas of red close to the bottom of our spectrogram, while mid-level areas of high frequencies would be seen as green areas close to the top of our spectrogram. And, of course, if the areas of color were close to the left of our spectrogram, it would mean that they occurred earlier in time than if they were located toward the right of it.

It can take a while to get used to visualizing sounds in this way, but, once you get used to it, it can become pretty intuitive. For some the garish colors can be a little distracting, so some software chooses (or gives the option) to substitute the full color spectrum for a monochrome interpretation instead. In this system, the amplitude is indicated by the brightness of the color. If the color scheme were green-based, then black areas would be zero amplitude, dark green would represent low amplitude, mid green would be mid amplitude, pale green would be high amplitude, and white would be maximum amplitude. I personally find this to be a little more intuitive, because we are dealing with mapping a single parameter of color (brightness) to a single parameter of sound (amplitude) rather than using a more-complex color change to represent an "up and down" change in amplitude. Your preferences might be different, and, as a result, you may find certain applications more useable, as the color scheme can vary from application to application. Below are the spectrogram displays of the different sounds that we showed waterfall plots for a few pages back.

At this point, though, what we have here is still an analysis tool. As it stands, we just have another way of looking at the frequencies contained within our audio, but things are about to get a whole lot more interesting. The great advantage with this kind of display is the fact that it is, in essence, a clever way of representing a three-dimensional sound in a two-dimensional space using position

FIGURE 15.4
Here we have the spectrograms for the same three sounds that we showed waterfall plots for above. You may find one or the other easier to understand or more informative, but, crucially, unlike waterfall plots, spectrograms allow us to actually make changes directly on the spectrograms themselves.

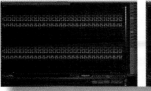

Editing in the Third Dimension

and color. While this in itself isn't anything particularly exciting, what this method opens up is the possibility to manipulate the sound by changing the color (or brightness) at any given point in order to change the amplitude. And if we add to that the fact that what we are dealing with here is essentially a graphical image, perhaps we can start to (literally) see the possibilities. In the next chapter, we will take a good look at just some of the ways in which we can use these spectral maps for editing purposes.

CHAPTER 16
Spectral Editing

INTRODUCTION

Given the complexity of the process and the sheer number of options, it would be impossible to fully explore this subject within the space limitations of this book. But it is such an important topic in advanced audio editing, and one that gives editing scope that just isn't possible in any other way, and as a result it deserves to be explored to at least a reasonable degree. With that said, let's get into things.

Spectral editing is a relatively new concept and one that is capable of being used in many different ways. At present there aren't a huge amount of options when it comes to software packages, and, for now at least, spectral editing capabilities are rarely a part of the feature set of DAW software. However, given the level of intricacy that this form of editing allows us, it wouldn't surprise me at all if one or more of the key players in the DAW world were to incorporate at least some basic spectral editing at some point in the near future. For now, though, we are limited to using stand-alone software to carry out our spectral edits.

Of these, perhaps the best known is iZotope RX/RX2, which is, for many, the only exposure (if any) that they will have had to spectral editing. Fortunately, in addition to being one of the most well known, it is one of the most advanced and complete in terms of feature set and work-flow. That's not to say there aren't other options, though. Adobe has Audition, Magix has Samplitude and Sequoia, and there is the excellent Spectro from Stillwell Audio, which goes one step further by actually being a plug-in rather than a piece of stand-alone software. Admittedly Spectro doesn't have some of the more-advanced features of the stand-alone software, but it is still an intriguing development in that it brings some spectral editing options inside your DAW (even if not natively) for the first time.

Tools of the Trade

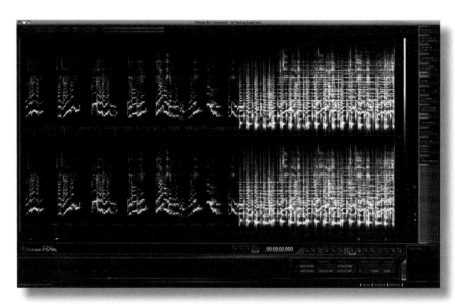

FIGURE 16.1
iZotope RX2 offers a large number of individual processes and tools that can be used to clean up and modify audio files. Spectral editing is quite a specialized area at present and is also quite processor-intensive, but the technology is bound to expand into other areas as processing power continues to increase, and it may even be a native editing option in your DAW one day alongside all the current tools.

TOOLS OF THE TRADE

Image-editing software has been around for many years, and, even at its most basic, it allows us, through the use of various tools, to select certain areas and change the color or brightness, "draw" onto the image using various "pens" and "brushes," and even erase parts of the image. The size and shape (and even intensity) of these tools can be varied to give anything from broad strokes to pixel accuracy, and, in the more-advanced packages, these tools can either have "hard" or "soft" edges. If we were to apply this editing paradigm to our "sound image," then using a tool that increased or decreased brightness in a certain area would be the same as boosting or cutting that frequency band using an EQ. The fundamental difference here is that we could apply this boost or cut to only a particular spatial part of our image. This means that it would be applied only to a particular frequency at a particular time. If we add to that the idea of "soft" edges to the tool, then it means that we would move from no change gradually through to maximum change, then gradually back to no change again, as we went from the very edges to the center of the area covered by our tool.

Earlier we spoke about the example of using a very precise EQ band and automating it from flat to a deep cut at the moment we wanted to remove an unwanted sound. We first had to identify the frequencies in that sound, and we then had to set up the EQ band and carry out the automation. Using a spectrogram with the kind of tools that we have been speaking about makes the whole process much easier. We can visually identify the offending area and then choose the appropriate tool, change the size of the area covered to be just big enough (in the frequency and time dimensions) to cover what we want to remove, and then apply the tool. It might be that the tool has a fixed "strength," and that, in order to obtain a greater reduction, we need to apply it several times, or, equally, and perhaps more commonly, there will be some kind of "strength" parameter associated with the tool.

We can select vertical columns which represent a full frequency-range "snapshot" over a particular period of time (the same kind of output that a spectrum analyzer would give if we use a short-enough time period), or we can select horizontal rows that represent the changes in amplitude of a specific range of frequencies (or a single frequency if we zoom in enough) over time. Each of these specific kinds of selections can prove very useful to use for different reasons. If we were trying to deal with short bursts of sound that we didn't want, such as a click from a damaged record, then we should consider the full-range/short-time selection, as this will ensure that whatever processing we do will affect only that short amount of time during which the click happens. But if we were looking to remove a mains hum from a guitar amp recording, then we could select specific frequencies (in this case 50 Hz or 60 Hz and the associated low-order harmonics) over the full duration of the recording, so that we would restrict any processing to just those particular frequencies.

There will most likely be other tools that you can use to make the selection as well. There will be the previously mentioned time-based selection and frequency-based selection tools, which will select in either the vertical or horizontal axis only, and there will be an extension of this that enables you to select a rectangular area of any size and proportion that will cover a limited range of frequencies for a limited time. There may also be "freehand" selection tools that allow you to click and drag to highlight the area that you want to process if it doesn't fit nicely into a rectangular selection (such as a curved or circular selection). Similar to this would be the "lasso" tool, which allows similar freedom of selection, but, instead of clicking and dragging over the area you want as if you were painting the selection, with this tool you use the mouse movements to draw around the edge of the area you want to select. These two tools offer the greatest amount of precision, and, for the most part, the areas that you want to select won't be exactly rectangular in nature; if you select a rectangular area to cover a small circular sound that you want to process, then you will be making change to frequencies that you didn't necessarily want to change.

Often there will be an option for you to hold down a modifier key and then draw additional areas to "cut out" of your initial selection. You could, for example, create a large, square selection and then hold down the modifier key while creating a smaller square within the selection that would then be removed. Or it may be that your freehand selection wasn't quite right, so, instead of having to redraw the whole thing, you could use the modifier key to erase just a part of it.

One final selection tool that is definitely carried over from the image-editing paradigm is the "magic wand" tool. In image-editing software, this tool is used to automatically select adjacent areas of the same (or similar, within a defined tolerance) color, and it is probably no surprise to learn that it does exactly the same thing here. Of course, given that we have already learned that the color (or brightness) represents the amplitude of a given frequency, it follows that this tool can be used to automatically detect and select nearby areas of the same frequency and amplitude. This is particularly useful if you have a long, sustained note, and you want to select all of it. If you use the magic wand and click on a part of the note, then the tool will automatically expand the selection to the rest of the note, given the similarity of color/amplitude and the proximity (same frequency).

Each of these tools serves a specific purpose and has its own place, and there will be times when you would use them all. Of course, having the tools to make the selection is only half the battle. Once you have your selections, you need to be able to do something with them.

CORRECTIVE SPECTRAL EDITING

Attenuation

Put simply, attenuation is reducing the level of sounds, or, in this case, the reduction in level of particular parts of the sound. This is probably the aspect of spectral editing that is the most immediate and familiar to people, and it is often used to remove imperfections from audio recordings. The examples that are always given are live recordings where somebody coughs, drops their keys, has a cell phone ringing in their pocket, and other things like that—in other words, random audio events that have found their way onto recordings of unrepeatable performances.

In practice, most basic spectral attenuation will be carried out simply by making a selection, choosing an attenuation amount, and then processing the file. There are rarely too many parameters associated with what is a comparatively simple operation. A control to widen the selection (in terms of the analysis, not the actual effect) might be present, and, by increasing this, the attenuation algorithm will take note of what the spectral content is around the edges of your selection, so that any abrupt changes can be avoided. There may also be, paired with this, a control that determines whether more emphasis should be placed on the spectral content before your selection or on that after the selection. Adjusting this parameter can prevent softening of transients if moved more in the "after" direction. Once these parameters have been set, the algorithm will reduce the level of the area within your selection but will also smooth off the edges based on the settings in the other parameters.

FIGURE 16.2 On the left we have a short section from a recording that has a couple of different problems. Around halfway through there is an unwanted noise in the background, and there is a loud click caused by record damage. On the right you can see the result of using different attenuation tools to remove the unwanted noise before moving on the deal with the click.

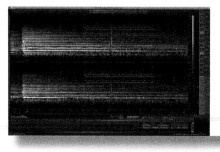 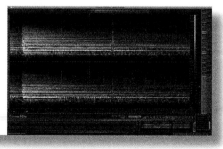

In order to try to help with potentially difficult situations, your spectral editing software may have different ways of dealing with the attenuation. These might well be grouped together under the heading of something like Spectral Repair, given that these tools are generally thought of as being ways of fixing unwanted glitches and sounds. Their use, in creative hands, does extend way beyond this, of course, but they were designed as corrective tools and are often labeled as such. They are generally more intelligent than the basic attenuation system, and, rather than just reduce the level of whatever is contained in your selection, they aim to completely replace it with other harmonic content.

These methods look at the surrounding spectral content, and then, based on the analysis, they fill your selected area with new information, completely replacing whatever was there in the process. The manner in which they choose to fill it can vary. Some algorithms simply look at the data either side of your selection and fill your selection with harmonic content to create a smooth transition from the "before" data to the "after" data. Other methods look at the surrounding data and try to find areas that are similar to your selection and fill your selection in this way. Other still are even more intelligent and actually analyze any strong harmonics (bright regions) on either side of your selection and will attempt to fill your region intelligently so that there is continuity of these harmonic regions.

The method that is most suitable really depends on the situation. If you are simply looking to remove a sound that is quite separate from the main content of the sound—perhaps a high-pitched cell phone ring in the background of a voice recording—then the attenuate method will probably suffice, as there isn't a great deal of other important harmonic information that you want to retain in that area. If, however, that same cell phone ring happened to be in the background of a solo piano recording, then you might find that the frequencies present in the cell phone ring were in the same area as some of the piano notes. In this case one of the more-complex approaches would probably give a far better result.

Copying and Pasting

Earlier in the book, we spoke about copy and pasting audio regions between different parts of a song. Although it is a common practice, it is not something that I feel gives the best solution, at least when large sections of multiple tracks are copied and pasted. It can just come across as very uninspiring, unless it is used quite sparingly. But there may be times when we simply don't have more than one chorus recorded, so, in that situation, we make use of what we have and use our little tricks to try to make it sound more interesting. In the case of spectral editing, however, I would not frown upon copying and pasting at all, simply because we are dealing with micro-sections of audio that would be impossible (if we do it right) to detect and to even know that anything had been changed. This is a very different concept than the macro copying we talk about in terms of the arrangement of the song.

Corrective Spectral Editing

To keep things consistent, let's continue thinking about our cell phone and piano dilemma. We have explored the possibility of removing individual frequencies from just that one section of the audio and have seen that, while it might be possible, it wouldn't be without its artifacts if the cell phone fundamental frequency happened to coincide with one of the fundamental frequencies of the piano notes. If that method doesn't work, then perhaps we can consider copying and pasting just a small section of audio—just long enough to cover the cell phone ring—from a different area of the piano performance. While there is nothing wrong with this in theory, in reality there are likely to be far too many variables to make this work effectively.

In order to get a smooth result, we would not only need the same notes (all of them) to be repeated somewhere else but they would also need to be played with the same (or a very similar) velocity, so that the tone was very similar. As well as this, there is the fact that even though the notes playing are the same, there could be the last remnants of the decay of preceding notes, which may be different between the two areas. All of the things can quickly conspire against you if you try to do a copy-and-paste job in a situation like this. And, worse still, we are considering just a solo piano. If there were a voice as well, then the chances of differences between the area with the cell phone and any possible candidates for copying and pasting would just increase exponentially.

So if this is all so unlikely and troublesome, then why did I bring up copying and pasting? The answer to that question lies in the fact that we are talking about three-dimensional spectral editing rather than the two-dimensional editing we were referring to during the previous copy-and-paste discussions. Because we are working with spectral editing, we don't need to copy the whole of the sound for a fixed amount of time. Instead we can just copy a part of the sound for a fixed amount of time. And that means we have much more flexibility when it comes to choosing what to copy and paste, because all we need to do is copy and paste a very small section that covers only the frequencies where the cell phone interrupts and only for the duration that the interruption happens.

Using the different tools that we discussed earlier, along with lots of auditioning to make sure that we have got all of the range of the cell phone ring covered but no more than we need to, we can then figure out what we need to locate in another section of the song that we can copy and paste. The easiest way to do this is to listen to the note and/or chord that is happening at the time the cell phone rings and then listen through the rest of the recording to see if there are any places where exactly the same note or chord is playing. It would be helpful if the dynamics and tone were similar, but, because of the nature of what we are doing, it isn't *quite* so crucial. It may be that there are a few of these, in which case you should note down the time location of them for later reference.

If, on the other hand, there are no other places in the recording where you get exactly the same combination of notes and chords, there may be places where the note (440 Hz, in our example) is present only with a different bass note or chord underneath it. Our two-dimensional editing would mean that this wasn't an option, but, because we are only going to be copying and pasting a very narrow frequency range, we might get away with it in a spectral editor.

The steps involved in copying the area will vary, and some spectral editors may not allow this copying and pasting. If this is the case, then you might still be able to do this by exporting the spectrogram as an image and then processing it with image-editing software before importing it back into the spectral editor. We have already mentioned the process of using an image editor (such as Photoshop) for spectral editing purposes, including some very avant-garde ideas, and more information is on the website, so take a look there if you want to know more, but for now let's assume that your editor does have a copy-and-paste facility.

The easiest option, if it is available, is to make the selection as we normally would and then simply drag the selection area along to one of the potential positions that we can copy from. If your spectral editor allows this, moving the selection area won't actually move the contents of that area but will, instead, move just the highlighted area to the new location. You might need to fine-tune the location in a frequency sense, and because of this it is often useful to identify some kind of marker on your selection area before you move it. You could, for example, note that the lowest point on your selection is located at about 390 Hz (you might need to zoom in to find this figure). Once you have this number, you can use it to fine-tune the position that you move the selection area to.

When you feel that it is in the right place, you should audition the selection, and, hopefully, you will have a very narrow range of frequencies that you can copy and paste and overwrite the cell phone ring with. Once you are happy that there is nothing unexpected in the area you are auditioning, you can go ahead and copy it. If there are other unexpected sounds, or at least ones that aren't present in the area you wish to replace, you should move on to the next possible location and try again, until you find one that will work.

Once the copying is done, the next step is to actually paste this data over the top of the unwanted sound. Most pastes of this nature are automatically positioned (in the time sense) at the current position of the playhead and (in the frequency position) at the same place vertically that they were copied from. You should move your playhead as close as you can to the left-most point of where your original selection was (it might be worth making a note of this too while you are noting down the lowest frequency) and then paste it. In essence that is all there is to it.

Now of course it would be misleading to suggest that this is going to work perfectly every time you do it. It isn't without its potential for problems, because you are changing very narrow frequency ranges in ways that are almost impossible to visualize, so this should be considered very much a last resort in

Corrective Spectral Editing

trying to fix problems like this. The simpler the content of the recording, the more likely you are to be able to do this without hearing any artifacts. But, oddly, the more complex the recording, the more likely you are to not notice any little artifacts, owing to the small differences being masked by the sound overall.

So now that we have looked at a few ways to eradicate unwanted noises, let's look at a few different situations where the unwanted noises aren't isolated incidents but occur throughout the recording.

Clicks and Crackle

Vinyl records, while preferred by many audiophiles, have the unfortunate property of being a physical medium that is prone to damage. Even the most well-looked-after records will, every time they are played, get progressively more and more damaged. It isn't hard to see why, because the device used to play them is a sharp object that will cause wear as it plays. This general degradation will cause a lack of fidelity over time, but in this case we aren't looking to correct that particular aspect. We are looking at the clicks and "crackle" that can occur when listening to a record.

There are a number of de-clicking plug-ins that can be very effective. They generally have few settings, and the settings that they do have are mainly related to the duration and the shape of the clicks as well as a general Threshold setting to make sure that they aren't removing too much. These can be a very quick and simple solution, and, if you have access to them, it is certainly worth giving them a try before you resort to the spectral editing methods.

Working within a spectral editor to remove these clicks and crackle is actually pretty straightforward. Many spectral editors or audio restoration packages (which are based on spectral editing anyway) will have dedicated sections for the removal and clicks and crackle. The parameters are generally very simple but differ slightly between de-clicking and de-crackling, so let's look at each one in turn.

De-clicking in a spectral editor generally has just a couple of parameters. There may well be an option to choose between digital and analog clicks, and this is used to change the shape used for the detection. Clicks caused in digital systems

FIGURE 16.3
Even the most drastic of clicks and pops can be dealt with, and the example on the left was removed by applying a de-clicking process followed by a spectral repair.

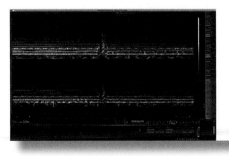 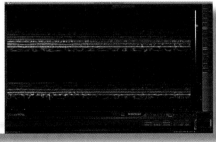

tend to be almost square in shape and are generally very short in duration (only a couple of samples). Analog clicks such as those caused by scratches on vinyl records are generally a little longer. Instead of this digital or analog choice, it may be that there is a simple Duration control instead, and this can be adjusted accordingly. If you are dealing with only a Duration control, then it is always wise to start on the shortest duration and increase from there in order to minimize the chances of the software removing things that it didn't need to.

The next control is likely to be a Strength or Threshold control. This is designed to adjust the sensitivity of the process, and, again, is there to try to prevent unnecessary removal of parts of the recording that don't need to be removed. In this case it is always better to start with the Threshold set as high as it will go and adjust downward from there, or, if it is a Strength control, start with this set at a minimum and increase.

The final control that is commonly found is an option to listen to just the clicks that are being removed. It might seem counterintuitive to listen to what you are actually removing rather than the end product with the removal done, but, in the case of any process like this, it this option can be very useful. It isn't useful so much to check that you are actually removing the clicks—that much will be obvious from listening to the processed signal—but more to check that you aren't removing other parts of the sound as well. Clicks will be conspicuous by their absence in the processed version, but you wouldn't necessarily notice the removal of other little details, as you aren't really listening for them. By listening to the actual audio that is being removed, it will become immediately clear if other things are being removed along with the clicks.

Once you have your settings, it is probably a very good idea to listen through the whole recording to make sure that the settings are appropriate, as there is no guarantee, particularly in the case of analog clicks, that they will all be the same in level and duration, so settings that are appropriate for one part of the recording may let a couple of clicks through later on. Depending on the length of the recording you are processing, this could be a hugely tedious process and one that could be prone to errors. In order to help with this, some of the editors have an option to search for similar events.

Because we are dealing with sound in a different way in a spectral editor, there are different ways of looking at the data that we have. Although the data is presented to us as a graphical representation, the actual data itself is numerical in nature. Each frequency band (the number is determined by the FFT analysis resolution) has an amplitude value for each sampling period or "window." While our eyes and brains are very adept at picking out visual patterns, computers in general are very adept at picking out numerical ones. As a result it is possible for you to select an area and for the software to look for other places in the file where a similar numerical pattern exists.

Your spectral editor may have an option to search for similar events, which will enable you to select a click event and then have the software search for other similar events rather than your having to listen through the entire track. If the software does have this facility, then it will likely have an adjustment for Tolerance or Similarity, which will allow you to vary the amount of similarity that is required for the software to detect it. If this is set too high, then only other areas with exactly the same spectral content will be found, and, if the recording is of acoustic instruments, it is very unlikely to find any exact copies. If the sounds were produced electronically, then there will be a greater chance of exact copies being found, but, even then, if there is any kind of processing (compression, reverb, delay, chorus, or flanging effects), then the chances are pretty slim. Equally, if the tolerance value is set too low, then the software might find areas of the recording that are only very superficially similar.

This could be a powerful tool to help us make sure that we have checked all the clicks on the recording, because it would be very easy to miss one if we were listening through, and even easier if we were just visually scanning the spectrogram. Once you had located a click that you wanted to remove, and prior to actually processing it with the de-clicker, you could make a selection around the click and then use this tool to find the locations of all the other clicks in the recording. Because of the nature of what we are trying to achieve here, it would make the most sense to have a slightly wider tolerance than absolutely necessary, because it would be better for the software to find things that weren't clicks, which you could then listen to and disregard, than for it to miss out on a click that is a little quieter or different in some way. As the tool locates each potential match, you can listen to it, and, if it is a click that you want to remove, you can quickly make a note of the time position, so that, when you have gone through the whole recording, you can go back to your de-clicking process, set up the parameters, and then start applying it to the clicks one by one, knowing that you have the time positions of all of them already noted down.

The other type of artifact that we might want to clean up from vinyl records is the previously mentioned crackle. This is a very different process than the de-clicking, because de-clicking tends to work on random (or at most periodic) events that tend to be very short in duration and quite loud in amplitude. Crackle, on the other hand, tends to be much more constant (even if it is made up of a large number of individual events) and at a much lower level. As a result the algorithm works on much more subtle audio events and allows very effective reduction of crackle in all but the most extreme cases. Unlike click removal, it is unlikely that you will be able to clearly see the crackle on the spectrogram, and as such it would be very difficult to achieve the reduction using any manual method.

The details of the algorithms used by the various companies that offer de-crackle solutions are, unsurprisingly, hard to come by. The best explanation that is available is that they are "based on psychoacoustic research and use multi-level decision algorithms"—whatever that may actually mean! Fortunately, with such sparse details on how the process works, the controls are generally very simple and based around a Strength control. This can take the form of a single parameter or a combination of Threshold and Amount controls, but the end result is the same whichever method is used: a gradually increasing effect in reducing the crackle.

Like the de-clicking processes, many de-crackle processes have an option to listen to the actual audio being removed, but this isn't quite as informative, simply because the clicks are usually isolated events, so it will be easy to tell if the settings are wrong. When listening to the crackle being removed, it is harder to tell if what you are listening to is just crackle or if there are some parts of the audio that you wanted to keep mixed in. It is still a useful option to have, but, in the case of de-crackle processing, it is perhaps more informative to actually listen to the processed end result rather than the crackle on its own.

Finally, some de-crackle processors also offer a degree of dynamic variation by way of a control that will vary the strength of the whole process according to the amplitude of the audio being processed. This offset can happen in either direction, so that the strength of the reduction can be either increased or decreased as the amplitude of the audio increases. This would allow you to, for example, have the crackle reduction set to be quite strong on quiet parts of the recording, where the crackle would be more noticeable, and then to reduce the strength in the louder parts, where there is a chance it will be masked by the sound of the recording.

Noise and Hum

With these two common issues with vinyl records out of the way, we can now move on to look at some other unwelcome visitors to our recordings and ways that we can deal with them. Keeping things in the realm of the actual recording media and system to begin with, let's take a look at the biggest nemesis of audio tape: tape hiss.

The actual recording medium of tape relies on a magnetized recording head to rearrange magnetic particles on the tape into particular patterns and formations, depending on the amplitude of the incoming signal. These patterns remain on the tape, until they are either re-recorded or until they are in the presence of a large, alternating current magnetic field. When the tapes are played back, these patterns are "read" by the playback head, and this signal is what forms your audio output. The magnetic particles that the tape is covered with will have variations

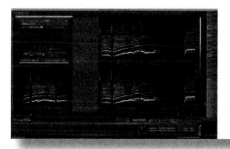
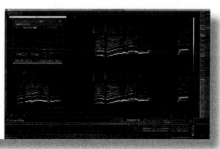

FIGURE 16.4
The intro to this Nina Simone song is an amazing performance, but it suffers a little from noise. While it's not going to ruin the performance, it is still a little distracting. With the help of RX2, this can be cleaned up very well, with almost no detrimental effect on the audio we wish to keep, as shown on the right.

in their size and thickness, and these variations will produce fluctuations in the magnetic fields used by the recording head to actually make the recordings. Upon playback these fluctuations manifest themselves as a high-frequency hiss. This problem is compounded each time a tape-to-tape copy is made, because the inherent tape hiss production in the recording system of the machine recording the copy is added to the high-frequency hiss already present in the "original" version. It only takes a few generations of copy for the hiss to reach unacceptable levels.

With spectral editing, however, we aren't limited only to tape hiss. Any kind of hiss from tape or recording equipment or consistent electrical or mechanical noise—in fact, any kind of constant and unchanging sound—can be processed and removed from recordings. The most obvious examples are, of course, tape hiss and "gain" hiss from preamps, mixers, compressors, etc., and most de-noiser processes will be used for this more than anything else. But the same tools can easily be used to remove mains hum, mechanical noise such as a motor in a video camera that has a built-in microphone, and even something like the noise from an air-conditioning unit.

Spectral de-noising works by first establishing a noise profile and then subtracting that from any audio you process. In essence it is simple enough, but it relies upon two things. Firstly, in order to get the best results, you need to have a section of your audio that is *just* the noise/hiss/hum that you want to remove. By analyzing the spectral composition of the sound you want to remove, the software becomes aware of which frequencies need to be reduced in which proportions. This could almost be viewed as a very complex EQ curve that we then apply to the audio in order to reduce the noise levels. Of course, this is a very simplistic view of what is happening, as the changes tend to be dynamic rather than static, so, in that sense, it is probably much closer to being a massively multiband compressor, where the individual Thresholds of each band are determined by the "noise" profile.

There will be a number of controls that determine exactly how this is carried out, which will vary from system to system, but, again, the process starts with a Strength control of some kind. As with the de-clicking and de-crackling processes, this can either take the form of a single Strength control, or it could be split into Threshold and Amount controls, but the adjustment of these controls will determine how strong the reduction amount is. In addition to these fundamental controls, you might also find controls for Attack and Release that determine, as you might imagine, how quickly or smoothly the de-noising starts to take place once the Threshold level has been crossed.

There may be additional controls to help make sure that transients are processed in error. We already know that transients generally involve large amounts of high-frequency energy, so it is quite possible that, after you have created your noise profile and adjusted your Threshold control, the majority of your audio passes through with only minimal de-noising going on, but, whenever there are particularly energetic transients, the de-noiser goes crazy and starts heavily compressing the high frequencies and making the transients sound unstable. In some systems this is done intelligently in the background without your having to intervene, whereas other systems have controls to allow you to adjust the sensitivity to your needs.

What all of these systems rely upon, though, is that the noise in question, whatever its source, remains consistent throughout the recording. Tape hiss should be pretty consistent in both level and spectral content, and hiss caused by high-gain levels when recording should be fairly consistent in spectral content at least, but may vary in gain, owing to things like fader movements or compression. The level changes may make it more difficult, but not necessarily impossible, to set the Threshold level accurately. The bigger problem comes if the actual spectral content of the hiss/hum changes during the recording. Things like air-conditioning units may change speed during a recording, which means that the associated noise profile would change. It might be possible to find an area of suitable "silence" during different parts of the recording and use different noise profiles for different sections, but you might not even be aware that a change has happened until a moment of silence, and by then it might be too late.

In order to counteract this, some systems have the ability to continuously monitor the audio and try to adapt the noise profile to any changes as the audio plays. These adaptive techniques are pretty good at coping with hiss and hum and other noises that change but don't change radically. But then in those cases, you would probably be very aware of the change in background noise and would (one hopes) be able to split the recording into different sections and create new noise profiles for each one, in order to increase the effectiveness of the processing.

The other potentially big problem is that all de-noising systems have some rather unpleasant artifacts associated with them. It is very hard to describe the effect in words, but some of the many terms used to describe the effect are "watery," "squirrely," "chirpy," and "warbly." While not immediately obvious, these words do tend to convey the effect that overzealous de-noising has. There is a very definite instability to the sound, as seemingly random softening of sounds happens along with very short-duration micro-resonances. The best way to understand the effect is to try it for yourself. If you create a noise profile in the way that is recommended but then turn the strength of the effect all the way up, it will soon become pretty obvious what those descriptions are referring to. Some de-noising systems have controls for "smoothing" that attempt to minimize the audible effects of this, but they will only go so far. The greater the strength of the de-noising process, the more chance of these artifacts, and the more smoothing you apply, the more noticeable the overall change in sound from your original to the processed version.

All in all, though, these de-noising systems can be very effective at removing all kinds of static (in a spectral sense) noise from the background of your recordings. They may not get rid of everything, as it might take some time to get everything set up correctly, but as long as you don't expect the impossible, then you might well find them very useful.

One particular kind of noise that can be dealt with by a de-noising process can, in addition, be dealt with in other ways. Mains hum can be a real problem when recording if you are using equipment that is wired in a certain way or if there are any kinds of loose wires, dry joints, or just general wear and tear. Fortunately mains hum usually comes in two flavors: 50 Hz and 60 Hz (depending on where you are in the world). Even better is the news that it is often pretty easy to deal with.

Mains hum has a number of different causes, but the end result is always the same. There is an audible frequency of either 50 Hz or 60 Hz (the same as the frequency of the AC mains electricity in your country) plus other harmonics. The harmonics will be simple multiples of the fundamental frequency and are easily calculable. As a result, mains hum can often be removed by an EQ with a good number of fully parametric bands, each with a very narrow bandwidth and set to 50 Hz, 100 Hz, 150 Hz, 200 Hz, and so on (or the 60 Hz equivalents). The difference with the spectral editor version of hum removal is really just a matter of being able to have an almost infinitely narrow bandwidth (depending on the resolution of the FFT analysis, of course) and by being able to use much higher gain reduction settings than a typical EQ. There is also the possibility that the spectral analysis may mean that the hum removal process may be able to automatically detect whether the hum is at 50 Hz or 60 Hz and then set up the frequencies of each of the bands needed automatically. In any case, having the option to carry out hum removal in the same places as you carry out de-clicking, de-crackling, and de-noising makes a lot of sense, even though, strictly speaking, spectral editing may not actually be needed for this particular part of the cleanup process.

The processes we have looked at here cover the vast majority of areas that spectral editing is used for, but there are a couple of additional uses that it can be put to that deserve a special mention. The first one tackles dynamic range issues, includes something that is quite a recent development in spectral editing, achieves something that is simply impossible using any other means, and may also be a massively helpful tool to have at your disposal. The second one, on the other hand, expands the realm of possibilities for spectral editing into almost unthinkable dimensions.

CREATIVE SPECTRAL EDITING

Compression

Compression is a fairly simple process to grasp. It is, as the name suggests, a way of squashing the dynamics of the sound. That's to say, you reduce the range of variation between the loudest and quietest signals in a recording. There are two distinct types of compression, though: downward compression (the most common type) and upward compression. At this stage we will just be looking at downward compression and the ways in which we can utilize spectral editing to allow us to do this in a far more precise and controllable way than with traditional compressors of the hardware or plug-in variety.

Spectral Editing CHAPTER 16

Downward compression reduces the dynamic range by reducing the level of the loudest peaks in a recording. If the loudest peaks were at −5 dB, and the lowest levels were at −45 dB, there would be a range of 40 dB. If we use compression to reduce the levels of the loudest peaks to, say, −15 dB, then the dynamic range (variation between loudest and quietest signals) would be reduced to 30 dB. This would also have the effect of reducing the apparent peak loudness of the track as well, so compression is often coupled with make-up gain to bring the peak levels back up to where they were. In our case that would mean applying a make-up gain of 10 dB. After this had been applied, the peak levels would, once again, be at −5 dB, but now the quietest levels would be at −35 dB, and the whole track would therefore seem louder. The only problem with this technique is that it can make those delicate transients sound squashed and unnatural.

Compressors, as a rule, will be set up and included as an insert effect (either in hardware or software), which means they will be active throughout the whole of the track. With the advent of plug-in compressors, it became much easier to bypass them or change the parameters in real time, so the process could be made more selective, but it wasn't without its risks. Any compressor works by changing the dynamic range, so, in most cases, bypassing a compressor will create quite a noticeable change in the feel of the audio passing through it. If the settings are very subtle, then this might not be too much of a problem, but a good majority of the time it will have a noticeable (at least) effect, so it isn't ideal.

A better solution might be to automate the Threshold level, so that it only compresses while the Threshold is lowered, and in this way it can be automated to affect only certain parts of the audio in a much more subtle way than simply bypassing. This might seem to be an ideal solution, and in many ways it is, but traditional compressors are full-range processors, and they will reduce the level of the whole frequency range of the audio, even if the peak is only in a very narrow band. If we were considering a mixed track, then a single loud kick drum could mean that the compressor kicks in, and the levels of *all* the sounds are reduced. To counteract this unwanted reduction across the whole frequency spectrum, we could use a multiband compressor. This would allow us to set up one of the bands to deal with the frequency range of only the sound we want to compress. If we combined this with automation of the Threshold level, then it would seem that we have a solution. The particularly loud kick drum in our example could have its own band on the multiband compressor, and the Threshold level could be automated down just before this offending sound and then back up again afterward, so as not to affect the rest of the recording.

We could, however, use our spectral editor to do much the same thing. We spoke earlier about the ability that we have to select only certain frequencies for a certain time period and process only those, and that is exactly what we would use the multiband compressor to do, so it makes sense that we can achieve a very similar result using our selection and attenuation tools. The method is simple enough, as all we have to do is find the particular sound we want to compress (which will probably be noticeably brighter than the surrounding areas or similar sounds nearby), make our selection, and then use the attenuate process to reduce the level by the amount that we want.

Using multiband compression has the advantage of being able to control the Attack and Release times of the compression, which means you have more control over the rate at which the level drops and then returns to normal. These two settings can be very important in getting the right effect with a compressor, but, at present at least, spectral editors have no way of doing the same thing for the attenuation process. It is possible to smooth the edges of the attenuation effect, but it does this equally on both sides of the attenuation, so that effective attack and release will be the same. This may be perfectly suitable, but there will be times when it would be better to be able to adjust them independently.

The major advantage that spectral editors have in doing this type of task is their much greater flexibility in choosing what will be attenuated. A kick drum, for example, will have a transient portion that will have quite a "clicky" attack that has a lot more mid-range and treble energy than the lower, "subby" sound that it ends up as. If you try to treat this with a multiband compressor, then you either need to set the band wide enough to make sure that you attenuate all the click and then risk it affecting other sounds at the end of the compression, or you need to set it narrow enough to avoid this, which would mean that the click part wouldn't be reduced in level by the compression. This fixed-frequency range of the band is equivalent to making a standard, rectangular selection in a spectral editor, which you can, of course, do. But the extra selection tools mean that you could select the entire frequency range of the start of the kick drum sound but then narrow the selection down as the kick drum progresses, in order to cover only the necessary lower frequencies. So what you lose to some extent in flexibility of the control of the level reduction amount, you gain in accuracy of the selection of the sound to be processed.

This approach may be a little heavy-handed for most compression tasks where hardware or plug-in compressors could be more immediate or even pleasing sonically, but, for those rare occasions when you might have just the odd thing to tidy up, perhaps a particularly loud plosive on a voice-over, and you are already in your spectral editor for other reasons, then it can be a very quick and simple fix without your having to resort to going outside of the editor for a simple task.

De-clipping

There is one other compression-related task that spectral editors are very adept at that can be very useful in a wide variety of situations. The process, commonly called de-clipping, involves processing a recording that has been recorded at too high a level, and as a result has audible distortion, because the dynamic range of the recording medium has been exceeded. This clipping can range from the subtle and not-too-disturbing (in the case of a mildly overloaded tape recording) to the harsh and downright nasty (in the case of an overloaded digital recording), but in either case there is a possible solution with the de-clipping process.

If you look at the illustration below, you will see two waveform displays and their associated spectral content. The images on the left represent a sound recorded without clipping, and the examples on the right are exactly the same sound but recorded at a much higher level, to the point where clipping has occurred. In the waveform display, you can clearly see that the peaks of the sound have been flattened out, because the natural "curve" of the waveform at that point would take it to a level above the maximum recording level. As a result the recording simply registers a maximum level for the entire duration that the waveform is above this maximum limit, and the resulting flat shapes occur. However, the spectral content shows us that this flattening out has actually resulted in a change of the relative balance of the harmonic content and the creation of harmonics that were present in the original sound. It is this change and these new harmonics that are commonly referred to as overdrive or distortion.

Obviously there are times and situations when this effect is actually desirable. Rock music, as we know it today, probably wouldn't exist without overdrive or distortion. And, on a more subtle level, mild overload (of the analog or modeled analog kind) can often add a subtle thickness and warmth to a sound. This is more commonly known as saturation and is the reason why there are an increasing number of plug-ins that claim to simulate the "warmth" of analog equipment (consoles, preamps, compressor, EQs, and even tape recordings). There is, of course, a fine line between saturation, overdrive, and distortion, and it's a very easy line to cross. It isn't at all uncommon for things to be recorded too "hot" (loud), and, up until fairly recently, with the advent of advance digital signal processing, there was nothing that could be done about it.

Fortunately that isn't the case anymore, and spectral editors, as well as some specialized plug-ins, give us the means to try to eliminate the distortions or, at the very least, reduce them. Once again we find that the actual manner in which this is achieved isn't divulged, but what we do know is that the audio surrounding the clipped areas is analyzed, and this analysis data is used to reconstruct the waveform as it originally would have been.

In use they are very simple, and the controls are usually limited to just some kind of Strength parameter that controls the amount of de-clipping that will occur. Behind the scenes this usually works as a Threshold of sorts, and any sounds above the Threshold level will be processed. As such, it is important to get this setting right. If the Strength/Threshold is set wrong, then either some of the clipped regions won't be processed or some audio will be that doesn't need to be, which could result in processing artifacts from trying to reconstruct undamaged audio.

One key thing to remember with all these de-clipping tools is that, when the missing peaks are reconstructed, they will just go over the maximum recording level again, as they did in the first place to create the clipping, so the overall output level of the process will need to be reduced. The actual amount of reduction will vary, depending on just how far over the limit the peaks originally would have been. Some de-clipping processes will automatically adjust the gain of the audio downward, while others will give you the option to control it yourself.

In either case it is important to remember that this gain change will happen, because if you were to select only a part of a file to be processed, then there would be an obvious gain drop during the selected area. You should, therefore, either select the whole file for processing—remembering that, if set correctly, only the clipped areas will be affected, so the rest of the audio will remain untouched—or, if you prefer, make sure that any gain change to the output of the de-clipper, if you can set it manually, is noted and a similar gain change applied to the unprocessed areas of the file to ensure consistency.

As useful as these tools are—and they are *very* useful—it is unreasonable to expect too much from them. You couldn't, for example, take the sound of a guitar recorded through a distortion pedal or an amp stack running at full tilt and get the original, clean guitar sound from it. That is beyond the capabilities of even the most advanced spectral repair. Similarly, you couldn't take a square wave from a synthesizer and "de-clip" it back down to a sine wave. The main reason for this is that the software needs non-clipped audio on either side of the clipped audio to actually carry out the reconstruction. A distorted guitar sound or a complex waveform from a synthesizer couldn't be cleaned up in this way, because there would be no "before" or "after" for the de-clipper to work from. But for that "almost perfect" take that was spoiled by a performance that was a little overenthusiastic in places or by a recording level set just a fraction too high, these tools could well be all you need to rescue that take and make it perfectly usable again.

CHAPTER 17
Applications of Audio Restoration

THE RECORDING STUDIO AND BEYOND

We have already seen how advances in plug-ins and software used commonly for ordinary recording, and production processes have made certain tasks in restoration much more accessible (such as the developments in Melodyne leading to Capstan). The two worlds of restoration and production/recording are, then, very closely linked, and this isn't really surprising, given the similarities in the technology and procedures involved. But, before we finish up the section, I thought it might be good to look more at the restorative and other uses of the tools and procedures that we have been talking about.

While there is much debate over whether digital recording will ever have the warmth or feel of analog recordings, what there is no question of is the longevity and durability of the format. Any particular storage medium, be it optical discs, hard drives, digital tape storage, solid stage storage, or any future medium, will have its limitations and life-span, but—and this is crucial—any digital audio can be copied from one place to another, from one medium to another, without any degradation in quality. This means that, with appropriate backup strategies in place, any recordings made today in a digital format will last indefinitely, and there will be no need of "restoration," at least not in the context that we think of restoration today. But, until we reach the point where all of our perishable and irreplaceable old analog recordings have been recorded to a digital format and preserved for all time, there is still a great demand for all our restoration tools.

Any restoration process will begin long before we ever even load up our spectral editor or image editor. In fact, it can begin long before we even load up our digital recorder. The first potential problem is in finding a machine, in optimum condition, to actually play back the recordings. Perhaps this isn't so much of a challenge in the case of vinyl records or tape recordings, but there are many recordings made on older media, such as wax cylinders and even magnetized wires, that are still in existence. Finding machines capable of playing these back could be a challenge in itself, but, given that the playback device is likely to be similar in age to the recording, it is unlikely it will be perfect. It might even be necessary to carry out some mechanical restoration of the playback device before actually doing any recording.

There may also be preparation to be done to the recording medium itself: anything from gentle cleaning to remove dust and dirt in the case of vinyl records to baking the reels of tape. This is exactly what it sounds like and involves heating the tape between 130 and 140 degrees Fahrenheit for anywhere between one and eight hours, which is dependent upon the tape size and width. Tapes that have been in storage for a long time and subject to humidity can suffer from what is known as "sticky shed." Magnetic tape has four layers: a back coat, a plastic base layer for strength (polyester or polyvinyl chloride), a binder, and the magnetic oxide layer. Over time the glue that holds these layers together can absorb moisture, and the bonds loosen, so that, if the tape were played, there is a good chance that the destabilized tape would simply break apart. The slow "baking" technique often re-strengthens the bond, giving the tape a much stronger bond and certainly enough for making several more passes to transfer to another medium.

So, once the playback device and medium itself are ready to go, then you have to make choices about which digital format to use. To put it simply, most people involved in restoration of this kind will go with the absolute highest quality that they can get for the master transfer. It doesn't really matter that "CD Standard" audio is recorded at sixteen bits and 44.1 KHz, because a higher quality (in both bit-depth and sampling frequency) can always be down-converted to this lower format if required. Twenty-four bits seems to be very widely accepted these days, so it would be a much better choice, but there are some thirty-two-bit recording systems available too that could be considered. And, to be honest, the higher the sampling frequency that you can record in, the better. It may be that you have to down-convert before you can use certain plug-ins or spectral editors, but having a higher sample rate "master" to work from will mean that the highest possible quality and fidelity is available for future processing systems that may become available.

Once the actual recording is done, the first stage of any restoration would be to try to stabilize and remove any pitch fluctuations inherent in the recording. Earlier in the book, we saw how Celemony's Capstan is very adept and dealing with these issues, and this, or a similar process, would be a good first port of call. If you are trying to clean up a recording that has wow or flutter or any similar pitch fluctuations, then it will just be that much harder than if you had one with a constant speed.

You could then work your way through all the different tools available, such as de-noising, de-clicking, de-crackling, and de-clipping, to try to clean up any unwanted artifacts before then addressing any other issues. Then any incidental noises could be removed (although this is more likely to be coughs and sounds of that nature, as cell phones weren't a problem fifty or more years ago). And at this point, the restoration process is largely completed.

Any of the many things that we could do beyond this point doesn't really come under the heading of restoration, as anything further would be making changes to the original rather than returning to its original state (the definition of restoration).

BEYOND RESTORATION

At this point, we move into the realm of improvements, which, in a musical sense, are often referred to as re-mastering. This can be something as simple as recovering some of the high or low frequencies that may have been lost (or at least not recorded very well) owing to the limitations of the recording medium or equipment, or it could be applied to an old multi-track tape master where individual tracks have been cleaned up, and then a new mixdown, using newer and more technically advanced equipment, is created.

As we have already said, for recordings made to digital systems either now, in the recent past, or in the future, the issues that we face in restoring shouldn't really be a problem. There may be an issue with noise or hum from a less-than-perfect recording session, and perhaps there might even be clipping issues, but we won't be faced with the task of restoring recordings that have been damaged owing to their age. Any issues that we have with future recordings will be issues that arose at the recording stage, and, as such, correcting them wouldn't really be considered to be restoration.

But in any case, the techniques that we use and the abilities that our current tools have allow us to make use of them in fields that aren't related to music or films/dialogue. There is a growing need for so-called forensic audio. In essence, this is a set of processes and techniques that are applied to audio recordings in the investigation of crime and the preparation of evidence. This can be anything from verifying the authenticity of recordings and establishing whether or not they have been edited previously to enhancing recordings and improving intelligibility and, most recently, techniques associated with forensic phonetics and voice biometrics.

Most specialist forensic audio facilities use specialized (or even custom made/programmed) equipment that is not only tailored specifically to their needs but also has been proven to be effective and, most importantly, cause as little change to the underlying audio as possible. Given that the results of forensic audio processing could well be crucial evidence in a trial, there needs to be no doubt at all that the results can be counted on. In addition, there are strict codes of practice, which means that every step of the process needs to be logged, so that it can be repeated and the results verified by a third party. As a result, there will need to be a degree of industry standard to what is being used. In many ways the technology available to the audio editors and recording engineers in the sound and film industries is more cutting-edge, but, until more investigation has been done to establish just how much of a change it creates in the underlying audio and how reliable it is in its results, it will be likely to remain out of the reach of a forensic audio facility.

Forensic audio analysis is also a hugely specialized field. There are very few dedicated facilities in comparison to the number of recording studios, but the demand is increasing constantly. As technology has improved and more and more can be reliably done to clean up audio recordings, there is an ever-widening range of possible applications. Calls to the emergency services, audio from mobile phones, CCTV cameras, faint voices in the background of other recordings, hidden-camera recordings, and even aircraft engine noise captured in the black box voice recorder of a plane shortly before an accident are all areas where forensic audio might be required. And while the equipment used might be different or specialized, many of the basic techniques that we have been discussing here are applicable.

One interesting twist is that the techniques we would use to remove a cough (or a cell phone ring, in the example we used) could also be used in reverse to isolate that cough. Therefore, if we could hear a faint conversation in the background of an audio recording, it might be possible for use to remove that conversation. Equally, though, it would be possible to try to invert the process and remove everything else. Sadly, it is much easier to remove a quiet sound from a louder recording successfully than it is to extract that same sound by removing the rest of the recording. It isn't the case that it is technically more difficult. On the contrary, it is pretty much the exact same process. The problem arises from the issue of perception. If we have a relatively quiet sound and we try to remove it, any residual parts of the sound that we don't manage to get rid of completely will get masked by the rest of the sounds that are present. If we managed to remove only the fundamental and first few harmonics of a single sound within a recording, then, even though higher-order harmonics remained, we would be unlikely to hear them as a distinct sound. If, on the other hand, we isolate those same few frequencies and then remove everything else, it would be quite difficult for us to actually get a good picture of what the sound was, because all the quieter higher-order harmonics that we couldn't pick out clearly are the ones that provide the majority of the definition, clarity, and, in the case of voices, intelligibility of the sound. Even the worst of recordings can be improved to a point, though, and if you are familiar with the ideas behind the processes as well as with the processes themselves, then you would have a good basis from which to build more experience with the more specialized tools used in forensic audio analysis.

One of the areas that seems to be growing the most, and which is outside of anything we have discussed so far in the book, is the subject of forensic phonetics and voice biometrics. These two techniques are meant to establish a person's identity through the unique tonal characteristics that each person's voice has. Not only does each person's voice have unique tonal properties (particular formants that occur as a result of the physical size and intricacies of an individual's vocal tract, mouth, and nasal cavity), but there are also very distinct speech patterns that we all have. The way we pronounce certain words, the placement and amount of stress we put on certain syllables, the musicality of pitch in our speech, the volume and speed that we speak at, our accent, and many other things are combined with the physical traits of our voice to form the basis of our voice biometric. It is in many ways analogous to a fingerprint, and there is a considerable amount of work in matching up voice recordings of suspects with recordings made at the scenes of crimes or over the telephone.

You as a forensic audio editor may be able to clean up and analyze recordings, pick up subtle variations, and do a really good job of isolating a single sound within a messy recording; and you may well be able to say, after looking at the audio analysis, that Person A has very strong formants at certain frequencies in his voice. However, the actual interpretation of that information moves out of audio and into the realms of linguistics, and a trained linguist would usually be involved at this point, as that person would have a far better understanding of the subtleties of the voice.

Although a forensic audio editor is faced with very strict (and for good reason) procedures and protocols, there will always be ways of combining those procedures that might be a little different from what is generally done in order to get the results that you need. Given the continual increase in possibilities of audio processing and the increase in quality and reliability of those processes, it is likely that there will be a growing demand for forensic audio in the future. It is not something that would suit everybody as a career choice, as it requires a very deep knowledge of audio theory and the tools you use, an almost-obsessive attention to detail, very good ears, and the ability to think outside the box when it comes to trying to solve some of the problems that you would be challenged with. If those are qualities that you have, then it might well be that forensic audio work could be an option for you in the future.

CHAPTER 18
Demixing

WHAT EXACTLY IS DEMIXING?

Demixing is very much a cutting-edge technology that has the rather lofty aim (or claim) of being able to take a finished stereo track and separate it out into its constituent parts. What this means in real terms is that you could take a recording of, for example, a live band, process it with the demixing software, and be left with individual tracks of drums, bass guitar, rhythm and lead guitar, keyboards, backing vocals, lead vocals, and whatever else might be in the track. At least, that's the idea. The technology and even the thinking behind it are still in their infancy, and it is, perhaps, hard to comprehend how such a thing would even be possible. Yet there are audio demos available for a few products that claim to offer exactly this ability. At present, the demos are mostly limited to more simple tasks such as vocal extraction in order to provide separate instrumental and a cappella versions of a song, but some of the demos are actually very convincing.

Perhaps a more realistic and ultimately achievable aim would be to not fully deconstruct a song into its constituent parts but rather to be able to extract certain elements from within that track and save them out as a separate file. Or, alternatively, we could remove certain parts from a recording that we didn't want there. Looking at it this way, it is just an extension of the spectral editing techniques we have already seen, where we can attenuate or even remove a particular isolated sound from within the context of an entire piece. This way demixing becomes more logically possible. The difficulty here, though, is that, rather than just being a one-off audio event that is quite different in terms of spectral content from the music around it (such as the often-used cough example), here we are looking to extract something that is more constant (presumably a melodic or chordal part) and is also musically related to, and therefore containing similar frequencies to, other musical parts playing simultaneously.

In either case, if this technology *does* work, even to a fraction of the potential proposed, then it would utterly revolutionize the way that we deal with audio recordings. But before we get ahead of ourselves and get too excited, let's just take a moment to find out the basics of how it all works to see if, at the present time at least, there is any hope of it doing what it claims to.

IS THAT EVEN POSSIBLE?

If you ask, "Does this demixing idea really work?" then the short answer is "Yes—with an 'if,'" and the long answer is "No—with a 'but.'" What I mean by that (semi-obscure pop-culture reference aside) is that it really depends on your expectations. Yes, it *does* work, in that it will allow you to remove certain sounds from a piece of music or, as in inverse operation, isolate certain sounds within a piece of music. However, this works only *if* you are prepared for those operations to not be 100% faithful to the original recording. In particular, sounds that have been extracted and isolated from within full pieces have a tendency to sound slightly "watery" and "washy." If you are expecting to be able to isolate an individual sound from within a mix and have it perfect, then you will be disappointed. I am sure that this is the eventual goal of software of this type but, for the time being, at the very least, it is beyond its capabilities in all but the most simplistic of situations.

This leads me on to the longer answer. No, it doesn't currently work as such, if we consider demixing to be the process of separating a mixed piece of music out into its component tracks, *but* that doesn't mean that it isn't a very useful tool to have. Nor does it mean that this technology will not grow in sophistication and mature over the coming years and eventually get much closer to the idealized goal. Although it seems unimaginable at the moment, that is probably how recording engineers from fifty years ago would have felt if you had mentioned the possibility of being able to automatically re-tune vocals and instruments to predefined musical scales in real time—or if you had said to recording engineers fifteen years ago (shortly after the release of AutoTune) that in the future you would be able to process polyphonic audio files and re-tune them, you probably would have been met with similar protestations that it simply wasn't possible.

Therefore people, myself included, are normally very careful about saying, "Oh, that will never happen" and instead tend to hedge their bets by saying something much more convoluted, along the lines of, "Well, with the current state of the technology, and factoring in current research, it is *unlikely* to happen in the near future—but you never know." So my longer answer to the question of whether demixing technology works is that, with the current state of the technology, and factoring in current research, it is *unlikely* to happen (at least to the extent that many people would like) in the near future—but you never know.

So we have established that it doesn't do quite as much as we might hope in its current (first-generation) incarnation, but there is still a lot that it can do. It is still surprising that the first reaction I usually get when mentioning this idea to people is one of utter disbelief. Indeed, this was my first reaction when I found out about it. I assumed that the audio demos presented were somehow faked and that the entire idea was nothing more than wishful thinking. But then

I thought about what Melodyne was already capable of, and it didn't seem so impossible after all. While Melodyne doesn't go as far as promising the ability to completely separate a track into its component parts, it does promise, and indeed deliver, the ability to be able to go into a polyphonic recording of a single instrument (or ensemble of similar instruments, such as a string quartet) and manipulate the pitch of individual notes within the recording. If you had a recording that wasn't too complex or dense, with relatively few instruments, and with each instrument working within a relatively narrow range, then you can coax this demixing functionality out of Melodyne with a little work.

Melodyne itself doesn't offer tools specifically designed for extraction or suppression of individual sounds, but it does have the ability to change the volume of an individual note after the analysis. So if you created a number of tracks and loaded the original audio file onto each track and then imported the audio into separate instances of Melodyne, it would be quite possible to choose to mute all the notes except the bass notes (for example) on the first track, mute all the notes except the vocal on the second track, all except the piano on the third track, and so on. Then, once you had split the audio out in this way, you would simply bounce each Melodyne-processed track to a new file, and you would have your demixed audio.

Of course the results wouldn't be perfect, and some audio files (simpler ones in general) would get better results than others, but the bottom line is this is the very first generation of software to offer anything like these capabilities, and I am absolutely sure that future developments in this area will only enhance the quality and improve the functionality of it. If you think back to how much of a revolution Antares AutoTune was when it first appeared on the scene, then you will have some idea of the potential impact of Melodyne Editor and specialized demixing software.

In some ways, this proof of concept with Melodyne goes some way toward convincing me that there might just be something to these claims made by the companies behind the demixing technology. What I am not convinced of at this

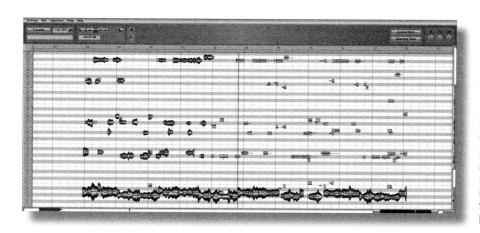

FIGURE 18.1
With Melodyne DNA, it is possible to change the volume of individual notes within a mixed piece. In the example above, we have muted all the notes in the second half that aren't part of the bass line.

stage is that the quality of such extraction would be high enough to justify the demixing name.

HOW DOES IT WORK?

The basic principle behind demixing software is that most sounds, at least in their steady-state portions, consist of a fundamental frequency plus additional harmonics whose frequencies are usually whole number multiples of the fundamental frequency and whose levels are proportional to the level of the fundamental. Yes there are exceptions to this, particularly in synthesized sounds (where the tonal qualities aren't limited by the physical properties of a sound-generating mechanism) and in transient portions of many acoustic sounds. Sibilant vocal sounds, for example, have a large noise component to the sound. Cymbals also don't tend to fall into the "fundamental plus harmonics" system as nicely as we might like. As a result these sounds would be much more difficult to isolate using the methods often used now. But, by and large, much of the content in any given recording will be of the "fundamental plus harmonics" nature and, therefore, is the real target of this type of software.

But even if we simplify things by dealing with only these parts of the audio, we are faced with a greater problem still. If we evaluate a piece of audio purely from a technical perspective (what frequencies occur at what times and with what amplitudes), then we will not be able to differentiate between a frequency of 440 Hz that is a fundamental frequency of a vocal note and a frequency of 440 Hz, occurring at exactly the same time, that is a second harmonic for a guitar note an octave below. Then we move up to 880 Hz, and we are faced with the same problem. Are the frequencies that we are hearing the second harmonics of the voice or the third harmonics of the guitar? In truth, it is probably a combination of the two and, mathematically speaking, no distinction is made. And this is where simple harmonic analysis breaks down.

When we listen to a piece of music, we can perceive the guitar and the voice. And not only can we perceive the guitar and the voice, but we can perceive the tone of each. Given that the tone of each is made up of different frequencies at different levels, it seems fair to say that we would hear the 440 Hz frequencies as both a second harmonic for the guitar and the fundamental for the voice. If we perceived this 440 Hz *only* as a fundamental of the voice, then the tone of the guitar would seem to change quite drastically while the voice was present. If we were considering a high-order harmonic that was at a very low level, then any lack of perception of that harmonic would probably lead to a relatively small perceived change in tone. But effectively removing the second harmonic could create a very big perceived difference. Equally, if we perceived the 440 Hz *only* as the second harmonic of the guitar, then our voice would be without a fundamental frequency, which could, at the least, make the voice sound very "thin," or, worse, could make us perceive the note as being an octave higher.

Yet neither of these things happens, and we clearly hear both sounds distinctly. This precise method by which we perceive sound is a matter of extensive research, and great advances are being made in all areas of psychoacoustics. Our greater understanding of this very complex subject has led to many of the advanced sound-manipulation technologies that we have today. And yet there is still something that seems, at least to the layman, to be beyond the scope of any algorithm. When we hear a piece of music, we know what a guitar sounds like, we know what a voice sounds like, and we know what a piano sounds like. There are variations among different guitars, different voices, and different pianos, of course, but our minds are attuned to the generic tonal qualities of each, and in many cases, we can identify what an instrument is even if the sound is heavily filtered, distorted, or otherwise compromised. Perhaps, this ability plays a part in our being able to hear sounds distinctly even when their harmonic content overlaps heavily.

This is further heightened by the fact that the tonal "fingerprint" of each sound within the piece will, at times, be separate from the other sounds. If we have a guitar and voice playing the same note (or notes separated by an octave) all the way through the recording, and they are never heard individually, then it would be more difficult for us to differentiate between the two. In fact, that is probably not the best example, because the voice will usually have distinct words that will allow us to separate it. Perhaps a guitar and a violin would be a better example. So if they were both playing the same notes at the same time all the way through the piece, we might find it hard to separate them in our minds. But if the melodies or musical parts diverge at any point, then our auditory system will quickly identify them as separate instruments and store their tonal fingerprint so that, when they cross paths again and play the same notes at the same time, we will have the ability to separate them based on what we know of the tone of each sound.

While demixing software may not (yet at least) have the interpretative abilities that our brains have, that's not to say that they don't have a degree of intelligence built into them. Many of the tools available to us now in demixing software (such as Sony SpectraLayers Pro and Prosoniq sonicWORX Pro and sonicWORX Isolate) will automatically select appropriate harmonics if we highlight a particular fundamental frequency. In doing so, they make it much easier for us to remove or isolate a sound, because we don't have to select each harmonic individually. Not only would this be time-consuming, but it could also be very difficult.

If we have a complex track, there could be a lot of harmonic information in a particular area, and if the harmonics that we would need to pick out are in a very crowded area, then it might not be easy for us to see them clearly. By calculating what those harmonics could be (based on the fundamental that we have selected), the software can search for these difficult-to-locate harmonics automatically. Some of these intelligent-selection systems allow us to click on any one of the harmonics, and both the higher harmonics and the sub-harmonics will be selected automatically. All that remains for us to do in this scenario is pick out the melody

What are the Applications?

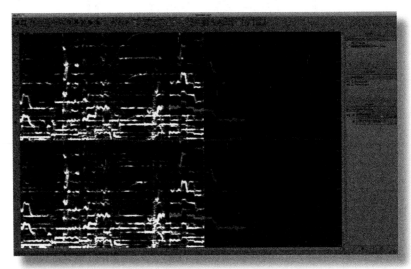

FIGURE 18.2
SpectraLayers Pro includes *very* sophisticated tools for picking out harmonics of a sound from among the entangled mess of frequencies. The left half of the screen shows the analysis of the track, and the right half shows the extracted vocal line. Of course, this isn't a perfect representation of the original vocal performance, as at least some of the detail in the voice is missing, but it is still impressive, given the very recent development of this technology.

we are trying to extract by locating just *one* of the harmonics. In many situations, this will still be the fundamental, because it is usually the highest in level and therefore most visible in a spectrogram, but, in cases where the fundamental may not be easily distinguishable from the other information surrounding it, having the option to pick out a different harmonic from a potentially less-crowded part of the audio spectrum to use as a guide can be very useful.

Once the selection has been made, the information can either be removed or extracted, as these processes are basically the inverse of each other. If you extract the sound, you will have it on its own. If you then phase-reverse the extracted sound and mix it with the original track, it will (assuming that both the original and the extracted part are phase-locked) remove the extracted part from the original audio. Whether the quality and integrity of this extracted sound will be sufficient depends very much on your intended use for it.

WHAT ARE THE APPLICATIONS?

To be honest, a simpler question to answer would be, What *aren't* the applications? If we consider a perfect demixing system, then the uses range from remastering and remixing (not to be confused with remixing in the sense of creating versions for alternate genres) old recordings to creating surround-sound mixes by being able to reposition single elements from within a stereo mix. There are also possibilities to correct timing and tuning errors on stereo recordings (such as live recordings) as well as rerecording individual parts while leaving the rest of the mix intact. These are all processes most likely to be done by a record label in order to reinvigorate older recordings and bring them to new markets and formats, and I can see the value in them as such.

But what would happen if the technology became accessible to everybody? These techniques could be used to create instrumental versions of songs for karaoke purposes, and they could equally be used to deconstruct tracks to give remixers access to individual parts of the mix (most likely vocals). It would mean that somebody who really liked a particular drum pattern could deconstruct the track and then sample only the drums. This gets us into a very gray area, which I will soon talk about more, as it has potentially far-reaching implications for the industry. But before we move on to that, let's look at what we are likely to be able to achieve with the state of the technology as it stands in its infancy.

First, we should make it clear that the actual applications really are no different between the current systems and the theoretical, ideal ones, but the degree to which some of these tasks could be completed is perhaps a little different. I don't believe that remastering from demixed tracks would be that much of a realistic proposition at this stage, simply because remastering is intended to improve the sound quality of the original recording. If the audio that we extract has a number of artifacts and perhaps loses definition and clarity, then, as much as recording equipment and technology has improved, we would be working with technically inferior tracks and trying to make a technically superior mix out of them, so any benefits gained from the new equipment would most likely be outweighed by the quality loss of the source files.

Remixing (as in re-balancing certain elements) I think is a much more realistic proposition at this stage. The biggest problem with extracting individual parts with this technology is the fact that a lot of detail will be left behind in the original file that should be in the extracted file. If we aim to use this extracted file in a new context, then we will lose that information forever. If, however, we will be keeping the original track and just changing the level of an individual part within that by extracting it and then changing the level, then the information that remained in the original file will still be there once we have boosted or cut the level (and perhaps even changed EQ or compression settings) of the extracted file. Technically it won't be perfect, because if we were to boost the level of an extracted lead vocal by 3 dB and then mix it back in, then the residual parts of the vocal that were left in after the extraction would now be 3 dB quieter than they should be relative to the main body of the vocal. This would change the tone, but it would still be much better than that information not being present at all, so it could be an acceptable compromise.

Equally, creating surround-sound mixes might be workable with the current quality of results, but it isn't quite as straightforward as with the simple re-balancing example above. While the residual harmonics would remain after we had extracted our part, there would be an additional complication in that they would remain in the main stereo sound field, while the main body of the sound could end up being positioned in the rear channels, and this spectral separation could be very disconcerting. If the extraction were a very good one, and there was little residual information left in the main stereo file, then we might well be able to reposition the extracted sound more freely among the surround channels. It would certainly be a matter of experimentation in seeing how good the extraction would be for differing parts or instruments and then basing the degree of freedom to move the sounds around on that quality.

One interesting possibility that wouldn't be dependent on having a perfect extraction and wouldn't have any associated spectral separation would be to extract a particular instrument and then use that extracted audio as an input to a reverb effect that fed into the rear channels. The fact that many reverbs have a tendency to slightly blur the sonic detail, owing to the accumulation of a large number of individual echoes, would help to gloss over any imperfections in the quality of the extracted sound, and the fact that only a single sound (or a select few) had been treated with this rear-channel reverb effect would mean that it sounded a lot more interesting than simply applying a "blanket" reverb to the whole mix and feeding that to the rear channels. This, I believe, is very much achievable, even with today's demixing software algorithms.

Moving on, we can see that the correction of tuning or timing on individual sounds within a mix using this technique is plausible, because, once again, the extracted sounds would be processed in some way and would then be mixed back into the track. As a result, any information missed in the extraction would still be in the original audio file. In this case, however, it could be that the residual information doesn't match perfectly with the newly corrected sound because of timing or tuning differences. How much of a problem this is would depend very much on the type of sound and the quality and completeness of the extraction. If only the very highest harmonic content were left in the original audio file, then it could work reasonably well, as the rest of the audio in the track would mask any differences in pitch or timing of the residual information.

In spite of all of these possibilities, I have a feeling that one of the main uses (in terms of the amount of use it would get) would be for vocal extraction or suppression. Being able to remove the lead vocal from a song to allow for the creation of a "karaoke" version or similar would be something that was very much in demand. There are often effects used on vocals, and it is hard to say how well the demixing software would deal with something like a reverb or an echo on a sound, but, in truth, I don't believe it would be a huge problem when looked at in context. These versions of songs are designed to have somebody resinging the lead vocal part, so any residual reverb or echo from the original vocal would simply sound like an effect added to the live vocal.

The other, related, use would be to extra a cappella vocals from songs. Constantly in demand by remixers and DJs the world over, a good a cappella recording is quite hard to find. Most remixers who are sent "stems" or a cappella vocals will not make them publicly available, as that would be breaking the trust of the labels that they are working for, and most of them are far too professional to do things like that. Nonetheless, there are a great many "studio" a cappella recordings out there on the Internet. And if an a cappella version of your favorite song isn't available, then there are many tried and tested techniques that you can use to try to create your own. Most of these are centered on phase reversal, although Mid/Side processing is used quite a lot as well. These techniques can often produce results that, although not perfect, are just about satisfactory for the intended use. There are, almost invariably, sounds carried through from the original song in some form, but these are often masked by the new music in the song that the a cappella extract is being used in.

These kinds of demixing tools could, if the quality of the extraction were good enough, replace all of that with a single step. And even if that weren't quite enough, the different techniques could be used in combination to try to find an ideal solution for any given track. The biggest problem that I can foresee with vocal extraction by using demixing software is the simple fact that, in its current form, the extraction is, under most circumstances, affected by artifacts. These artifacts tend to make the extracted vocal sound just—not right. It is often variously described as "watery," "washy," "phasey," "soft," and many other things, none of which are complimentary. If we were talking about a "backing" instrument that wasn't especially prominent in the mix that we just wanted to extract, do a little tidying up on, and then put back in its original place, then we could probably not worry hugely about those artifacts. But vocals are meant to be front and center, and prominent in the mix. And if we take those vocals out of the track they were in and place them in a new musical setting, then we can't even retain that lost information in our original track. It will be gone for good, and the artifacts will remain.

So, in summary, there are a number of uses for demixing technology as it stands today, but almost invariably these are centered on using demixing to pull out one sound, clean it up, tidy it up, perhaps change the sound slightly with additional processing, and then either put it back in the same track or use it in another work. For it to really fulfill its potential, I think we would need to see a reasonable level of improvement on the detection and extraction algorithms. Only then do I see it really deserving the name "demixing."

But if that *were* to happen, that would put us in a very sticky situation legally and morally.

LEGALITIES OF POSSIBLE USES

I need to start this section with a disclaimer that I am not qualified to offer legal advice, and nothing in the following section should be taken as legal advice. The information here is merely intended as an overview of the potential legal matters that may require you to obtain such advice.

Sampling is very much a way of life now. While the levels of sampling in music creation seem to have dropped considerably from their heyday in the late 1980s, it still happens a great deal, but there seems to be a lot more respect with the process now, and, for the most part, it is done legally and legitimately, and appropriate royalties are paid. What would happen, though, if people were able to sample recordings in a far less obvious way and thereby potentially avoid the source of the sample ever being discovered? I think that, even with good morals in the music industry in general, it would happen a great deal more than it does now while being declared a great deal less.

One of the things that define a sample is the "wholeness" of it. It is a brief section of a whole record, and, in many ways, that is often the appeal of it. Samples can be used to bring an instantly recognizable hook into a song. Of course the musical content in the sample can be re-created, and this often happens, but unless the recreation is a really amazing one, it will often lack the magic of the original sample. This "wholeness" of the sample also means that they are often quite easily detectable, and therefore, intellectual property issues are fairly easy to establish and enforce.

Now let's imagine a world where our hypothetically perfect demixing software exists and think about the possibilities for creative sampling. If we can separate out a track into its component parts, then we no longer need to sample a whole track. We might sample only something from the rhythm guitar track, or perhaps the bass track. We may sample only a snare drum that we thought sounded especially fantastic. If we were still in the pre-demixing era, then this wouldn't be possible: the individual sounds would, to a large extent, be protected by the "wholeness" of the sample.

As I have just said, a lot of the appeal of samples now is that they add a hook or a "vibe" or a feeling *because* of that wholeness, so, in that sense, would people even want to sample individual elements from a track if they could, when it is the entire track that gives the feel that they are looking for? In all honesty, I have to say that the answer to that would be a resounding *yes*. You could say that people currently sample the whole track because it has the "feel" that they want. That may be true, but I think it also has a lot to do with the fact that that is all they *can* sample. If that limitation were removed, then I am sure there would be a number people who still wanted to sample whole tracks to capture a particular feel, but, equally, I think there would be large numbers of people who might want to just capture, for example, a guitar part without the rest of the band.

So somebody samples a guitar riff without the bass and drums, it would still be recognizable, and therefore the rules governing its copyright would be easily enforceable, right? Well, yes, assuming that it wasn't changed in any way. But if you can extract an individual instrument, and you can then shift the notes and timing around (polyphonically, no less), then maybe time-stretch a little, EQ slightly, and maybe add some other effects, then pretty soon the sound is far enough away to be difficult to necessarily recognize as a sample and potentially even hard to verify that it was a sample in legal cases.

FAIR USE

Without exception, the best course of action to take is to request clearance for the use of any samples from the owner of the sound recording copyright and also publishing clearance from all publishers (or writers directly, if they are unpublished). There does exist, however, something called "fair use," and I should at least discuss that briefly in its relation to sampling. Fair use is defined as the right to copy a portion of a copyrighted work without permission because your use is for a limited purpose, such as for educational use in a classroom or to comment on, criticize, or parody the work being sampled.

In assessing the validity of any claim of fair use, there are three main factors used by courts:

- The amount of the work that was sampled
- Whether the material was transformed in some way
- Whether significant financial harm was caused to the copyright owner

As I have previously stated, I am not qualified to offer specific legal advice on these matters, but, from a layman's viewpoint, the shorter the sample, the greater the likelihood that fair use could be proven. In addition, the more processing/editing has taken place to the sample (and therefore the less it sounds like the work that has been sampled), the greater the likelihood of fair use being proven. And, finally, and obviously, if you use a sample and release a work commercially, and you sell five copies, then there is more likelihood of fair use being proven than if you released the work and it sold 1 million copies, because, in the former case, there would be little financial loss to the copyright holder. Once again, though, the safest option is always to seek sample clearance with any form of sampling. If you feel that your usage might come under fair use because the sample is short or virtually unrecognizable, then you should definitely seek legal advice as to whether fair use might apply. But always do this *before* commercially releasing any material, as the penalties for copyright infringement, although different in each country, are without exception quite serious.

MORAL ISSUES

But what of the moral issues? There are arguments from both sides here, and both are understandable and compelling. On the one hand, it is arguable that if you were to extract a guitar part from a song, then you changed a little of the timing and some of the notes, then time-stretched and applied EQ and effects, wouldn't the end result be different enough to not justify it being a "sample"? If the guitar sound was re-amped, for example, and the melody and tempo changed completely, then surely it wouldn't sound anything like the original and therefore wouldn't be "stealing"! From a publishing point of view, I imagine that it wouldn't be clear-cut, because you have in effect created a new composition and merely played that composition on a particular sound. From a sound-recording perspective, things are much easier to see. Regardless of what you have done to the sound, regardless of how much post-production you have done and how much of an effort you have made to change the sound, it still remains a sound recording made by somebody else to which you have never been granted a license to use or had usage rights assigned to you. In that sense, it is cut and dried: it would still be a sample, and therefore usage would need to be approved ahead of time and fees (and advance and/or royalties) would need to be paid.

Studio time could well have been paid to record the sound that you extracted and changed, and time and effort led to the skills of the player who played it, the engineer who recorded it, and the producer who crafted the sound. All of those people put their skills into the recording you sampled from, and they should be compensated if their work is used, albeit in a rather different form than it was originally created in. I think that, morally, these people should be recognized.

All of this does beg the question, however, of why you would go to all that trouble to extract a sound from a particular recording, then change the melody, change the timing, maybe change the EQ or add effects, when you could just create a part from scratch that has a very similar sound without any legal or moral complications. After all, if you are not going to get the recognition of the full sample from the listeners, then where is the benefit in using the sample? Perhaps this will be enough of a deterrent to prevent people from going to all that trouble to create something from a sample when they could possibly create their own version just as easily. If you hear a synth sound that you like on a record and can program a very similar sound and play your own melody with it, then why jump through hoops to do the same thing using the extracted synth sound? This doesn't apply, of course, if you hear a guitar sound that you like and aren't a guitar player yourself (or don't have access to one). In this situation, I can see temptation rearing its ugly head.

IT'S NOT ALL BAD

To end this chapter, I do see one very beneficial use of demixing software, but even this would need careful consideration as to whether or not it was legal. I, for one, listen to some pieces of music and think to myself that I would love to be able to listen in and hear the part individually just to get a better idea of how it was all put together—to listen to the intricacies of the individual parts that make up the whole from a musical point of view, of course, but also from a production perspective. Sometimes, no matter how hard you listen, there are just too many other things going on for you to really isolate what is making up that delicious little sound you can hear running through the back of the chorus that really makes the chorus come alive. I can imagine demixing software being highly useful and relevant as a teaching tool. According to our definition of fair use, this should be legally allowed, at least for personal education. Whether or not that fair use would be granted if, for example, a college were to use demixing software to extract elements from a well-known song to teach their students about certain elements of music production is perhaps less clear. It's understandable to protect the rights of the copyright holders, of course, but it would be a shame if this potentially great use of the technology would be outlawed from the very start.

CHAPTER 19
Thinking Outside the Box

SOUND DESIGN

Now that we have come to the end of the book, and we have covered techniques from basic and routine through to the more complex and specialist, there is one more thing that I would like to discuss. I have deliberately concentrated on actual editing tasks throughout the book. Right at the beginning, I defined the role of the editor as that of trying to get the best possible version of a particular audio file. "Best" in this sense can mean a lot of things, but, most importantly, it means making it as close to what the client (who could be yourself) needs in order to fulfill its purpose. We may, along the way, have strayed from that path a little, having been distracted by something shiny hiding in the undergrowth, but we always made it back on to that path. But, now that our journey is complete, let's allow ourselves to stray a little further, away from the confines of the "best possible version" definition of audio editing, and take a brief moment to consider what other applications we could find for the techniques that we have discovered here.

The most obvious extension to our editing techniques is to consider them in the context of not just polishing and refining audio to improve it, but rather to take it, twist it, turn it inside out, and make something totally new from it. The realm of sound design is often, in recent times at least, associated more with synthesis than with creative audio manipulation, but there is much to be said for using audio files as a basis for sound design. That doesn't mean that the results are any better than those that are purely synthetic in nature, but it can make a difference, depending on the required usage of the sound.

In order to really understand what we can do in this area, we first need to look at what sound design is used for. Historically we could trace the idea of sound design back to the late 1920s, when Jack Donovan Foley started working on creating sound effects to be added to the dialogue recordings for the ground-breaking "talkie" *The Jazz Singer*. Audio recording for film was such a new concept at the time, and the technology was so limited, that the on-set recordings purely captured dialogue and required any additional sound effects to be added later. Foley was pioneering in this regard, so much so that the whole art of creating these sounds effects for film adopted his name and is now generally known as Foley.

Perhaps surprisingly, many of the same techniques are used some eighty years later, and a great deal of Foley work today involves the use of physical objects being chopped, squashed, rubbed, dropped, and otherwise abused in the name of re-creating the sounds either not possible to capture while filming or better left to Foley artists. Many of the sounds that we associate with certain actions in films aren't exactly the same as their real-life counterparts, but, ironically, the Foley sounds have become so widely used and accepted that, were we to use the real recorded sounds, they might seem strangely out of place.

Now, of course, this isn't really what we would consider sound design in a contemporary context, but the principles of creating sounds, rather than simply recording them, for the purposes of either adding a sense of realistic detail to a picture or making sounds that create an atmosphere and have no absolute relationship to any on-screen action is common to Foley's original work and sound design (for picture) today. While Foley is largely limited to recording the physical manipulation of tangible objects, the tools and abilities that we have to manipulate audio today go far beyond anything conceivable even a couple of decades ago, and, not surprisingly, sound design is becoming more important and more valuable every day.

USING AND ABUSING AUDIO EDITING TECHNIQUES

One of the main reasons why good sound design is essential lies in its ability to create a mood or feeling. Unfortunately there is no bullet-point list for how to create a certain feeling or mood. A good sense of the emotional effect of different kinds of music and sounds is a prerequisite, but, in terms of the process, it is very open, and there are no rules. And even if there were, just consider this for a moment: a vast majority of live bands in the world today will have a guitarist who will use, either all the time or just occasionally, some kind of overdrive/crunch/distortion on their guitar sound. It has become synonymous with rock music and all its derivatives, is widely used in other forms of music and yet, ultimately, it is distortion of one form or another, which, in almost every aspect of recording, from circuit design to mixer gain structure to mastering, is considered a bad thing and something to be avoided if at all possible. So technology and practices have been developed that give us superb clarity and freedom from distortion, and then the very thing that all of this technology was designed to avoid becomes a rock 'n' roll staple.

So there really are no rules, no limits, and no preset formulas for success. One very fortunate side effect of the digital/software world is that you aren't going to blow anything up or damage or destroy any equipment by using extreme settings or using things in the way that they weren't intended. This freedom to really experiment means a much greater range of possibilities, and it encourages experimentation. Many of the techniques described in this book could be used or abused, even things as simple as copying and pasting or comping.

Thinking Outside the Box CHAPTER 19

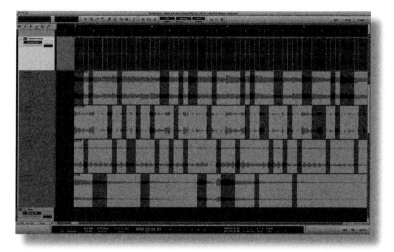

FIGURE 19.1
Four unrelated audio files place into layers and then comped at random may sound like a recipe for disaster, but it can bring some very inspiring results. If you use the Snap function, then the result will always have a strong rhythmic component to it, but the timing and melody of the underlying files could be at a completely different tempo, which can create subtle polyrhythmic effects.

Sometimes there can be unexpected sounds hidden within very unassuming and otherwise uninspiring audio files. Cutting and pasting between random or unrelated sounds, or swipe-comping a collection of unrelated audio files, can lead to some very interesting results. This could be combined with actual dialogue or relevant sound effects and might give the impression of some kind of "random data stream" or glitch that could be very effective in creating a "dream state." If the edits all had fairly long cross-fades between them, then, depending on the sounds used, of course, the feeling might be one of a floating randomness, but if the edits were all very instantaneous with little or no cross-fading, then it would create a greater sense of pace, urgency, and perhaps even panic. By punctuating the random sequence with intelligible words or sounds that are relevant to the on-screen action, you could bring the mood from the entirely abstract into a slightly disjointed reality.

Another very quick and easy way to create otherworldly sounds is the use of extreme time stretching or pitch-shifting. While stretching or shifting beyond a certain amount will almost always go beyond the realms of retaining the original character of the sound, such methods can be used to create very atmospheric effects that still retain an element of the original sound. One very well-known use of this is the so-called PaulStretch effect that uses spectral smearing to create sounds that are stretched by large multiples of their original length. Many examples exist on YouTube, including possibly the best example of this, which takes a Justin Bieber track and converts it into a shimmering soundscape that you would never believe came from the place that it did. If you want to have a listen, do a search for "Justin Bieber 800 percent slower" on YouTube, and you will be able to hear the effect. While you cannot hear any distinct words or musical melodies from the original song, there is still that connection to it in terms of key and (very) gradually evolving musicality. Used with dialogue (and perhaps at the smaller end of the stretching scale) you could create very ghostly voices, which could be used very effectively.

FIGURE 19.2
Stretching sounds will almost always create artifacts that we ideally don't want, but, if we take that stretch to extreme measure (800% or more), then the artifacts can actually take on a quality all their own. The image above shows the effect on the waveform of stretching a file by 800%.

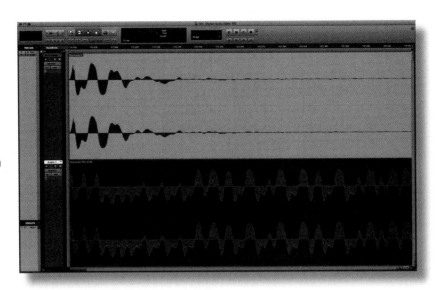

I think that the greatest amount of sonic manipulation (and fun) can be had with spectral editing, and it is here that you can really start messing with sounds in ways that retain the essence of the sound you are working on but change it in ways that are often hard to describe but almost always very effective. As a simple example, let's take a sound, any sound, and load it into a spectral editor. Then, using the attenuation tools, you highlight a particular harmonic from the sound, and you reduce its volume to zero (or at least substantially). How different does it sound? It's hard to say, really. If it is one of the higher-order harmonics, then it will not sound that much different, but if it is one of the lower-order harmonics, then it would be noticeable at the very least. This in itself isn't exactly cutting-edge sound design, but if you selectively remove quite a few of the harmonics, and not necessarily in any particular pattern, then you will notice the sound changes quite substantially. This will work best on harmonically complex sounds, simply because you have more harmonics to play with and more combinations and permutations to try. If you remove most (or even many) of the harmonics, then you might end up with something very weak and uninteresting. The key here is knowing when enough is enough. You will want to change the sound—to make it substantially different but still recognizable for what it is. If you were so inclined, you could take normal Foley effects for footsteps, jangling keys, running water, or whatever happened to fit with the image that you were working to, and create a surreal world where even-numbered harmonics didn't exist! I realize that this would be a ridiculous thing to do, but it is just one example of the possibilities.

I have mentioned using Photoshop (or a similar image editor) and a piece of software called PhotoSounder to effect all kinds of weird and wonderful spectral editing techniques, and I strongly urge you to take a look at the many possibilities that are possible both in a traditional spectral-editing sense and in a more off-the-wall sense. You could, for example, take a very "normal" sound and give it a sinister edge by slightly stretching the harmonic scale so that the second harmonic

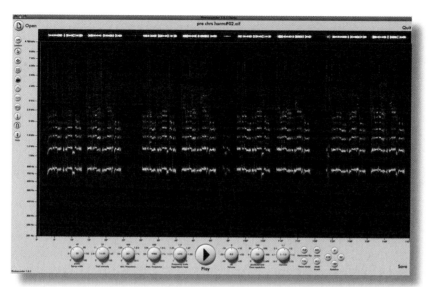

FIGURE 19.3
PhotoSounder is an amazing tool if you are looking for something truly outside the box when it comes to audio editing. It isn't a tool for everybody, and it certainly takes a bit of getting used to, but if you put the effort in, you can get effects that I can't imagine being possible any other way.

isn't exactly double the frequency of the fundamental. If the fundamental is 300 Hz, then, instead of being 600 Hz, maybe the second harmonic is 606 Hz, the third is 909 Hz instead of 900Hz, the fourth is 1212 Hz, and so on. While sounds like this would be quite unusable in a musical context, in the context of spot effects they can be very effective for adding a sense of unreality and unease.

Similarly, you could apply this type of spectral stretching or compression in a time-variable way, so that a sound starts with all the harmonics compressed to a single frequency, and then, over a period of time, the harmonics would spread out and resolve to their original positions. This starting point could be the fundamental frequency from which all the harmonics spread upward, and this would have the sonic effect of starting as a sine wave (almost) and ending up as the final sound. It isn't something that would be useful in that many situations, but it is certainly one for the "How did they do that?" list.

How about using spectral vocoding to start with one sound, then applying a graphical fade-in to another sound that is layered with the first (on a multiply layer) to provide a vocoded mid-section to the sound, and then having the first sound fade out (with a graphical fade) to leave only the second sound remaining? Because of the vocoding, this sounds very different to a traditional cross-fade and could be used in conjunction with other processes to create natural sounds that have a synthetic edge or attack to them. Once again, if used subtly, this could add just a faint air of something being advanced or futuristic without necessarily having to resort to full-on sci-fi sounds.

Another potentially interesting thing to do is to have a sound and then bounce a copy of it with distortion on. If you then load up the two layers in Photoshop, you can subtract the clean sound from the distorted sound, and this will give you the difference between the two. Or you could do the same thing with completely random pairs of sounds. It would be misleading of me to say or even imply that the results from these kinds of abstracted graphical manipulations are predictable, but that is one of the things that I find most appealing about them. When I want a controlled and (to a large extent) predictable process, I have a large range of tools and techniques available to me. This applies not only to audio editing but also to recording, sound design, mixing, production, arranging, and pretty much every audio-related area in use today. But every now and then it can be fun to do things out of sheer randomness and just see what happens.

Many times the results will be utterly unusable, and other times they will be unusable but interesting, and you will hear something in the result that leads you to go back a step and try something similar but slightly different. And then there will be times where the result is a polar opposite to what you expected, but nonetheless is scary, inspiring, beautiful, strange, ominous, endearing, epic, tragic, flowing, twisted, or just downright weird! And at times like those, you can thank the cosmic forces of Fate and Serendipity—and also perhaps give a tip of the hat in thanks to those equally scary, inspiring, beautiful, and strange programming wizards who have made it possible to manipulate audio in the ways that we can today.

Index

Note: Boldface page numbers refer to figures.

A
absolute pitch 202
AC tape bias 19
Action 169
additive synthesis 198
Algorithm controls 254
alternate rhythms 203
ambient noise control 121
Ampex 6
analogue circuitry 239
Antares AutoTune 95, 301
APTrigga (apulSoft) **180**
ARA (Audio Random Access) plug-in 259
Arrange Window 59
Arrow tool 41, 213
attack transients 135
attenuation 278–9
audible editing, visual editing *vs.*: adjusting plug-in settings 30; clear timing reference 31; DAWs 32; human feeling 31; track arrangement 30; zero-crossing points 31
audible thump 26
Audio Bend 231; algorithm 214; button 145, 169, 215; panel 214, 231; processing engine 145
audio editing 12–13; corrective editing *vs.* creative editing 15–17; DAW 9–11; definition 1–3, 5, 311; digital tape and hard disk–based 6–9; editor role 20–1; restorative editing 17–19; tape 5–6; techniques 312–16
audio manipulation 311
Audio Pool option 234
audio restoration: applications of 293–7
Audio section 256
Audio Warp 173
AudioSuite plug-in 212, 256; *TimeShift* 256
Auto-Tune (Antares) **241**, 241–3; effect 253
automatic nondestructive edits 75

automatic stretching 221
automation 13

B
baking technique 294
balance control 184
bars and beats format 154
beat detective **167**, 212
beat-mapping: applications 164; bars and beats format 154; bouncing 156; creative usage 159–61; destination tempo 154; DJs, tempo variations 155; inconsistent tempo 160–1; live performances 160; power of **154**; quality limitations 155–6; smearing effect 155; time stretching algorithm 155–6; transient markers creation 153–4, 159
Bend Markers 231–2, **232**
Bend Panel 169
Bounce In Place 166
bouncing 5–6
Browser panel 191
built-in drum replacement tool 185, **186**

C
C-Lab/Emagic 11
cappella vocals 306
Capstan 294
Capture Selection 167
Catch Range 173
CD Standard 294
Celemony Melodyne 95
Celemony's Capstan 19–20, **20**
Cents shift 254
Cher Effect 241
Classic option 207
click-track tempo 34
clicks and crackle **282**, 282–5
Clip Conform option 166, 168
Clip Gain, Pro Tools 78–80, **79**
Clip Separation option 166, 168
color-coding tracks 113

compiling: Cubase 108–11; DAW software in 91, **93**; double track vocal 94; *Duplicate Tracks* 111; gaps/overlaps 96, **97**; *layers* concept 105–7, **106**; Logic 99–102, **100**; master track 92; name them/color them **92**; performance emotion/delivery 90; performance grade 91–3; pitch adjustments 94–6; Pro Tools 102–5, **104**; Pro Tools 898; Quick Swipe Comping 97–8; Studio One 105–8; Swipe comping **98**, **110**; timing changes 96–7; tonal matching 93–4; vocal comping 90; vocalists 89–90; voice tone 90
comping 6
compression 288; advantages 290; compressors 289; downward 289; multiband 289–90
compressors 75, 289
Conform button 168
conform to tempo 228
Convert Regions to New Sampler Track dialogue box 165
copying and pasting 279–82
correct pitch 250
correction map 19–20
corrective editing processes 6
corrective editing *vs.* creative editing 15–17; elastic audio 15; MIDI sequencing 17; pitch correction 17; vocal performance 17; whole choruses 16
Create Groove button 174
Create MIDI Notes 193
Create Warp Markers 235
cross-fading 200
Crossfade Length box 168
Cubase 215–18; audio 11; compiling 108–11; drum replacement 192–4; elastic audio 233–7; fade shapes 64–7; level control 82–4; multi-track

Index

Cubase (*continued*)
comping 130–2;
pitch-shifting 260–2;
Time Stretch window in 217;
time-stretching 215–18;
transient detection 147–50, **148**
Curve Kind buttons 65
cutting, copying, pasting, and moving: audible thump 26; chorus 26; click-track tempo 34; complexity and time-consuming 26–7; costs to avoid 27; Cubase 43–6, **44**; focus on quantization grid 27; listener's ability 28; locator 26; Logic 35–7, **37**; manual work 34; multi-track tape recording 26; multiple-track editing *see* multiple-track editing; potential problems 33–4; Studio One 40–3, **41**; tempo variations 34; tightness 27–8; timing reference 27; tone difference 34; visual editing *vs.* audible editing 30–1; zero-crossing edits *see* zero-crossing edits
cutting and pasting 12; *see also* cutting, copying, pasting, and moving

D

Damped Spline Interpolation 65
DAW software 179, 275; Deck software 10; FFT Window 10; graphical user interfaces 9–10; MIDI sequencing software 11; music sequencing, platform 10; Pro Tools 187; Soundstream system 9; SoundTools software 10
de-clipping 290; rock music 291; Strength parameter 291; tools 291–2; waveform display 291
de-crackle processes 285–6
de-noising 286–8
decay parameter 136
Default Settings box 212
Define Bars 216
Delay: parameter 187; setting 254
demixing: Antares AutoTune 301; application 304–7; definition 299; fair use 308–9; fundamental plus harmonics system 302; legal permit 310; legalities 307–8; Melodyne 301, **301**; moral issues 309–10; principle 302–4; SpectraLayers Pro 303, **304**; tonal fingerprint 303; tools 307
Destination: box 208; value 208
destructive edits 72–3
destructive *vs.* nondestructive edits: files/regions uasge 71; safeguards levels 70; Undo command 70; untouched original copy 70
Detection: controls 81; section 167
Detune control 254
Digidesign 10
Digital Performer 698
digital recording 196
digital tape and hard disk–based, editing: Alesis ADAT recorder 8; CD format 6–7; copy and paste sounds 7; DASH digital multi-track systems 6–7; noise level 7; *non-linear editing* process 8; RADAR system 7–8; recording 7; S-VHS format cassettes 8; storage space 9; wobbling effect 7
Direct Note Access (DNA) technology 243
Disable Elastic Audio 145
"divide down" technique 239
double triggers **186**
drag 59; and drop approach 221
dream state 313
drums 254; algorithm 214; comping 114
drum replacement: Cubase 192–4; logic's built-in 185–7; plug-ins 180–4 *see also* SPL DrumX-changer; Pro Tools 187–90; recycled loops, using 177–9; Studio One 190–2; tools 179–84
Drum—Groove Agent ONE from 175
duplicate audio files **223**
Duration control 283
Dynamic Velocity 193

E

Edit Hitpoints button 148
Edit Smoothing option 166, 168
editor role: audio editing 20; mixing and production 20; planning 21; time-stretching and pitch correction 21
elastic audio 168; algorithms 209, 220; auto-conform and follow tempo 223–4; automatically re-quantize audio files **222**; Cubase 233–7; definition 219; duplicate audio files **223**; enabled track 209; engine 255; events 169; functionality **220**; Logic 225–8; plug-in selector 229; Pro Tools 228–31; quantization of audio files **222**; regular time-stretching 221–2; Studio One 231–3
Elastic Properties dialogue 210
Élastique algorithm 215; *Efficient* 215; *Pro* 215; *Pro Format* 215
Eltro Information Rate Changer 239
Eltro system 244
End Bar 166
Enhanced Resolution option 167
Eraser tool 42
Event Operations dialogue 168
Event Operations Transpose dialogue 256
Event Sensitivity control 143–4, **144**
Exchange section 175
expander **120**, 120–2
exponential fade curves 50, **50**
EXS-24 sampler 165–6

F

Fade Flags 62
fade shapes: cross-fades 54; Cubase 64–7; DAWs and 51–2, **52**; decay characteristics 53; drum and percussion sounds 53; exponential 50; housekeeping edits 52; linear 48–9; logarithmic 49; Logic 58–60, **59**; periodic patterns 56–7; phase cancellation 54–6; preset curves 47; Pro Tools 60–2, **61**; S-curve 50–1; sound characteristics, change in 52; Studio One 62–4, **63**
fades: and cross-fades 47; tool 58–9
fair use 308–9
FFT analysis and waterfall plots **270**; complex sound breakdown 269; dynamic versions 270–1; harmonic content 271; pseudo-three-dimensional space 270; snapshot 269–70; static displays 270
File Tempo box 212
Fill And Crossfade option 168
first harmonic 198
First Selected Track option 193

Index

Flex: algorithms 164; markers 164, 226–8; mode 58, 125, 142, 164–5, 225–7; time editing 225
Flexed audio files 225, **227**
flutter 18–19
Foley, Jack Donovan 311–12
Foley artists 312
forensic audio 295; analysis 296; editor 297
Formants section 211
frames, transient detection 137
Free option 207
freehand selection tools 277
frequency-dependent gating 119–20
full frequency-range snapshot 277
full-range/short-time selection 277
fundamental frequency 198

G

General option 231
global offset 181
Grabber cursor 230, **230**
gramophone 18
granular synthesis 199–201; advantages 201
granular techniques **200**, 245
Grid mode 212
Groove: *Agent* **174**, 193; *Agent ONE* 192; *Clipboard* 169; *Panel* 170, 190, **191**; *Template* 164–5; *Template Extraction* 167; *Template Extraction* option 166
Group: button 175; editing 131–2

H

hard-disk recording 8
hard-tuning effects **241**
Harmonic Correction button 254–5
harmonizer: H910 240; H949 240; H3000 240–1
hi-hat transients **134**
High Emphasis option 167
Hitpoints 147–9, 192
hyper-cardioid microphones 115

I

image-editing software 276
inconsistent tempo 160–1
individual notes 138
Inspector panel 62, 164, 187, 212, **213**, 254; *Quantize* parameter 227
Instrument option 185–6
integer multiplication of length 200
iQ Mode 173
iZotope RX/RX2 275, **275**

J

Justin Bieber 800 percent slower 313

K

karaoke 304, 306
Keyboard Commands Focus Mode 188
kick drum *see* snare drum
kick replacement 191

L

lasso tool 277
latency 180–1
layers concept 105–7, **106**
legal permit, demixing 310
length-based stretches 204
Length in Samples option 208
level control: Cubase 82–4; definition 69; destructive *vs.* nondestructive edits 69–72; Logic 76–8, **77**; Pro Tools 78–80; Studio One 80–2; types 72–5
linear fade curves 48, **48**
Linear Interpolation 65
live performances 160
live recordings 18
locator 26
logarithmic fade curves 49, **49**
Logic Pro 9 98
Logic's built-in drum replacement tool 185; *see also* drum replacement
Low Emphasis option 167

M

macro time-strech 204
magic wand tool 277–8
magnetic tapes 294
Make Groove Template 164
manual comping method 113–14
Mark of The Unicorn (MOTU) 11
Melodyne (Celemony) 242–4, **243**, 258, **259**, 260, 301, **301**
Micro-Pitch Curve 261–2
micro time-strech 204
MIDI file: advantages 189; MIDI controllers 179; note velocity 179; timing offset 177; track 189
Mini Browser 194
Minimum Length control 81
Mix control 254
Mode: control 254; option 185, 207
Monophonic: algorithms 210, 256; section 211
MPEX 216
multi-sampled patches 163

multi-track comping: ambient noise control 121; color-coding tracks 113; Cubase 130–2; direct snare sound 117; drum comping **114**; expander **120**, 120–2; *Flex Mode* 125; frequency-dependent gating 119–20; Group Editing 131–2; *hyper-cardioid* microphones 115; instruments 113; lanes combination 130–1, **131**; Logic 122–5, **124**; manual comping method 113–14; noise gates 118–19, **119**; phase cancellation 117–18; *Phase-Locked Audio* check-box 124; phase-locked groups **129**; Pro Tools 125–7, **126**; room mics 115; *Scissors* tool 123; source sound **116**; Studio One 127–30; swipe comping 115; timing cleaning up 117–18; *Type* of group 125–6; unchanging waveform 116; Unveil 121, **121**
multi-track drum recording 178
multi-track tape 5–6; master 295; recording 26
multiband compression 289–90
multiband compressor 267, **267**
multiple-track editing: audible ghost 32; drum kit recording 32; live-band recording 33; live instruments 33; multi-mic recording 32; noise gates 32; phase-aligned zero crossing principle 32
"munchkinization" effects 257
musical mode check-box 234
Mute tool 42–3

N

New MIDI Track option 193
noise: gates 118–19, **119**; and hum **285**, 285–8
non-automation level control 76
non-musical/rhythmic audio clips 204
Non-Q 173
normalization 72, **72**
Nyquist frequency 19

O

Object Selection tool 43–4, 109, 217
Operation heading 167
Options Panel 231

Original value 208
OSC 10

P

Pad Edit 194
PaulStretch effect 313
periodic patterns 56–7
phase cancellation 117–18, 180; phase alignment 54; sine waves 54, **54**; sound, frequency content 56
Phase-Locked Audio check-box 124
phase-locked groups **129**
PhotoSounder 314, **315**
pitch 256; change 195; correction plug-ins 253; modulation 242; parameter 194; section 211, 257; tracking 19; variant 215
Pitch & Warp button 261
pitch adjustments: DAWs 95; pitch-shifting plug-ins 95; timing adjustments 96; tools 95; vocal tuning 94
pitch-manipulation software tools 138–9
Pitch Map (Zynaptiq) 243
Pitch Shifter II plug-ins 253
pitch-shifting 239, 313; analogue 239–40; approaches to 244–5; computer-based 241–4; Cubase 260–2; digital 240–1; formants 247–50; Logic 253–5; ProTools 255–7; size and 245–6; Studio One 257–9; uses of 250–2
playback rate 202
plug-in processing 189
Poly Fast algorithm 216
Poly Musical algorithm 216
Polyphonic algorithms 210, 256
polyphonic audio files 243, **262**
Polyphonic/Rhythmic modes 211
polyrhythm 203
Pre-Roll and *Post-Roll* controls 81
Prelisten option 185
Preview button 65, 211, 257
Preview Quality option 216
Pro Tools 187–90, **188**, 209–12; Clip Gain 78–80, **79**; compiling 102–5, **104**; DAW software 187; drum replacement 187–90; elastic audio 228–31; fade shapes 60–2, **61**; *Grabber* tool 38; level control 78–80; multi-track comping 125–7, **126**; preferences dialogue 211; *Selector* tool 38–9; *Separation Grabber* tool 39; Smart tool 39, **40**; time-stretching 209–12; transient detection 142–5; *Trimmer* tool 38
Processed Length option 211
Processed Tempo display 211
Processing tab 211
production/recording 293
Project Clipboard option 193
Propellerhead's Recycle software **156**

Q

quantization of audio files 222
Quantize Grid options 169
Quantize panel 171, 190
Quantize parameter 164
Quantize Pitch control 261–2
Quantize Value 232; box 170
Quick Swipe Comping 97–102

R

Random Access Digital Audio Recorder (RADAR) system 7–8
random data stream 313
Randomize control 169, 173
Range control 254
Range Selection tool 45
Range tool 41–2
recording studios 293–5
Recycle-like beat slicing 166
recycled loops 177–9
recycling: Attack times 162; creative usage 161–2; extended usage 158–9; fades 158; Propellerhead's Recycle software **156**; slices **157**, 157–8; tempo and timing 157; transient markers 156
Relative Threshold option 185
remastering 305; *see also* demixing
remixing 305
Rename Preset 174
Render button 211, 257
Resolution controls 167
Response control 253
restoration 293, 295–7
restorative editing: analog recordings 18; Celemony's Capstan 19–20, **20**; correction map 19–20; deteriorated over time 17–18; flutter 18–19; live recordings 18; pitch and tempo 19; pitch tracking 19; speed fluctuation 18–19; tape reels 18; vinyl records 18; wow 18–19
Restore button 66
Resulting Length sections 216
rhythm 163–4
Rhythmic algorithms 210
room mics 115
Root controls 253
"round robin" sample playback 183

S

S-curve: in cross-fading 51; Type 1 50–1, **51**; Type 2 51, **51**
sample-based track 229
Sample Editor 76–7
saturation 291
Save Preset button 174
Save to Disk 168
Save to Groove Clipboard 168
sawtooth wave 198
Scale controls 253
scrubbing 134
Segments button 260
Selection section 167
Selector tool 38–9, **40**, 60, 79, 104
Semi Tones shift 254
sensitivity control 182
Sensitivity option 167
Separate at Selection command 169
Shift control 257
Show Lanes button 108
silence, audio regions 80–1
sine wave oscillators 198
Sizing Applies Time Stretch 44; version 217
Sizing Moves Contents 44
slice 177–8
SMPTE 27, 208
snap 12
Snap mode 232
Snap Timebase control 231
snare drum 177, 179, 203; clean 178; straight pattern 178; volume control 179
Solo algorithm 214
Solo Fast algorithm 216
Solo Musical algorithm 216
Song Setup button 231
Song tempo 231
sonic identity 199
sonic manipulation 314
sound: algorithm 214; design 311–12
Soundstream 10
source file 181
spectral analysis **197**

Index

spectral editing: attenuation 278–9; clicks and crackle 282–5; compression 288–90; copying and pasting 279–82; DAW software 275; de-clipping 290–2; freehand selection tools 277; full frequency-range snapshot 277; full-range/short-time selection 277; iZotope RX/RX2 275, **275**; lasso tool 277; magic wand tool 277–8; noise and hum **285,** 285–8; notes and chords 280–1; spectrogram 276; tools of trade 276–8
Spectral Repair 279
SpectraLayers Pro 303, **304**
spectrograms **272**; advantage 272–3; *amplitudes* 272; monochrome interpretation 272; spectral editing 276; topographical maps 271–2
Speed display 211
speedup method 213
SPL DrumX-changer **182**
Spline Interpolation 65
Split at Grid command 43
Split tool 42, 45
square wave 198
Standard algorithm 216
Start Bar 166
static nondestructive edits 73–4
Steinberg 11
sticky shed 294
Store button 65
Straighten Pitch control 262
Stretch audio files 231
stretching sounds 314
Strip Silence function 81
Studio One: compiling 105–8; cutting, copying, pasting, and moving 40–3, **41**; drum replacement 190–2; elastic audio 231–3; fade shapes 62–4, **63**; Inspector Panel in **213**; level control 80–2; multi-track comping 127–30; pitch-shifting 257–9; time-stretching 212–15; transient detection 145–7, **146**
Sub-Beats 167
surround-sound mixes 305
Swing slider 168
swipe comping **98, 110,** 115
Synchro Arts Vocalign 34
synchronization 207

T

Tab to Transient mode 187–8
tape reels 18
tape variant 215
TC/E Plugin box 211–12
TCE algorithm 211
TCE Factor box 210
tempo: and timing 157; variations 34
Tempo and *Musical Mode* 234
tempo-based stretches 203–4
Tempo Change 208
Tempo in BPM value 216
Threshold control 267, 283
3D editing: automated EQ 268; bass guitar recording 266; FFT analysis and waterfall plots *see* FFT analysis and waterfall plots; loud thump 266, **266**; lower-order harmonics 268–9; multiband compressor 267, **267**; spectrogram method *see* spectrograms; Threshold control 267; waterfall plot 269
throat 249
tick-based tracks 229
tightness 27–8
Time and Pitch Machine 207, 253–4, **254**
time-based stretches 202–3
Time section 256
Time Shift 257; AudioSuite plug-in **211**
time-specific files 209
Time stretch 212; audio 219
Time Stretch Ratio sections 216
time-stretching 195–7; additive approach 198–9; algorithm 155–6; algorithms 196; audio **196**; Cubase 215–18; difficulties 197–8; filling in gaps 204–6; granular approach 199–201; length-based stretches 204; Logic 207–9; manual 206–7; Pro Tools 209–12; Studio One 212–15; tempo-based stretches 203–4; time-based stretches 202–3; tools 139; use of 201–2
Time variant 215
Timebase Selector 229
TimeShift option 212
Timestretch control 215
Timestretch mode 231
timing: cleaning up 117–18; discrepancy 177; map 167

Timing Offset 187
tonal fingerprint 303
tonal matching: compiling 93–4; definition 85; EQ band 85–6; EQ plug-ins matching 86, **87**; response time 86; smoothing parameter 86
tone difference 34
tools of trade 276–8
track/channel automation 73–4
Track heading 215
Track View Selector 169, 230
transient detection: advanced level-based detection 136–7; applications and limitations 138–40; attack transients 135; conventional waveform display 134–5; Cubase 147–50, **148**; decay parameter 136; drum hit characteristics 137; frames 137; hi-hat transients **134**; individual notes 138; Logic 140–2, **141**; *nonperiodic* components 139; pitch-manipulation software tools 138–9; Pro Tools 142–5; purpose of 133; scrubbing 134; sensitivity control 136; simple level-based detection 135; spectral analysis 133; spectral detection techniques 137–8; Studio One 145–7, **146**; time stretching tools 139; transient markers *see* transient markers
Transient Editing tool 141
Transient Editing Mode button 140
transient markers 165; in audio region 140; beat-mapping 153–4, 159; Logic **141**; recycling 156; Sample Editor 142; sensitivity control 140; timing references 139
transient peak probability 137
Transient section 211
Transpose 258; control 257
Transposition and *Harmonic Shift* button 254
Trigger Note drop-down box 187
Trimmer tool 211
tube resonances 247–8
Tune 258
tuning/timing correction 306

U

Units box 211
Unveil 121, **121**

Index

V

variable nondestructive edits 74–5
VariAudio **251**, 260
Varispeed algorithms 210
Velo knob 192
Velocity Mode 193
vinyl records 18, 284
visual editing *vs.* audible editing:
    adjusting plug-in settings 30;
    clear timing reference 31;
    DAWs 32; human feeling 31;
    track arrangement 30;
    zero-crossing points 31
vocal comping 90
Vocals 254
vocoded mid-section 315

Voice button 194
voice tone 90
volume dynamics over bar 179
volume envelope 73–4, **74–5**;
    region **81**, 81–2
Volume Handle 83–4, **84**

W

Warp Marker 173, 230, 235, **236**
waterfall plot 269
waveform 195; displays **197**
wow 18–19

X

X-Form algorithms 210

Z

zero-crossing edits 206; modern music production 30; phase-aligned 28–9, **30**; phase alignment and 30; pop sound 29; sounds stages 29; speaker oscillation 29; speaker position 28; steady waveform pattern 29–30; vibrating guitar string 29; waveform direction (phase) 29
Zynaptiq 243